Advances in Disease Diagnostics and Pathogen Biocontrol of Horticulture Crops

Guest Editors
Graciela Dolores
Ávila-Quezada
Facultad de Ciencias
Agrotecnológicas
Universidad Autónoma
de Chihuahua
Chihuahua
Mexico

Irasema Vargas-Arispuro
Coordination of Food Sciences
Research Center for Food and
Development (CIAD)
Hermosillo
Mexico

Editorial Office
MDPI AG
Grosspeteranlage 5
4052 Basel, Switzerland

This is a reprint of the Special Issue, published open access by the journal *Horticulturae* (ISSN 2311-7524), freely accessible at: https://www.mdpi.com/journal/horticulturae/special_issues/SH53K2N3H5.

For citation purposes, cite each article independently as indicated on the article page online and as indicated below:

Lastname, A.A.; Lastname, B.B. Article Title. *Journal Name* **Year**, *Volume Number*, Page Range.

ISBN 978-3-7258-4379-4 (Hbk)
ISBN 978-3-7258-4380-0 (PDF)
https://doi.org/10.3390/books978-3-7258-4380-0

Cover image courtesy of Graciela Avila-Quezada

Contents

About the Editors

Graciela Dolores Ávila-Quezada

Graciela Dolores Avila-Quezada was born in Chihuahua, Mexico, and currently serves as a research professor at the Autonomous University of Chihuahua. She has authored more than 100 scientific articles published in peer-reviewed journals indexed in Scopus and Web of Science. She holds an H-index of 19 on Scopus and a Level 2 recognition from the National System of Researchers in Mexico. She is an associate editor of the Mexican Journal of Plant Pathology. Her work has earned her several prestigious recognitions, including the 2023 National Award for Agricultural Parasitology "Maestro Ignacio Hernández Olmedo" in Mexico and the 2011 "Esther Orozco" Outstanding Woman in Science Award in Chihuahua. She has served as President of the Mexican Society of Phytopathology (SMF) and has exhibited a dedication to strengthening the connection between science and agricultural producers. She actively participates in scientific outreach initiatives, including public events, radio programs, and training programs aimed at empowering women in science. Additionally, she has led the UACH-2021-02 Plant Protection Disciplinary Group at the Autonomous University of Chihuahua.

Irasema Vargas-Arispuro

Irasema Vargas-Arispuro was born in Guaymas, Mexico, and currently works as a research professor at the Center for Food and Development Research (CIAD). She has been a member of the Mexican Academy of Sciences since 2014. She holds Level 3 recognition from the National System of Researchers in Mexico. She is an associate editor of the Mexican Journal of Plant Pathology. Since 2021, she has led a strategic project for the agroecological transition of corn cultivation nationwide. She actively participates in training programs for corn producers to facilitate the adoption of sustainable agricultural practices and clean technologies, helping to mitigate the negative effects of climate change and promoting more resilient and efficient farming systems in the use of natural resources such as water and soil nutrients.

Preface

Horticultural agriculture is currently facing significant challenges that compromise crop productivity and quality, as well as the sustainability of agricultural systems globally. Plant pathogens such fungi, bacteria, viruses, or nematodes, represent a threat that, historically, has been combated through the intensive use of chemical pesticides. However, this strategy has generated adverse environmental consequences, pathogen resistance, and risks to human health, which have prompted an urgent search for more sustainable alternatives.

In this context, the use of biocontrol agents and the development of accurate diagnostic strategies emerge as key tools to transform the way plant health is managed in horticultural crops. This Special Issue of *Horticulturae* brings together a variety of research that responds to this need, highlighting innovative approaches to disease diagnosis and control using environmentally friendly, effective, and locally adapted methods.

The articles presented here explore a diversity of cases, from the epidemiology of specific diseases such as downy mildew in grapevines or gray mold in pecan nuts, to the experimental and validated use of microorganisms such as *Trichoderma asperellum, Stenotrophomonas maltophilia* or *Exiguobacterium acetylicum* for the control of phytopathogens. In addition, aspects such as the molecular characterization of viruses in legumes, the impact of the use of certified seeds, and integrated agricultural practices to reduce the spread of diseases are addressed.

This Special Issue also highlights the holistic approach of this research, which seeks to mitigate diseases and to strengthen the resilience of agricultural systems, improve plant nutrition, and reduce environmental impact. As the agriculture of the future is redefined, biocontrol and sustainable plant health are emerging as fundamental pillars for achieving fairer, safer, and more environmentally friendly agricultural production.

Graciela Dolores Ávila-Quezada and Irasema Vargas-Arispuro

Guest Editors

horticulturae

Editorial

Special Issue: "Advances in Disease Diagnostics and Pathogen Biocontrol of Horticulture Crops"

Graciela Avila-Quezada [1,*] and Irasema Vargas-Arispuro [2]

1 Facultad de Ciencias Agrotecnológicas, Universidad Autónoma de Chihuahua, Chihuahua 31350, Mexico
2 Centro de Investigación en Alimentación y Desarrollo, AC. Carretera Enrique Astiazarán Rosas 46, Col La Victoria, Hermosillo 83304, Mexico; iris@ciad.mx
* Correspondence: gdavila@uach.mx

1. Introduction

Horticultural crops play an important role in global food security and the economy [1]. However, they face persistent threats from various pathogens—fungal, bacterial, viral [2], and insect-borne [3]—that impact yield, quality, and sustainability. Traditionally, reliance on chemical pesticides has led to environmental concerns [4] and resistance development in pathogens and pests [5], highlighting an urgent need for eco-friendly and sustainable disease management strategies.

In line with the United Nations' Zero Hunger goal [6], which aims to end hunger, achieve food security, improve nutrition, and promote sustainable agriculture, this Special Issue compiles research on disease diagnostics and the use of biocontrol agents to manage horticultural crop diseases. By advancing sustainable plant protection, these studies contribute to the development of resilient agricultural systems capable of supporting global food needs.

2. Overview of Published Articles

Carrera et al. (Contribution 1) analyzed the epidemiology of downy mildew (*Plasmopara vitícola)* in grapevines during a two-year study in northwestern Spain, evaluating two bioclimatic zones. Temperature, humidity, and wind speed were identified as key variables influencing disease dynamics. These findings emphasized the importance of region-specific disease forecasting for improved disease management.

Zheng et al. (Contribution 2) reported the first occurrence of gray mold (*Botrytis cinerea)* on pecan fruits and leaves in China. Molecular and pathogenicity analyses confirmed its identity and laid the groundwork for future disease control research in nut crops.

Huang et al. (Contribution 3) explored *Exiguobacterium acetylicum* SI17's potential as a biocontrol agent against litchi downy blight. Although the bacterium did not antagonize *Peronophythora litchii,* a genomic analysis revealed several secondary metabolite genes contributing to plant defense, suggesting significant promise for pre-harvest biocontrol.

Expósito-Goás et al. (Contribution 4) focused on bean common mosaic virus (BCMV) in protected geographical indications (PGIs). The PGI "Faba de Lourenzá" faced challenges from BCMV due to uncertified seeds. Field strategies such as virus-free seedlings, rogueing, and intercropping were tested, revealing effective control methods and highlighting the need for farmer education and off-site seed production.

Mrkvová et al. (Contribution 5) used high-throughput sequencing to analyze the virome in Fabaceae species, including red clover, peas, and beans. The genetic characteriza-

 https://doi.org/10.3390/horticulturae11050557

tion of Slovak bean yellow mosaic virus isolates provided insights into host susceptibility and virus–host interactions in pea genotypes.

Liu et al. (Contribution 6) studied strawberry root rot control with *Trichoderma asperellum*, where the isolate CMT10 controlled *Neopestalotiopsis clavispora* through multiple biocontrol mechanisms, including competition, hyperparasitism, and metabolite production. It also promoted plant growth, showing potential as a sustainable alternative to fungicides.

Shanmugaraj et al. (Contribution 7) is about collar rot management in tomato. In this study, among 20 *T. asperellum* isolates, A10 exhibited the strongest antagonism against *Agroathelia rolfsii* and the highest enzyme activity. In vitro and greenhouse assays confirmed its effectiveness in reducing collar rot and producing antimicrobial compounds.

Tomar et al. (Contribution 8) investigated biological control of the cabbage pest *Pieris brassicae* (an invasive pest affecting cole crops, with rising resistance to current control methods) using *Steinernema carpocapsae* and its symbiotic bacterium *Xenorhabdus nematophila*. Lab tests showed high larval mortality, highlighting the potential of these bacteria for integrated pest management, pending further field trials.

Osman et al. (Contribution 9) studied the biocontrol of onion white rot using the bacteria *Stenotrophomonas maltophilia* and *Serratia liquefaciens* which significantly suppressed *Stromatinia cepivora* infection and enhanced the plant growth and yield, aided by volatile metabolite production and plant defense activation.

Finally, Tyagi et al. (Contribution 10) reviewed the use of biocontrol agents in sustainable horticulture. Their review examined the roles of fungal, bacterial, and viral biocontrol agents, the environmental and genetic factors affecting their performance, and the challenges and opportunities in integrating these agents into mainstream horticultural practices.

3. Conclusions

This Special Issue presents recent advancements in understanding and managing horticultural diseases through innovative diagnostic tools and eco-friendly biocontrol approaches. The studies included emphasize the importance of region-specific strategies, molecular identification, and the integration of biocontrol agents into sustainable agriculture. Collectively, these findings underline the potential of biocontrol methods to reduce the dependency on chemical pesticides, protect biodiversity, and enhance crop productivity. Future efforts should focus on scaling these solutions under field conditions, improving the formulation of biocontrol microorganisms, and integrating them into holistic crop management systems.

Author Contributions: Conceptualization, writing—original draft preparation, writing—review and editing, G.A.-Q. and I.V.-A. All authors have read and agreed to the published version of the manuscript.

Funding: This research received no external funding.

Conflicts of Interest: The authors declare no conflict of interest.

List of Contributions:

1. Carrera, L.; Fernández-González, M.; Aira, M.J.; Espinosa, K.C.S.; Otero, R.P.; Rodríguez-Rajo, F.J. Airborne *Plasmopara viticola* Sporangia: A Study of Vineyards in Two Bioclimatic Regions of Northwestern Spain. *Horticulturae* **2025**, *11*, 228. https://doi.org/10.3390/horticulturae11030228.
2. Zheng, X.-R.; Huang, X.-X.; Peng, J.-F.; Gafforov, Y.; Chen, J.-J. Occurrence of *Botrytis cinerea* Causing Gray Mold on Pecan in China. *Horticulturae* **2024**, *10*, 1212. https://doi.org/10.3390/horticulturae10111212.

3. Huang, S.; Lv, X.; Zheng, L.; Guo, D. *Exiguobacterium acetylicum* Strain SI17: A Potential Biocontrol Agent against *Peronophythora litchii* Causing Post-Harvest Litchi Downy Blight. *Horticulturae* **2024**, *10*, 888. https://doi.org/10.3390/horticulturae10080888.
4. Expósito-Goás, S.; Pinacho-Lieti, L.G.; Lago-Pena, F.; Cabaleiro, C. Epidemiology and Management of Bean Common Mosaic Virus (BCMV) in Traditional *Phaseolus vulgaris* L. Landraces within Protected Geographical Indications. *Horticulturae* **2024**, *10*, 699. https://doi.org/10.3390/horticulturae10070699.
5. Mrkvová, M.; Kemenczeiová, J.; Achs, A.; Alaxin, P.; Predajňa, L.; Šoltys, K.; Šubr, Z.; Glasa, M. Molecular Characteristics and Biological Properties of Bean Yellow Mosaic Virus Isolates from Slovakia. *Horticulturae* **2024**, *10*, 262. https://doi.org/10.3390/horticulturae10030262.
6. Liu, P.; Yang, R.; Wang, Z.; Ma, Y.; Ren, W.; Wei, D.; Ye, W. Biocontrol Potential of *Trichoderma asperellum* CMT10 against Strawberry Root Rot Disease. *Horticulturae* **2024**, *10*, 246. https://doi.org/10.3390/horticulturae10030246.
7. Shanmugaraj, C.; Kamil, D.; Kundu, A.; Singh, P.K.; Das, A.; Hussain, Z.; Gogoi, R.; Shashank, P.R.; Gangaraj, R.; Chaithra, M. Exploring the Potential Biocontrol Isolates of *Trichoderma asperellum* for Management of Collar Rot Disease in Tomato. *Horticulturae* **2023**, *9*, 1116. https://doi.org/10.3390/horticulturae9101116.
8. Tomar, P.; Thakur, N.; Sidhu, A.K.; Laskar, B.A.; Hashem, A.; Avila-Quezada, G.D.; Abd_Allah, E.F. The Isolation, Identification, and Insecticidal Activities of Indigenous Entomopathogenic Nematodes (*Steinernema carpocapsae*) and Their Symbiotic Bacteria (*Xenorhabdus nematophila*) against the Larvae of *Pieris brassicae*. *Horticulturae* **2023**, *9*, 874. https://doi.org/10.3390/horticulturae9080874.
9. Osman, H.E.M.; Nehela, Y.; Elzaawely, A.A.; El-Morsy, M.H.; El-Nagar, A. Two Bacterial Bioagents Boost Onion Response to *Stromatinia cepivora* and Promote Growth and Yield via Enhancing the Antioxidant Defense System and Auxin Production. *Horticulturae* **2023**, *9*, 780. https://doi.org/10.3390/horticulturae9070780.
10. Tyagi, A.; Lama Tamang, T.; Kashtoh, H.; Mir, R.A.; Mir, Z.A.; Manzoor, S.; Manzar, N.; Gani, G.; Vishwakarma, S.K.; Almalki, M.A.; Ali, S. A Review on Biocontrol Agents as Sustainable Approach for Crop Disease Management: Applications, Production, and Future Perspectives. *Horticulturae* **2024**, *10*, 805. https://doi.org/10.3390/horticulturae10080805.

References

1. Khan, M.M.; Akram, M.T.; Janke, R.; Qadri, R.W.K.; Al-Sadi, A.M.; Farooque, A.A. Urban horticulture for food secure cities through and beyond COVID-19. *Sustainability* **2020**, *12*, 9592. [CrossRef]
2. Xu, J.; Zhang, N.; Wang, K.; Xian, Q.; Dong, J.; Chen, X. Exploring new strategies in diseases resistance of horticultural crops. *Front. Sustain. Food Syst.* **2022**, *6*, 1021350. [CrossRef]
3. Mani, M. Pest management in horticultural crops under protected cultivation. In *Trends in Horticultural Entomology*; Springer: Singapore, 2022; pp. 387–417.
4. Lamichhane, J.R.; Dachbrodt-Saaydeh, S.; Kudsk, P.; Messéan, A. Toward a reduced reliance on conventional pesticides in European agriculture. *Plant Dis.* **2016**, *100*, 10–24. [CrossRef] [PubMed]
5. Venkatesan, T.; Chethan, B.R.; Mani, M. Insecticide resistance and its management in the insect pests of horticultural crops. In *Trends in Horticultural Entomology*; Springer: Singapore, 2022; pp. 455–490.
6. United Nations. The Global Goals. Zero Hunger. End Hunger, Achieve Food Security and Improved Nutrition and Promote Sustainable Agriculture. 2025. Available online: https://www.globalgoals.org/goals/2-zero-hunger/ (accessed on 18 May 2025).

horticulturae

MDPI

Article

Airborne *Plasmopara viticola* Sporangia: A Study of Vineyards in Two Bioclimatic Regions of Northwestern Spain

Lucía Carrera [1], María Fernández-González [1,*], María Jesús Aira [2], Kenia C. Sánchez Espinosa [1], Rosa Pérez Otero [3] and Francisco Javier Rodríguez-Rajo [1]

1 Department of Plant Biology and Soil Sciences, Faculty of Sciences, University of Vigo, 32004 Ourense, Spain; lcarrera@uvigo.gal (L.C.); keniacaridad.sanchez@uvigo.gal (K.C.S.E.); javirajo@uvigo.es (F.J.R.-R.)
2 Department of Botany, Faculty of Biology, University of Santiago de Compostela, C/Lope Gómez de Marzoa, s/n, 15782 Santiago de Compostela, Spain; mariajesus.aira@usc.es
3 Phytopathological Station Areeiro, Subida a la Robleda, s/n, 36153 Pontevedra, Spain; rosa.perez@depo.gal
* Correspondence: mfgonzalez@uvigo.es; Tel.: +34-988368912

Abstract: Downy mildew, caused by *Plasmopara viticola*, is one of the most destructive diseases affecting grapevines, particularly in areas with bioclimatic conditions that favor its development, such as northwestern Spain. This study examined the presence of *P. viticola* sporangia in three vineyards located in two distinct bioclimatic regions (ultra-oceanic and subcontinental) over two consecutive years (2023 and 2024) using Hirst-type volumetric samplers. The relationship between *P. viticola* and grapevine phenology, along with meteorological variables, was analyzed to help develop effective strategies for managing this disease. Spearman correlation analysis showed that temperature was the most influential variable in all vineyards. However, water-related variables (relative humidity and precipitation) showed stronger correlations in the ultra-oceanic vineyard, which also had the highest sporangium concentrations. Principal Component Analysis revealed that sporangium concentrations in the ultra-oceanic region were most strongly associated with relative humidity and wind speed. In contrast, sporangium concentrations in the subcontinental vineyards were more closely related to temperature, dew point, and wind speed (in Alongos), as well as wind speed (in Alongos and Cenlle). PCA results clearly differentiated the two bioclimatic zones. These findings provide valuable insights that can improve downy mildew management in vineyards in northwestern Spain.

Keywords: *Plasmopara viticola*; sporangia; vineyard; aerobiology; phenology

1. Introduction

The grapevine (*Vitis vinifera* L.) is known for its significant agricultural and economic value in various geographical regions [1], with Spain being the world's leading wine exporter by volume and the third by value [2,3]. The northwestern region of the country has over 33,200 hectares of vineyards, with the vast majority of wine production centered in the provinces of Ourense and Pontevedra [4]. The bioclimatic conditions in this area are characterized by high humidity, abundant rainfall, and moderate temperatures, all propitious to the development of downy mildew (*Plasmopara viticola*) on grapevines, one of the most devastating diseases affecting these crops [5–7]. Downy mildew epidemics result in both direct and indirect yield losses, including reduced photosynthetic activity in affected leaves, the premature defoliation of vines, and the rotting of inflorescences, shoots, and clusters [8].

https://doi.org/10.3390/horticulturae11030228

The causal agent of this disease is the oomycete *Plasmopara viticola* (Berk. & M.A. Curtis) Berl. & De Toni, which was brought to Europe in 1878 from North America through rootstock of American wild grapevines resistant to phylloxera [9]. *P. viticola* is an obligate parasite belonging to the family Peronosporaceae, the order Peronosporales, and the class Peronosporomycetes [10]. During the vegetative period of grapevine development, *P. viticola* completes its life cycle, which is governed by oospores and sporangia. In early spring, zoospores released from macrosporangia, formed by the germination of oospores, are dispersed by rain and wind onto new leaves and shoots, eventually reaching the grape clusters. New zoospores are produced through asexual reproduction, continuously infecting fresh tissues throughout the growing season [11]. Moreover, evidence shows that the survival of the inoculum in plant debris allows it to germinate in successive crops, complicating the eradication of the disease [12]. All grapevine varieties are susceptible to this disease, although the degree of susceptibility varies among them [13,14]. However, genetic diversity analysis has revealed distinct patterns between populations in different geographical regions, highlighting the influence of historical introductions and local environmental conditions [15].

The traditional and most effective method for controlling downy mildew is the periodic application of copper-based formulations, although other types of phytosanitary products, operating either on contact or systemically, have been developed [16]. However, the European Union has established restrictions on the prolonged use of these phytosanitary products due to their potential to promote the emergence of pathogen-resistant strains [6], as well as their negative impact on soil and water quality, which can ultimately affect human health [17–19]. There is therefore an increasing need for research into the alternative management of cryptogamic diseases that minimizes the use of these products [20]. Cultivation practices, such as planting grapevines in open, well-ventilated layouts rather than compact ones; managing cover vegetation; employing specific pruning techniques; and ensuring the appropriate handling of pruning residues, represent some alternatives for disease prevention [21,22].

Given that *P. viticola* sporangia are the main focus for monitoring infection, aerobiological studies that continuously detect their presence and correlate it with meteorological variables are a useful tool for detecting and reducing the spread of the disease [23,24]. Various studies in Spain have described these relationships; however, they differ between years and regions. There are wine-making regions where studies on the detection of *P. viticola* sporangia in the air are limited, such as the region of the Rías Baixas [13,25–27]. The aim of this study was to evaluate the influence of meteorological variables on the aerial concentration of *Plasmopara viticola* sporangia in two different bioclimatic regions of northwest Spain during the grapevine vegetative cycle.

2. Materials and Methods

2.1. Location and Climatic Characteristics of the Study Area

The study was conducted in three vineyards in northwestern Spain: Areeiro (42°24′13″ N, 8°40′18″ W), Alongos (42°20′03″ N, 7°57′48″ W), and Cenlle (42°18′56″ N, 8°06′03″ W) (Figure 1). This region is strongly influenced by maritime conditions, which diminish as one heads inland toward the east and the distance from the coast increases [28].

Areeiro is located in the Rías Baixas region and biogeographically belongs to the Eurosiberian region. Many of the vineyards are situated close to the coast, creating a microclimate with high humidity, mild temperatures, and frequent rainfall. The traditional cultivation system is in elevated training trellises, structures that lift the vines off the ground to prevent diseases induced by humidity. These systems also provide optimal sun exposure and ventilation. The climate is Atlantic oceanic, with mild winters and precipitation throughout the year, falling to a minimum in the summer [29].

Figure 1. Location of the study vineyards: Areeiro, Alongos, and Cenlle.

Alongos and Cenlle are located in O Ribeiro, Ourense Province, in the Miño River basin, a region characterized by rugged terrain with fertile valleys and gentle hills. Biogeographically, their location is in a transitional zone between the Eurosiberian and Mediterranean regions [30]. The landscape is dominated by vineyards extending across terraced slopes, utilizing traditional cultivation systems that maximize the use of the hilly terrain and optimize solar exposure. The climate is oceanic with a Mediterranean influence, with mild, wet winters and warmer, drier summers compared to the coastal areas [28].

2.2. Phenological and Aerobiological Study

This study was carried out during the active grapevine growing seasons in 2023 and 2024, focusing on phenological research and sampling *P. viticola* sporangia in the vineyards using aerobiological methods.

For the phenological study, the most characteristic grapevine varieties of each wine-producing region were selected: Albariño in Areeiro and Treixadura in Cenlle and Alongos. The Biologische Bundesanstalt, Bundessortenamt und Chemische Industrie (BBCH) scale proposed by Meier [31] was followed. In each vineyard, 20 plants were selected to monitor six phenological stages (PS): PS 0 (bud development), PS 1 (leaf development), PS 5 (inflorescence emergence), PS 6 (flowering), PS 7 (development of fruit), and PS 8 (ripening of berries). To determine these stages, weekly visits were made to each vineyard, with two visits per week during the flowering stage (PS 6). The start date of each phenological stage was recorded as the date when 50% of the studied plants were observed in that stage.

Fungal propagule samples were collected using Lanzoni VPPS-2000 model volumetric collectors (Lanzoni Manufacturing, Bologna, Italy) [32], placed 2 m above ground level to avoid obstruction by vine growth. The samplers were calibrated to a flow rate of 10 L/min. For sample processing, the authors applied the Spanish Aerobiological Network protocol [33]. Air samples were trapped in a cylindrical drum covered with Melinex tape previously coated with a 2% silicone solution. In the laboratory, this tape was divided into seven sections, and microscope slides were prepared using glycerinated gelatin with 1% basic fuchsin as the preparation medium. Aerobiological preparations were analyzed under a light optical microscope (OLYMPUS BX50F, Tokyo, Japan) at 400× magnification. *P. viticola* sporangia were identified as hyaline, ovoid structures, wider at the apex, with lengths ranging from 17 to 40 μm and thicknesses from 10 to 20 μm. The results were expressed as daily sporangium concentration per cubic meter of air (sporangia/m^3) using the terminology proposed by Galán et al. [34].

2.3. Meteorological Data

The meteorological data used in this study were recorded by a HOBO meteorological station (ONSET HOBO USB Micro Station Data Logger) located at each vineyard. The daily meteorological variables recorded were average temperature (AvgT, °C), maximum temperature (MaxT, °C), minimum temperature (MinT, °C), relative humidity (RH, %), and dew point (Dp, °C). Precipitation (Rain, L/m^2) and wind speed (Ws, km/h) data were obtained from the nearest meteorological stations in the MeteoGalicia network (http://www.meteogalicia.gal/, accessed on 28/11/2024).

To classify the regions where the vineyards are located (Rías Baixas and O Ribeiro), the Currey continentality index [35] was used. This index is calculated based on thermal amplitude and latitude:

$$Kcu = Am/(1 + (\phi/3)) \tag{1}$$

where:

Kcu = Currey index;
Am = thermal amplitude;
ϕ = latitude.

2.4. Statistical Analysis

To evaluate the degree of association between meteorological variables and the concentrations of *P. viticola* sporangia up to 7 days prior, the non-parametric Spearman correlation test was performed. Three periods were considered for the analysis: the 2023 season, the 2024 season, and both seasons combined. The significance levels for the confidence intervals were set at 95% ($p < 0.05$) and 99% ($p < 0.01$).

Additionally, a Principal Component Analysis (PCA) was conducted to identify differences in sporangium concentrations between bioclimatic regions and to analyze the influence of all variables on sporangium concentrations in each area. Statistical analyses were performed using IBM SPSS Statistics 24 software.

3. Results

3.1. Grapevine Phenology Cycle and P. viticola Concentrations

The duration of the grapevine growing cycle in all vineyards was similar in both years, with an average duration of 170 days, the exception being Alongos in 2023, where it was shorter, lasting 142 days. Considering both growing seasons, the average duration was longer in Cenlle (177 days) and Areeiro (175 days) than in Alongos (159 days). In 2024 (179 days), the growing cycle was longer than in 2023 (161 days) in all three vineyards (Table 1).

The duration of the phenological stages showed variations across all locations and years (Table 2). In all study vineyards, the longest phenological stage was fruit development (PS 7), lasting 74 days in Areeiro, 72 days in Alongos, and 68 days in Cenlle. The next longest stage was the ripening of berries (PS 8), which lasted 64 days in Areeiro, 51 days in Alongos, and 60 days in Cenlle, using the average value for both growing seasons.

The *P. viticola* total concentration showed notable differences between the different vineyards and sampling years. In 2023, Areeiro yielded 53,447 sporangia/m^3, whereas

Alongos yielded 2483 sporangia/m^3 and Cenlle 1581 sporangia/m^3 (Table 2). In 2024, there was a considerable decrease compared to the previous year, with a 37% reduction at Areeiro, 47% at Alongos, and 71% at Cenlle (Table 2).

Table 1. Dates and duration of the phenological study.

	Study Period		
Vineyard	**Areeiro**	**Alongos**	**Cenlle**
Start date 2023	24 March	28 March	14 March
Duration (days)	167	142	173
Harvest	7 September	17 August	3 September
Start date 2024	13 March	19 March	12 March
Duration (days)	182	168	181
Harvest	11 September	3 September	9 September

Table 2. Start date of the phenological stages (PS 0—bud development, PS 1—leaf development, PS 5—inflorescence emergence, PS 6—flowering, PS 7—development of fruit, and PS 8—ripening of berries) in the three vineyards in 2023 and 2024, the sum of *Plasmopara viticola* daily concentrations (SSIn, sporangia/m^3) over each phenological stage, the peak maximum daily concentration (Peak, sporangia/m^3), and the total sporangium concentration in each study period (Total, sporangia).

	2023			2024		
	Start Date	**SSIn**	**Peak**	**Start Date**	**SSIn**	**Peak**
			Areeiro			
PS 0	24 March	14	5	13 March	21	5
PS 1	2 April	30	4	2 April	57	23
PS 5	19 April	1013	205	17 April	1149	484
PS 6	18 May	20,925	5604	28 May	3811	629
PS 7	7 June	22,002	1474	10 June	13,333	1158
PS 8	26 July	9463	1028	30 July	15,241	1402
Total		53,447			33,612	
			Alongos			
PS 0	28 March	14	13	13 March	85	25
PS 1	3 April	31	5	31 March	166	43
PS 5	16 April	85	9	15 April	205	22
PS 6	13 May	11	4	26 May	83	11
PS 7	30 May	1854	455	7 June	587	117
PS 8	19 July	490	67	21 July	196	56
Total		2483			1321	
			Cenlle			
PS 0	14 March	109	22	12 March	120	18
PS 1	5 April	110	48	5 April	89	16
PS 5	22 April	168	23	20 April	57	5
PS 6	20 May	23	5	31 May	39	5
PS 7	7 June	808	110	13 June	110	22
PS 8	25 July	364	83	23 July	48	5
Total		1581			462	

In terms of temporal distribution, *P. viticola* sporangia were recorded at all phenological stages in both years and all vineyards (Table 2 and Figure 2). The highest concentrations in Areeiro were recorded during the flowering (PS 6) and fruit development (PS 7) stages in 2023. However, in 2024, peaks were observed during the fruit development stage and the ripening of berries (PS 7 and 8, respectively). In Alongos and Cenlle, in both growing seasons, the highest concentration of sporangia was observed during the fruit development stage (PS 7). In general, the days with the highest concentration of *P. viticola* sporangia in the three vineyards were mainly recorded during the fruit development and ripening stages (PS 7, PS 8).

In 2023 in Areeiro, several daily peaks were recorded: 7 June (1045 sporangia/m^3), 13 June (1474 sporangia/m^3), and 7 and 24 July (1193 and 1454 sporangia/m^3), all during the fruit development stage (PS 7). Additionally, between 28 May and 1 June, 17,303 sporangia/m^3 were recorded, representing 33% of the total annual count in just five days. However, the maximum peak in Areeiro occurred during the flowering stage (PS 6) on 29 May 2023, with 5604 sporangia/m^3. In 2024, in the same vineyard, several days with values over 1000 sporangia/m^3 were also recorded during both PS 7 and PS 8.

In the Alongos vineyard, the maximum daily concentrations were lower, with notable peaks of 455 sporangia/m^3 on 23 June 2023 and 117 sporangia/m^3 on 9 July 2024, both during the fruit development stage (PS 7).

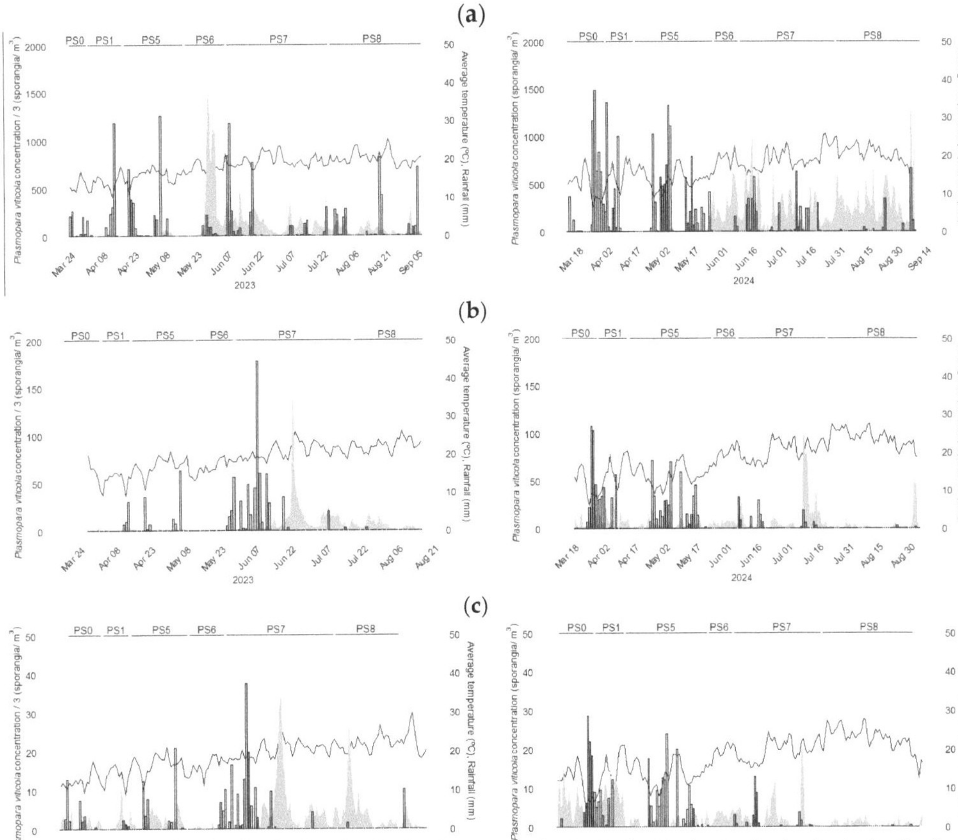

Figure 2. Phenology (upper lines: PS 0—bud development, PS 1—leaf development, PS 5—inflorescence emergence, PS 6—flowering, PS 7—development of fruit, and PS 8—ripening of berries), *Plasmopara viticola* daily concentrations (gray area), average temperature (lines), and rainfall (bars) during the years of study. In order to facilitate the comprehension of the graphics, *Plasmopara viticola* sporangium concentrations were divided by three in the 2023 graphs. (**a**) Areeiro 2023–2024, (**b**) Alongos 2023–2024, and (**c**) Cenlle 2023–2024.

3.2. Meteorological Parameters During the Study

The meteorological conditions in the three vineyards showed marked location-related differences. The continentality index for Areeiro (Kcu = 0.43) classifies this area as ultra-oceanic due to its lower thermal amplitude. The inland area, where Alongos and Cenlle are located, is classified as subcontinental, with Kcu values of 1.5177 and 1.6511, respectively.

Analysis of the meteorological data showed that the average and minimum temperatures did not exhibit significant differences among the three locations over the study period. However, the maximum temperature varied between 26 and 27 °C in Alongos and Cenlle, while in Areeiro, it ranged between 23 and 24 °C (Table 3). On the other hand, all climatic variables contributing to ambient humidity were higher in Areeiro compared to the other two vineyards. The average precipitation over both years was 475 L/m^2 in Areeiro, compared to 247 L/m^2 in Alongos and 321 L/m^2 in Cenlle. Similarly, the average relative humidity was 83% in Areeiro, in contrast to 70% in Alongos and 75% in Cenlle. In terms of the average dew point, Areeiro recorded 14 °C, while Alongos and Cenlle were approximately 1 degree lower. Wind speed was also higher in Areeiro compared to that in Alongos and Cenlle (3.7 km/h in Areeiro vs. <1 km/h in Alongos and Cenlle). When comparing both growing seasons, 2023 was slightly warmer and less rainy than 2024 in all three vineyards (Table 3).

Table 3. Meteorological variables (MaxT °C—maximum temperature, MinT °C—minimum temperature, AvgT °C—mean temperature, RH %—relative humidity, Dp °C—dew point, and Ws Km/h—wind speed) and total precipitation (Rain—L/m^2) in the three vineyards (2023 and 2024).

		MaxT	MinT	AvgT	RH	Dp	Ws	Rain
Areeiro	2023	23.87	13.39	18.07	83.53	14.43	3.55	360.60
	2024	23.10	12.43	17.12	83.19	13.41	3.82	589.80
Alongos	2023	27.50	11.86	19.11	69.22	12.47	0.54	204.20
	2024	26.88	11.76	18.60	71.37	12.52	0.68	290.20
Cenlle	2023	26.20	12.16	18.35	74.50	12.95	1.05	255.00
	2024	26.06	11.62	17.84	75.27	12.66	2.30	321.30

Analysis of the meteorological data averaged by phenological stage (Table S1) reveals that the highest average temperatures occurred during the summer months (PS 8), ranging from 19.98 °C in Areeiro to 23.81 °C in Alongos, in both cases during the 2024 growing season. The lowest minimum temperature values were recorded during the leaf development stage (PS 1), ranging from 5.23 °C in Alongos to 8.52 °C in Areeiro in 2023. As far as rainfall is concerned, in 2023, all three vineyards recorded the highest precipitation during the fruit development stage (PS 7), with values over 100 L/m^2. In 2024, precipitation was higher during the bud development stage (PS 0), with 79.73 L/m^2 in Alongos and more than 100 L/m^2 in Areeiro and Cenlle.

3.3. Meteorological Influence on *P. viticola* Sporangia

Among the meteorological variables studied, temperature had the greatest influence on the *P. viticola* sporangium concentration in three vineyards, regardless of the study period (Tables 4 and S2).

In particular, the average temperature from the seven preceding days in Areeiro (r = 0.626, 2024) and the same-day temperature in Alongos (r = 0.627, in 2023) showed the strongest correlations. Similar r values were also observed for the dew point, with mostly positive correlations. In general, the influence of variables related to water (precipitation and relative humidity) was weaker, showing both positive and negative correlations at different periods. Similar relationships were observed between the detected concentrations and wind speed (Tables 4 and S2).

Analysis of the correlation results by vineyard revealed that in the ultra-oceanic area (Areeiro), the mean temperature generally exhibited stronger correlations than the maximum or minimum temperatures, with positive correlations in all study periods. Overall, the r value for both the maximum and mean temperatures increased when considering the influence of previous days, reaching its peak seven days before. The temperature on previous days had a greater influence than the temperature on the day of measurement on the *P. viticola* presence in the air. By contrast, the relationship with the dew point on previous days was weaker than that on the same day. Precipitation showed a negative correlation when analyzing the 2024 data, with r values increasing when considering the influence of previous days. Relative humidity correlated positively with sporangium concentration in 2023. However, in 2024, significant correlations were negative.

In the subcontinental area (Cenlle and Alongos), temperature was also the most influential variable, with correlation results consistently stronger in Alongos than in Cenlle. What was notable about the temperature influence in the transition zone was that it showed a positive correlation in 2023 and a negative one in 2024. After temperature, the dew point and wind speed were the factors exhibiting the highest r values. With regard to wind speed, a similar pattern was observed in both vineyards: a negative influence in 2023 and a positive one in the preceding days in Alongos in 2024. By contrast, relative humidity and precipitation had significantly weaker correlations with the *P. viticola* sporangium concentration. Relative humidity always showed a positive influence, while rainfall had variable effects.

Table 4. Spearman correlation test results comparing *Plasmopara viticola* concentration (sporangia/m^3) and the meteorological variables recorded (MaxT—maximum, MinT—minimum, and AvgT—average temperature (°C); Rain—rainfall (mm), Ws—wind speed (km/h); RH—relative humidity (%); and Dp—dew point (°C)) (signification level: $p < 0.01$ **, $p < 0.05$ *). Only the values of the correlations with the highest r values for each variable are shown.

	Areeiro		Alongos		Cenlle	
	2023	2024	2023	2024	2023	2024
MaxT	0.433 ** MaxT_5	0.498 ** MaxT_7	0.467 ** MaxT	−0.277 ** MaxT_3	0.289 ** MaxT	−0.229 ** MaxT_5
MinT	0.604 ** MinT_1	0.618 ** MinT_7	0.544 ** MinT_7	−0.250 ** MinT_7	0.288 ** MinT_1	−0.199 * MinT_3
AvgT	0.559 ** AvgT_4	0.626 ** AvgT_7	0.627 ** AvgT	−0.278 ** AvgT_3	0.347 ** AvgT	−0.222 ** AvgT_5
Rain		−0.318 ** Rain_7	−0.224 ** Rain	0.249 ** Rain_2	−0.203 ** Rain	0.202 * Rain_7
RH	0.314 ** RH_1	−0.158 * RH_3	0.194 * RH_7	0.243 ** RH_1	0.200 ** RH_1	0.222 ** RH_7
Dp	0.677 ** Dp_1	0.632 ** Dp_7	0.603 ** Dp_7	−0.243 ** Dp_3	0.361 ** Dp_1	−0.170 * Dp_3
Ws	−0.192 * Ws_1	0.243 ** Ws_4	−0.396 ** Ws_7	0.216 ** Ws_5	−0.308 ** Ws_1	

When the correlations were compared across both seasons (2023 and 2024), a general trend toward a decrease in the significance and strength of the correlations with meteorological factors became evident. In the ultra-oceanic area, correlations with temperature and dew point (both on the same day and up to seven before days) remained significant and positive. Relative humidity and precipitation on the same day and previous days showed less influence when combining seasons than when the years were analyzed separately. In the vineyards in the subcontinental area, maximum temperature only showed significant correlations with sporangia on the same day. Taking the dataset for each location separately, in Alongos, most correlations were positive and significant, with minimum temperature, mean temperature, and dew point maintaining their significance, although with lower r values. However, in Cenlle, for these same variables, correlations were only present on the same day and the preceding day, and for dew point, for the two preceding days. Relative humidity lost its strength when both years were considered together, although the correlation remained positive. With regard to leaf humidity, significant correlations were only found in Areeiro one and seven days before. Wind speed from four and seven days prior had a positive influence in Areeiro, while in Cenlle, it had a negative influence across all the evaluated days.

3.4. Principal Component Analysis

To determine the relationship between the concentrations of *P. viticola* sporangia recorded in the three vineyards, a principal component analysis was conducted. Principal component 1 (PC1) and principal component 2 (PC2) accounted for approximately 74% of the variance in the data. PC1 grouped the sporangium concentrations from the Alongos and Cenlle vineyards, while those from Areeiro were placed in PC2 (Figure 3).

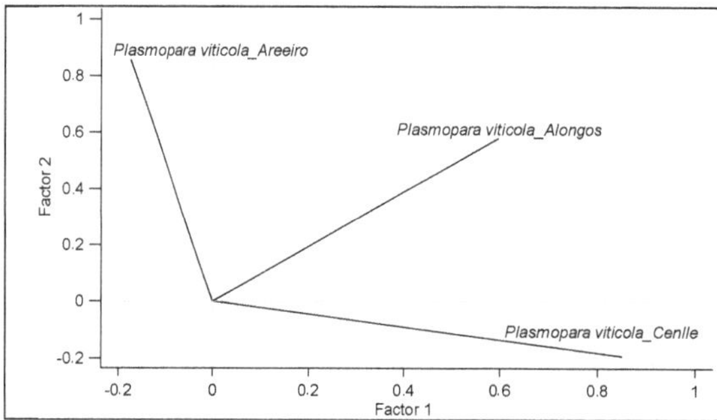

	Principal Components	
	1	2
Eigenvaule	1.136	1.073
Percent of variance	37.866	35.770
Cumulative percentage	37.866	73.636
*Plasmopara viticola*_Cenlle	0.849	−0.193
*Plasmopara viticola*_Alongos	0.595	0.578
*Plasmopara viticola*_Areeiro	−0.174	0.855

Figure 3. Results of the principal component analysis applied to *Plasmopara viticola* concentrations in the vineyards of Areeiro, Alongos, and Cenlle. Factor 1 = Principal Component 1; Factor 2 = Principal Component 2.

In order to identify the meteorological factors that most influence the variation in *P. viticola* sporangium concentrations across the different vineyards, three additional PCAs were performed, one for each study area. The PCA for Cenlle grouped the data into three principal components, while the PCA for Areeiro and Alongos grouped the data into two components. In both Areeiro and Alongos, the two components accounted for a similar percentage of the data variance, close to 70%. In contrast, the three components from the Cenlle PCA accounted for 83.3% of the data variance (Figure 4).

With regard to the distribution of meteorological variables in the different principal components, it was observed that the *P. viticola* concentration in the ultra-oceanic area (Areeiro) was more strongly associated with component 2, which also grouped relative humidity, rain, and wind speed (Figure 4). Meanwhile, the sporangium concentration in Alongos was more closely associated with component 1, where temperature (maximum, minimum, and average), dew point, and wind speed (with a weaker and negative association) were highlighted. Finally, the sporangium concentration in Cenlle and wind speed were grouped in component 3. These results are consistent with the findings from the Spearman correlation analysis when considering both years (2023 and 2024) together.

Finally, when we took into account the grouping of the meteorological variables analyzed with the PCA, it was observed that the maximum, minimum, and average temperatures, as well as the dew point, were grouped in the same component across all three vineyards, always in the component that accounted for the highest variability (Figure 4). Precipitation was found in the second component, both in the ultra-oceanic vineyard and in the subcontinental vineyards. In the latter, relative humidity was grouped with precipitation in all study areas (Figure 4).

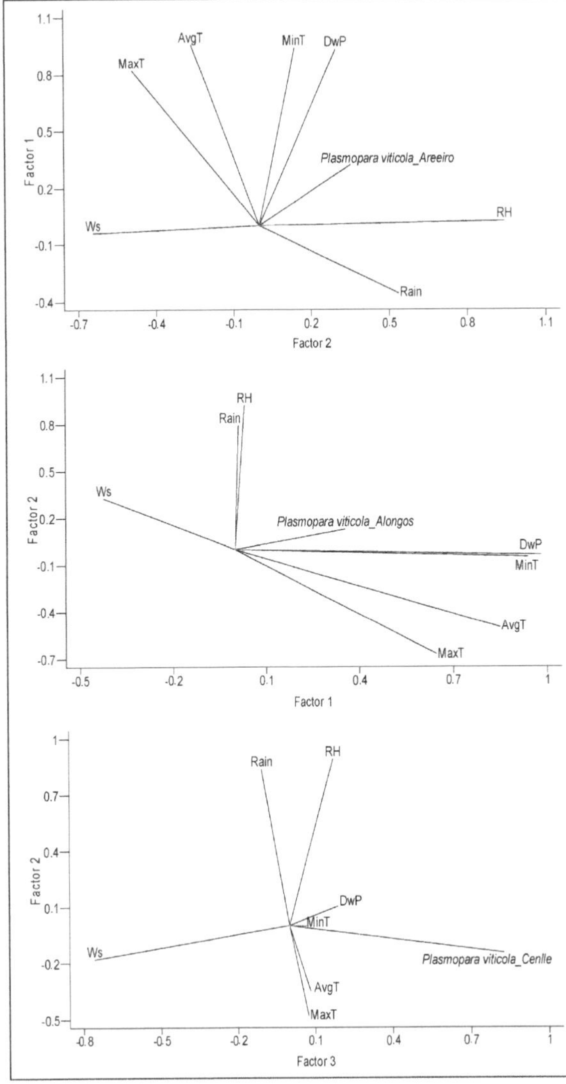

	Areeiro	
	PC 1	PC 2
Eigenvaule	3.611	2.072
Percent of variance	45.135	25.896
Cumulative percentage	45.135	71.031
Plasmopara viticola	0.320	0.353
MaxT	0.819	−0.491
MinT	0.938	0.136
AvgT	0.953	−0.264
Rain	−0.356	0.535
Ws	−0.040	−0.643
RH	0.023	0.940
Dp	0.931	0.297

	Alongos	
	PC 1	PC 2
Eigenvaule	3.892	1.674
Percent of variance	48.645	20.927
Cumulative percentage	48.645	69.572
Plasmopara viticola	0.354	0.126
MaxT	0.645	−0.671
MinT	0.937	−0.049
AvgT	0.847	−0.500
Rain	0.010	0.792
Ws	−0.423	0.324
RH	0.032	0.916
Dp	0.980	−0.038

	Cenlle		
	PC 1	PC 2	PC 3
Eigenvaule	3.850	1.684	1.131
Percent of variance	48.128	21.048	14.134
Cumulative percentage	48.128	69.176	83.310
Plasmopara viticola	0.069	−0.147	0.823
MaxT	0.822	−0.480	0.075
MinT	0.958	−0.003	0.065
AvgT	0.923	−0.349	0.082
Rain	−0.082	0.834	−0.107
Ws	−0.137	−0.180	−0.759
RH	−0.138	0.888	0.170
Dp	0.961	0.103	0.187

Figure 4. Principal component analysis (PCA) results for the three locations: Areeiro, Alongos, and Cenlle. Factor 1 = PC 1, Factor 2 = PC 2, and Factor 3 = PC 3. MaxT—maximum temperature, MinT—minimum temperature, AvgT—average temperature, Dp—dew point, Rain—rainfall, RH—relative humidity, and Ws—wind speed.

4. Discussion

Knowledge of grapevine phenology over several seasons is essential for ascertaining the characteristics of wine-growing regions. This information enables the effective planning of agricultural tasks and the application of phytosanitary products [36,37]. Furthermore, aerobiological studies of these crops help to optimize harvest yields and to improve the quality of the final product [38].

The main factors that influence the vegetative activity of the vine and the development of its various phenological stages are meteorological in nature, although agricultural management also plays a role [39,40]. The average duration of the grapevine cycle in this study is consistent with that in other studies carried out in the northwest Iberian Peninsula [5,26]. The shortest vegetative cycle throughout the study period was observed in 2023 at the Alongos vineyard. This difference was mainly due to a shortening of PS 8 by 11–14 days compared to that in Cenlle and Areeiro. This may have been caused by temperature variations, which are a key factor for grapevine development in temperate regions [25,41]. In this regard, maximum and average temperatures during this stage were higher in Alongos, and other variables related to humidity were lower. This may have influenced earlier ripening, since the heat dries the soil and stops fruit growth [42].

The meteorological conditions in the three vineyards favored the presence and spread of *P. viticola* sporangia in the air during the vine growth cycle. For downy mildew to develop in vineyards, climatic conditions are crucial in determining the prevalence and severity of the disease [5]. During spring rains, oospores germinate. These oospores (or the zoospores produced) are transported by rain or wind to the lower leaves, where they penetrate with the help of free water and initiate disease. Once the mycelium develops, sporangiophores emerge through the stomata, releasing sporangia that are carried by the wind and rain to other healthy plants [43]. Oospore germination requires soil temperatures around 12–13 °C and moisture, while the conditions for *P. viticola* sporulation include a minimum temperature of 13 °C, an optimal temperature of 19 °C, relative humidity higher than 95%, and around 4 h of darkness [44]. In both the vineyard located in the ultra-oceanic area (Areeiro) and the two in the subcontinental area (Cenlle and Alongos), the average relative humidity was above 69%, and the average temperature ranged between 17.12 °C and 19.11 °C, conditions propitious to the presence of the pathogen. However, the sporangium concentration was significantly higher in Areeiro, which distinguishes this vineyard from the others based on the PCA results. This variation may be attributable to the higher average values of relative humidity, accumulated rainfall, and wind speed in Areeiro. Furthermore, the Albariño variety cultivated in Areeiro is more susceptible to downy mildew than the Treixadura variety found in Cenlle and Alongos. Boso et al. [45] conducted a study where they confirmed that the Albariño variety grown in Rías Baixas is more susceptible than the Treixadura variety from O Ribeiro. This could be due to Albariño's compact bunches, which can retain moisture and hinder ventilation, while Treixadura bunches are looser.

In Areeiro, the peaks of *P. viticola* sporangium concentration took place during flowering in 2023 and during the ripening of berries in 2024, coinciding with the most critical stages for the development of downy mildew [46]. In this same vineyard, where the highest sporangium concentration in the entire study was recorded (29 May 2023, during GS6), rainfall, daily relative humidity above 85%, and daily average temperatures ranging between 15.99 °C and 17.47 °C were recorded over the three preceding days. The secondary peaks were also characterized by rainfall in the preceding days and daily average temperatures above 18.00 °C, although relative humidity was slightly lower (around 77%). In Alongos and Cenlle, both the maximum concentration peak and the secondary peaks occurred during stage 7 (development of fruit) in both years. In all cases, there were episodes of rainfall, daily relative humidity above 65%, and daily average temperatures exceeding 17.00 °C. These meteorological conditions may have favored the sporulation and viability of the sporangia and thus the increase in their concentrations in the air [47]. In fact, the viability of these asexual propagules is a critical biological and epidemiological trait influencing the development and spread of the disease [48].

Meanwhile, a comparison of concentrations between years shows that although the accumulated rainfall and average relative humidity were generally lower in 2023 than in 2024, the slight increases in temperature (maximum, minimum, and average) in 2023 might have positively influenced the presence of *P. viticola* sporangia in the air. Indeed, the correlations demonstrated that temperature was the meteorological variable that most influenced the sporangium concentrations in the vineyards, especially in Alongos and Cenlle, when analyzing the influence of all variables in the PCA. These findings are consistent with those reported by [26] in the O Ribeiro wine-growing region of northwestern Spain, where variations in the positivity/negativity of correlation were also observed between different study years. For this same region, Bracero et al. [49] demonstrated that the temperatures on preceding days have a greater influence on sporangium concentrations in the air than those on the day of sampling. This finding acts as an aid for the timely prediction of infection in plants, allowing for the efficient use of phytosanitary products.

Among the variables related to humidity, dew point had one of the strongest influences on *P. viticola* sporangium concentration, both in the Spearman correlation analysis and in the PCA for the three vineyards. This aligns with results from other studies in the southern Iberian Peninsula [49]. The positive relationship is due to the fact that dew point generates the necessary humidity on the leaf for zoospores to penetrate the stomata and simultaneously promotes the release of sporangia from the sporangium stalks [50]. This humidity can also be generated by rainfall and the relative humidity on previous days, which generally correlated positively in this study. In Areeiro, the bioclimatic area with the highest average relative humidity values, the PCA revealed that this variable was the one most contributing to sporangium concentrations. In this regard, other authors have pointed out that sporangia are only released during rainfall, which is why they are abundant during these periods [5,51]. By contrast, same-day rainfall decreased sporangium concentrations due to the effect of atmospheric washout [52]. This same effect may explain the negative correlations with both variables in the ultra-oceanic area, where rainfall is more frequent.

Wind speed was not always correlated with the sporangium presence, nor in the same way. In other studies in northwestern Spain, it has been shown that no correlation exists between these variables over the course of the preceding days, although when a correlation is detected, it is positive [5,27]. The PCA confirmed that wind speed played an important role in the dispersion and detection of sporangia in all the vineyards. Cortiñas et al. [26] found that both variables were present in the same component but in opposite directions. Wind disturbance, while aiding in the dispersion of propagules, also promotes mixing and dilution, which may be a likely cause for the inverse relationship in the years when it was detected [53].

The detection of airborne *P. viticola* sporangia is valuable for identifying infection events. Information about the presence of sporangia, combined with an understanding of the meteorological factors that promote their development and, therefore, the spread of the disease, is essential, as it facilitates disease control. In fact, controlling the disease requires a significant amount of plant protection products [54]. Copper compounds, which are commonly used in the vineyards examined in this study to combat the spread of *P. viticola*, are also widely employed in viticulture to reduce sporangium concentrations to acceptable levels [44]. These treatments help to control the spread of the pathogen, thereby minimizing crop damage. Although numerous alternatives to copper-based treatments have been investigated to control *P. viticola*, the issue remains unresolved [55], and copper continues to be one of the most commonly used treatments against downy mildew, despite its environmental impacts due to soil accumulation [56]. However, the use of these products could be reduced by combining aerobiological techniques with phenological studies, enabling the more efficient and precise application of plant protection products [26,27,38].

This is a preliminary study, and more years of research are needed to develop predictive models for sporangium concentrations that would allow for the development of an early warning system. With additional data, more reliable results can be achieved, enabling the adaptation of predictive models to specific bioclimatic zones. This would facilitate early decision-making based on weather forecasts, leading to more effective and sustainable disease management. Consequently, the overall chemical treatment load can be reduced and the environmental impact minimized.

5. Conclusions

Grapevine diseases, such as downy mildew, can diminish or even destroy the harvest of a season, especially in regions where the climatic conditions are favorable for the development of these diseases. The present aerobiological study of *P. viticola* sporangia emphasizes the significance of climatic conditions in vineyards in the northwest of Spain, particularly the role of temperature and rainfall in the days preceding sporangium detection. These factors had a slight but distinct impact on the spread of downy mildew in the two studied bioclimatic zones.

Furthermore, the use of aerobiological samplers that allow for the detection of sporangium concentration, which causes fungal diseases in the vine, is very helpful for implementing an effective integrated pest management strategy. This enables the early detection of the disease, providing enough time for prompt intervention to halt the development of the disease and, therefore, its undesirable effects on the final production.

Supplementary Materials: The following supporting information can be downloaded at: https://www.mdpi.com/article/10.3390/horticulturae11030228/s1, Table S1: Main meteorological parameters (MaxT—maximum temperature, MinT—minimum temperature, AvgT—average temperature, Dp—dew point, Rain—rainfall, RH—relative humidity, and Ws—wind speed) averaged by phenological stage (PS 0—bud development, PS 1—leaf development, PS 5—inflorescence emergence, PS 6—flowering, PS 7—development of fruit, and PS 8—ripening of berries) for the studied grapevine seasons (2023 and 2024); Table S2: Spearman correlation test results comparing Plasmopara viticola concentrations (sporangia/m^3) to recorded meteorological variables (MaxT—maximum, MinT—minimum, and AvgT—average temperature (°C), Rain—rainfall (mm), Ws—wind speed (km/h), RH—relative humidity (%), and Dp—dew point (°C)) (significance level: $p < 0.01$ **, $p < 0.05$ *).

Author Contributions: Conceptualization, M.F.-G. and F.J.R.-R.; methodology, M.F.-G. and M.J.A.; formal analysis, L.C., M.F.-G., M.J.A., K.C.S.E., R.P.O. and F.J.R.-R.; investigation, L.C., K.C.S.E. and R.P.O.; data curation, L.C. and M.F.-G.; writing—original draft preparation, L.C.; writing—review and editing, L.C., M.F.-G., M.J.A., K.C.S.E., R.P.O. and F.J.R.-R.; visualization, L.C. and M.F.-G.; supervision, M.F.-G., M.J.A. and F.J.R.-R.; funding acquisition, F.J.R.-R. All authors have read and agreed to the published version of the manuscript.

Funding: This research received no external funding.

Data Availability Statement: The raw data supporting the conclusions of this article will be made available by the authors on request.

Acknowledgments: Lucía Carrera received a Predoctoral Grant PRE2021-098872, financed by the Ministry of Science and Innovation. Kenia C. Sánchez Espinosa is a beneficiary of the Predoctoral Grant PREUVIGO-23 from the University of Vigo, Spain. This work was supported by the Project "Sustainability of vineyard production: decrease of external inputs, enhance soil biodiversity and increase crop performance" (reference: PID2020-116764RB-I00). The Xunta de Galicia (Spain) provided financial support through recognition as a Competitive Reference Group (ED431C 2023/19, GI-1809 BIOAPLIC), and GISA/BV2 research group.

Conflicts of Interest: The authors declare no conflicts of interest.

References

1. Alston, J.M.; Sambucci, O. Grapes in the World Economy. In *The Grape Genome*; Cantu, D., Walker, M.A., Eds.; Springer: Cham, Switzerland, 2019; pp. 1–24.
2. Fraga, H. Viticulture and Winemaking under Climate Change. *Agronomy* **2019**, *9*, 783. [CrossRef]
3. INFOVI. Información Del Sector Vitivinícola. Available online: https://www.mapa.gob.es/es/agricultura/temas/producciones-agricolas/vitivinicultura/ (accessed on 10 December 2024).
4. Organización Interprofesional del Vino de España (OIVE). *La Relevancia Económica Del Sector Vitivinícola en Galicia*; Organización Interprofesional del Vino de España (OIVE): Madrid, Spain, 2023.
5. Fernández-González, M.; Rodríguez-Rajo, F.J.; Jato, V.; Aira, M.J. Incidence of Fungals in a Vineyard of the Denomination of Origin Ribeiro (Ourense-North-Western Spain). *Ann. Agric. Environ. Med.* **2009**, *16*, 263–271. [PubMed]
6. Rhouma, A.; Hajji-Hedfi, L.; Atallaoui, K.; Chihani-Hammas, N.; Godwin Okon, O. Biology, Diversity, Detection and Management of *Plasmopara viticola* Causing Downy Mildew of Grapevine (*Vitis* spp.). *Asian J. Mycol.* **2024**, *7*, 37–50.
7. Clippinger, J.I.; Dobry, E.P.; Laffan, I.; Zorbas, N.; Hed, B.; Campbell, M.A. Traditional and Emerging Approaches for Disease Management of *Plasmopara viticola*, Causal Agent of Downy Mildew of Grape. *Agriculture* **2024**, *14*, 406. [CrossRef]
8. Caffi, T.; Legler, S.E.; González-Domínguez, E.; Rossi, V. Effect of Temperature and Wetness Duration on Infection by *Plasmopara viticola* and on Post-Inoculation Efficacy of Copper. *Eur. J. Plant Pathol.* **2016**, *144*, 737–750. [CrossRef]
9. Rumbou, A.; Gessler, C. Genetic Dissection of *Plasmopara viticola* Population from a Greek Vineyard in Two Consecutive Years. *Eur. J. Plant Pathol.* **2004**, *110*, 379–392. [CrossRef]
10. Mycobank. Fungal Databases, Nomenclature & Species Banks. Available online: https://www.mycobank.org/ (accessed on 11 December 2024).
11. Valleggi, L.; Carella, G.; Perria, R.; Mugnai, L.; Stefanini, F.M. A Bayesian Model for Control Strategy Selection against *Plasmopara viticola* Infections. *Front. Plant Sci.* **2023**, *14*, 1117498. [CrossRef] [PubMed]
12. Armijo, G.; Schlechter, R.; Agurto, M.; Muñoz, D.; Nuñez, C.; Arce-Johnson, P. Grapevine Pathogenic Microorganisms: Understanding Infection Strategies and Host Response Scenarios. *Front. Plant Sci.* **2016**, *7*, 382. [CrossRef] [PubMed]
13. Boso, S.; Alonso-Villaverde, V.; Gago, P.; Santiago, J.L.; Martínez, M.C. Susceptibility to Downy Mildew (*Plasmopara viticola*) of Different *Vitis* Varieties. *Crop Prot.* **2014**, *63*, 26–35. [CrossRef]
14. Gaforio, L.; Cabello, F.; Muñoz Organero, G. Evaluation of Resistance to Downy Mildew in Grape Varieties Grown in a Spanish Collection. *Vitis J. Grapevine Res.* **2015**, *54*, 187–191.
15. Gouveia, C.; Santos, R.B.; Paiva-Silva, C.; Buchholz, G.; Malhó, R.; Figueiredo, A. The Pathogenicity of *Plasmopara viticola*: A Review of Evolutionary Dynamics, Infection Strategies and Effector Molecules. *BMC Plant Biol.* **2024**, *24*, 327. [CrossRef]
16. Puelles, M.; Arbizu-Milagro, J.; Castillo-Ruiz, F.J.; Peña, J.M. Predictive Models for Grape Downy Mildew (*Plasmopara viticola*) as a Decision Support System in Mediterranean Conditions. *Crop Prot.* **2024**, *175*, 106450. [CrossRef]
17. Ortega Rioja, P. Assessment of Different Strategies to Reduce Environmental Fungicide. Ph.D. Thesis, Universitat Politècnica de Catalunya, Barcelona, Spain, 2024.
18. Silva, V.; Mol, H.G.J.; Zomer, P.; Tienstra, M.; Ritsema, C.J.; Geissen, V. Pesticide Residues in European Agricultural Soils—A Hidden Reality Unfolded. *Sci. Total Environ.* **2019**, *653*, 1532–1545. [CrossRef] [PubMed]
19. Llamazares De Miguel, D.; Mena-Petite, A.; Díez-Navajas, A.M. Toxicity and Preventive Activity of Chitosan, *Equisetum arvense*, Lecithin and Salix Cortex Against *Plasmopara viticola*, the Causal Agent of Downy Mildew in Grapevine. *Agronomy* **2022**, *12*, 3139. [CrossRef]
20. Pertot, I.; Caffi, T.; Rossi, V.; Mugnai, L.; Hoffmann, C.; Grando, M.S.; Gary, C.; Lafond, D.; Duso, C.; Thiery, D.; et al. A Critical Review of Plant Protection Tools for Reducing Pesticide Use on Grapevine and New Perspectives for the Implementation of IPM in Viticulture. *Crop Prot.* **2017**, *97*, 70–84. [CrossRef]
21. Fernandes de Oliveira, A.; Serra, S.; Ligios, V.; Satta, D.; Nieddu, G. Assessing the Effects of Vineyard Soil Management on Downy and Powdery Mildew Development. *Horticulturae* **2021**, *7*, 209. [CrossRef]

22. Mesguida, O.; Haidar, R.; Yacoub, A.; Dreux-Zigha, A.; Berthon, J.Y.; Guyoneaud, R.; Attard, E.; Rey, P. Microbial Biological Control of Fungi Associated with Grapevine Trunk Diseases: A Review of Strain Diversity, Modes of Action, and Advantages and Limits of Current Strategies. *J. Fungi* **2023**, *9*, 638. [CrossRef]
23. Hasanaliyeva, G.; Si Ammour, M.; Yaseen, T.; Rossi, V.; Caffi, T. Innovations in Disease Detection and Forecasting: A Digital Roadmap for Sustainable Management of Fruit and Foliar Disease. *Agronomy* **2022**, *12*, 1707. [CrossRef]
24. Hui, W.; Shuyi, Y.; Wei, Z.; Junbo, P.; Haiyun, T.; Chunhao, L.; Jiye, Y. Modeling the Dynamic Changes in *Plasmopara viticola* Sporangia Concentration Based on LSTM and Understanding the Impact of Relative Factor Variability. *Int. J. Biometeorol.* **2023**, *67*, 993–1002. [CrossRef] [PubMed]
25. Martínez-Bracero, M.; González-Fernández, E.; Wójcik, M.; Alcázar, P.; Fernández-González, M.; Kasprzyk, I.; Rodríguez-Rajo, F.J.; Galán, C. Airborne Fungal Phytopathological Spore Assessment in Three European Vineyards from Different Bioclimatic Areas. *Aerobiologia* **2020**, *36*, 715–729. [CrossRef]
26. Rodríguez, J.A.C.; Fernández-González, E.; Fernández-González, M.; Vázquez-Ruiz, R.A.; Aira, M.J. Fungal Diseases in Two North-West Spain Vineyards: Relationship with Meteorological Conditions and Predictive Aerobiological Model. *Agronomy* **2020**, *10*, 219. [CrossRef]
27. Fernández-González, M.; Piña-Rey, A.; González-Fernández, E.; Aira, M.J.; Rodriguez-Rajo, F.J. First Assessment of Goidanich Index and Aerobiological Data for *Plasmopara viticola* Infection Risk Management in North-West Spain. *J. Agric. Sci.* **2019**, *157*, 129–139. [CrossRef]
28. Blanco-Ward, D.; García Queijeiro, J.M.; Jones, G.V. Spatial Climate Variability and Viticulture in the Miño River Valley of Spain. *Vitis J. Grapevine Res.* **2007**, *46*, 63.
29. Lorenzo, M.N.; Taboada, J.J.; Lorenzo, J.F.; Ramos, A.M. Influence of Climate on Grape Production and Wine Quality in the Rías Baixas, North-Western Spain. *Reg. Environ. Change* **2013**, *13*, 887–896. [CrossRef]
30. Fernández-González, M.; Lara, B.; González-Fernández, E.; Rojo, J.; Pérez-Badia, R.; Rodríguez-Rajo, F.J. *Pinus* Pollen Emission Patterns in Different Bioclimatic Areas of the Iberian Peninsula. *Forests* **2021**, *12*, 688. [CrossRef]
31. Meier, U. Growth Stages of Mono and Dicotyledonous Plants. In *BBCH Monograph*, 2nd ed.; Federal Biological Research Centre for Agriculture and Forestry: Bonn, Germany, 2001; ISBN 9783955470715.
32. Hirst, J.M. An Automatic Volumetric Spore Trap. *Ann. Appl. Biol.* **1952**, *39*, 257–265. [CrossRef]
33. Galán, C.; Cariñanos, P.; Alcázar, P.; Domínguez, E. *Manual de Calidad y Gestión de La Red Española de Aerobiología*; Servicio de Publicaciones de la Universidad de Córdoba: Córdoba, Spain, 2007; ISBN 9788469063545.
34. Galán, C.; Ariatti, A.; Bonini, M.; Clot, B.; Crouzy, B.; Dahl, A.; Fernandez-González, D.; Frenguelli, G.; Gehrig, R.; Isard, S.; et al. Recommended Terminology for Aerobiological Studies. *Aerobiologia* **2017**, *33*, 293–295. [CrossRef]
35. Currey, D.R. Continentality of Extratropical Climates. *Ann. Assoc. Am. Geogr.* **1974**, *64*, 268–280. [CrossRef]
36. Mariani, L.; Failla, O.; Dal Monte, G.; Facchinetti, D. IPHEN: A Model for Real Time Production of Grapevine Phenological Maps. In Proceedings of the Climate and Viticulture Congress, Zaragoza, Spain, 10–14 July 2007; pp. 272–278.
37. Verdugo-Vásquez, N.; Acevedo-Opazo, C.; Valdés-Gómez, H.; Araya-Alman, M.; Ingram, B.; García de Cortázar-Atauri, I.; Tisseyre, B. Spatial Variability of Phenology in Two Irrigated Grapevine Cultivar Growing Under Semi-Arid Conditions. *Precis. Agric.* **2016**, *17*, 218–245. [CrossRef]
38. Cortiñas, J.A.; Fernández-González, M.; Vázquez-Ruiz, R.A.; Aira, M.J.; Rodríguez-Rajo, F.J. The Understanding of Phytopathogens as a Tool in the Conservation of Heroic Viticulture Areas. *Aerobiologia* **2022**, *38*, 177–193. [CrossRef]
39. Santos, J.A.; Malheiro, A.C.; Karremann, M.K.; Pinto, J.G. Statistical Modelling of Grapevine Yield in the Port Wine Region under Present and Future Climate Conditions. *Int. J. Biometeorol.* **2011**, *55*, 119–131. [CrossRef] [PubMed]
40. Vasylenko, O.; Kondratenko, T.; Havryliuk, O.; Andrusyk, Y.; Kutovenko, V.; Dmytrenko, Y.; Grevtseva, N.; Marchyshyna, Y. The Study of the Productivity Potential of Grape Varieties According to the Indicator of Functional Activity of Leaves. *Slovak J. Food Sci.* **2021**, *15*, 639–647. [CrossRef] [PubMed]
41. Parker, A.; de Cortázar-Atauri, I.G.; Chuine, I.; Barbeau, G.; Bois, B.; Boursiquot, J.M.; Cahurel, J.Y.; Claverie, M.; Dufourcq, T.; Gény, L.; et al. Classification of Varieties for Their Timing of Flowering and Veraison Using a Modelling Approach: A Case Study for the Grapevine Species *Vitis vinifera* L. *Agric. For. Meteorol.* **2013**, *180*, 249–264. [CrossRef]
42. Leeuwen, C.; Van Barbe, J.; Garbay, J.; Gowdy, M.; Lytra, G.; Plantevin, M.; Pons, A.; Thibon, C. *La Madurez Aromática Puede Ser El Tipo de Madurez Con Mayor Impacto En La Tipicidad Del Vino Tinto. Parte II: Factores Del Terroir y Prácticas de Manejo Que Afectan La Madurez Aromática*; IVES Technical Reviews, Viña & Vino: Villenave d'Ornon, France, 2023; pp. 1–2.
43. Agrios, G.N. *Plant Pathology*, 3rd ed.; Academic Press: New York, NY, USA, 2012.
44. Gessler, C.; Pertot, I.; Perazzolli, M. *Plasmopara viticola*: A Review of Knowledge on Downy Mildew of Grapevine and Effective Disease Management. *Phytopathol. Mediterr.* **2011**, *50*, 3–44.
45. Boso, S.; Gago, P.; Santiago, J.L.; Fuente, M.D.; La Martínez, M.C. Factors Affecting the Vineyard Populational Diversity of *Plasmopara viticola*. *Plant Pathol. J.* **2019**, *35*, 125–136. [CrossRef] [PubMed]

46. Fernández-González, M.; Escuredo, O.; Rodríguez-Rajo, F.J.; Aira, M.J.; Jato, V. Prediction of Grape Production by Grapevine Cultivar Godello in North-West Spain. *J. Agric. Sci.* **2011**, *149*, 725–736. [CrossRef]
47. Kennelly, M.M.; Gadoury, D.M.; Wilcox, W.F.; Magarey, P.A.; Seem, R.C. Primary Infection, Lesion Productivity, and Survival of Sporangia in the Grapevine Downy Mildew Pathogen *Plasmopara viticola*. *Phytopathology* **2007**, *97*, 512–522. [CrossRef] [PubMed]
48. Hong, C.F.; Scherm, H. A Spectrophotometric Approach for Determining Sporangium and Zoospore Viability of *Plasmopara viticola*. *J. Phytopathol.* **2020**, *168*, 297–302. [CrossRef]
49. Martínez-Bracero, M.; Alcázar, P.; Velasco-Jiménez, M.J.; Galán, C. Fungal Spores Affecting Vineyards in Montilla-Moriles Southern Spain. *Eur. J. Plant Pathol.* **2019**, *153*, 1–13. [CrossRef]
50. Caffi, T.; Legler, S.E.; Rossi, V.; Bugiani, R. Evaluation of a Warning System for Early-Season Control of Grapevine Powdery Mildew. *Plant Dis.* **2012**, *96*, 104–110. [CrossRef]
51. Magyar, D.; Frenguelli, G.; Bricchi, E.; Tedeschini, E.; Csontos, P.; Li, D.W.; Bobvos, J. The Biodiversity of Air Spora in an Italian Vineyard. *Aerobiologia* **2009**, *25*, 99–109. [CrossRef]
52. Albelda, Y.; Rodríguez-Rajo, F.J.; Jato, V.; Aira, M.J. Concentraciones Atmosfericas De Propagulos Fungicos En Viñedos Del Ribeiro (Galicia, España). *Boletín Micológico* **2005**, *20*, 1–8. [CrossRef]
53. Roy, S.; Gupta Bhattacharya, S. Airborne Fungal Spore Concentration in an Industrial Township: Distribution and Relation with Meteorological Parameters. *Aerobiologia* **2020**, *36*, 575–587. [CrossRef]
54. Dagostin, S.; Schärer, H.J.; Pertot, I.; Tamm, L. Are There Alternatives to Copper for Controlling Grapevine Downy Mildew in Organic Viticulture? *Crop Prot.* **2011**, *30*, 776–788. [CrossRef]
55. De La Fuente, M.; Fernández-Calviño, D.; Tylkowski, B.; MMontornes, J.; Olkiewicz, M.; Pereira, R.; Cachada, A.; Caffi, T.; Fedele, G.; De Herralde, F. Alternatives to Cu Applications in Viticulture: How R&D Projects Can Provide Applied Solutions, Helping to Establish Legislation Limits. In *Grapes and Wine*; IntechOpen: London, UK, 2022. [CrossRef]
56. Komárek, M.; Čadková, E.; Chrastný, V.; Bordas, F.; Bollinger, J.C. Contamination of Vineyard Soils with Fungicides: A Review of Environmental and Toxicological Aspects. *Environ. Int.* **2010**, *36*, 138–151. [CrossRef] [PubMed]

horticulturae

MDPI

Article

Occurrence of *Botrytis cinerea* Causing Gray Mold on Pecan in China

Xiang-Rong Zheng [1], Xiao-Xiao Huang [1], Jin-Feng Peng [1], Yusufjon Gafforov [2,3] and Jia-Jia Chen [1,*]

[1] College of Landscape Architecture, Jiangsu Vocational College of Agriculture and Forestry, Zhenjiang 212400, China; zhengxr@jsafc.edu.cn (X.-R.Z.); xiaoxiaohuang@jsafc.edu.cn (X.-X.H.); pjf912530@163.com (J.-F.P.)

[2] Central Asian Center for Development Studies, New Uzbekistan University, Tashkent 100007, Uzbekistan; yugafforov@yahoo.com

[3] Mycology Laboratory, Institute of Botany, Academy of Sciences of Republic of Uzbekistan, Tashkent 100125, Uzbekistan

* Correspondence: jiajiachen@jsafc.edu.cn

Abstract: Pecan (*Carya illinoinensis*), a globally economically significant dried fruit and woody oil tree, faces significant challenges in production and nut quality due to the rampant outbreak of severe fungal diseases. From 2020 to 2021, an extensive occurrence of a disease resembling gray mold was observed on the leaves and fruits of pecan trees in Jiangsu Province, China. Upon isolation from symptomatic samples, *Botrytis cinerea* was identified through morphological analysis and phylogenetic studies of the *G3PDH*, *HSP60*, and *RPB2* gene sequences. Furthermore, pathogenicity tests conclusively attributed the gray mold disease observed on pecan leaves and fruits to *B. cinerea*. To the best of our knowledge, this is the first time that *B. cinerea* has been reported on pecans. These findings thus provide a basis for further research on the management of pecan gray mold.

Keywords: pecan; gray mold; *Botrytis cinerea*; pathogenicity

1. Introduction

Carya illinoinensis, better known as pecan, is a significant species among dried fruit and woody oil trees [1]. Its kernels constitute a superb source of healthful and nutritious food [2]. In recent years, there has been a substantial expansion in pecan cultivation in China, fueled by the escalating demand for nut production. However, this rapid growth may lead to an increased prevalence of diseases within these newly established plantations.

The Sclerotiniaceae genus *Botrytis* includes significant plant pathogens that inflict gray mold on a broad spectrum of economically vital fruits, crops, and ornamentals [3]. Gray mold often occurs as leaf blight [4], flower blight [5], or fruit rot [6], posing significant challenges to agriculture. However, reliance solely on morphological characteristics like conidial spore size and shape is inadequate for the definitive identification of Botrytis species, necessitating the integration of molecular methodologies [7]. Multigene phylogenetic analyses leveraging housekeeping genes, such as RNA polymerase II (*RPB2*), heat shock protein 60 (*HSP60*), and glyceraldehyde-3-phosphate dehydrogenase (*G3PDH*), have emerged as informative tools in accurately distinguishing *Botrytis* species [8].

In June 2020, a widespread occurrence of a disease exhibiting symptoms akin to gray mold was observed on the leaves and fruits of pecan trees in Jurong, Jiangsu Province, China. Approximately 45% of pecan trees suffered from this disease (200 trees were surveyed), significantly hindering the growth of young trees and compromising nut production. Notably, to our current understanding, there is no research regarding the pathogenicity of Botrytis species causing gray mold disease on pecan in China. Therefore, the objectives of the present study were to isolate, identify, and characterize the pathogenic fungi causing gray mold disease on pecan by means of virulence tests, multi-locus phylogenetic analysis, and morphological studies.

2. Materials and Methods

2.1. Field Survey and Fungal Isolation

A field survey was carried out from September to December 2021 in five pecan orchards in Nanjing and Jurong, Jiangsu Province. In each orchard, 25 plants in each of four corners (southwest, southeast, northwest, and northeast) were arbitrarily selected, and the economic loss was estimated for each orchard. In total, 21 diseased samples bearing gray mold were collected. Each sample was composed of five leaves/fruits from the most severely infected trees within one hectare of the orchard, and sampling was performed on approximately 40 hectares. All the samples were transported in an ice box to a laboratory for further analysis.

Leaf tissues (3×3 mm^2) at the border of lesions were surface sterilized (5% NaClO for 45 s, 75% ethanol for 45 s), then placed on a potato dextrose agar (PDA) plate supplemented with ampicillin (100 μg/mL) and incubated at 25 °C in the dark. Emerging fungal hyphae were transferred to fresh PDA and pure culture was obtained by single-spore isolation [9]. All fungal isolates were primarily identified by colony morphology and the ITS sequences.

2.2. Molecular Characterization

Total genomic DNA was isolated from approximately 7-day-old PDA cultures by gently scraping the mycelium with a sterile scalpel and following the manufacturer's protocol for the Fungi Genomic DNA Extraction Kit (Solarbio, Beijing, China). The ITS, RPB2, HSP60, and GAPDH were amplified and sequenced using the primer pairs ITS1/ITS4 [10], RPB2-F/RPB2-R, HSP60-F/HSP60-R, and G3PDH-F/G3PDH-R [11], respectively. PCR was performed in an Eppendorf Nexus Thermal Cycler (Eppendorf, Hamburg, Germany) with a total volume of 50 μL, containing 2 μL of each primer (10 pmol/mL), 4 μL of DNA template (100 ng), 25 μL of 2xEasyTaq PCR SuperMix (Transgen, Beijing, China), and 17 μL of double-distilled H_2O. PCR amplification settings were conducted following previous studies [12]. The PCR products were verified by agarose gel electrophoresis (1.3%), and the positive amplicons were purified and sequenced in both directions by Sangon Biotech (Shanghai) Co., Ltd., Shanghai, China. The quality of the obtained DNA sequences and the contig assembly were performed using Bioedit software v. 7.0.9. All sequences derived in this study have been deposited in GenBank. Sequences with high similarities were downloaded from NCBI and selected for phylogenetic analysis (Table 1).

Table 1. Strains of the *Botrytis* spp. studied in this study and GenBank accessions of the sequences generated.

Species	Strain [a]	Origin	GenBank Accession [b]		
			G3PDH	*HSP60*	*RPB2*
Botrytis aclada	MUCL8415	Germany	AJ704992	AJ716050	AJ745664
B. allii	MUCL403	The Netherlands	AJ704996	AJ716055	AJ745666
B. byssoidea	**MUCL94**	USA	AJ704998	AJ716059	AJ745670
B. californica	X503	USA	KJ937068	KJ937058	KJ937048
B. calthae	MUCL2830	USA	AJ705001	AJ716062	AJ745673
B. caroliniana	**CB15**	USA	JF811584	JF811587	JF811590
B. cinerea	**MUCL87**	The Netherlands	AJ705004	AJ716065	AJ745676
B. cinerea	JSAFC 2180 [c]	China, leaf	*PP955196*	*PP955207*	*PP955218*
B. cinerea	JSAFC 2181	China, leaf	*PP955197*	*PP955208*	*PP955219*
B. cinerea	JSAFC 2186	China, fruit	*PP955198*	*PP955209*	*PP955220*
B. cinerea	JSAFC 2187	China, fruit	*PP955199*	*PP955210*	*PP955221*
B. cinerea	JSAFC 2188 [c]	China, fruit	*PP955200*	*PP955211*	*PP955222*
B. cinerea	JSAFC 2190	China, leaf	*PP955201*	*PP955212*	*PP955223*
B. cinerea	JSAFC 2208	China, leaf	*PP955202*	*PP955213*	*PP955224*
B. cinerea	JSAFC 2213 [c]	China, leaf	*PP955203*	*PP955214*	*PP955225*
B. cinerea	JSAFC 2214	China, leaf	*PP955204*	*PP955215*	*PP955226*
B. cinerea	JSAFC 2223	China, fruit	*PP955205*	*PP955216*	*PP955227*
B. cinerea	JSAFC 2296 [c]	China, fruit	*PP955206*	*PP955217*	*PP955228*
B. convoluta	MUCL11595	USA, The Netherlands	AJ705008	AJ716069	AJ745680
B. croci	MUCL436	The Netherlands	AJ705009	AJ716070	AJ745681

Table 1. *Cont.*

Species	Strain [a]	Origin	GenBank Accession [b]		
			G3PDH	*HSP60*	*RPB2*
B. deweyae	**CBS 134649**	UK	HG799521	HG799519	HG799518
B. elliptica	BE0022	The Netherlands	AJ705010	AJ716071	AJ745682
B. fabae	**MUCL98**	Spain	AJ705014	AJ716075	AJ745686
B. fabiopsis	BC-2	China	EU519211	EU514482	EU514473
B. ficariarum	**CBS 176.63**	Belgium	AJ705015	AJ716076	AJ745687
B. galantina	MUCL435	The Netherlands	AJ705018	AJ716079	AJ745689
B. gladiolorum	MUCL3865	The Netherlands	AJ705020	AJ716081	AJ745692
B. globosa	MUCL444	Belgium	AJ705022	AJ716083	AJ745693
B. hyacinthi	MUCL442	The Netherlands	AJ705024	AJ716085	AJ745696
B. narcissicola	MUCL2120	Canada	AJ705026	AJ716087	AJ745697
B. porri	**MUCL3234**	–	AJ705032	AJ716093	AJ745704
B. prunorum	**Bpru-1.9**	Chile	KP339985	KP339999	KP339992
B. pseudocinerea	X004	USA	KJ796651	KJ796655	KJ796647
B. pyriformis	**SedsarBC-1**	China	KJ543484	KJ543488	KJ543492
B. ranunculi	**CBS178.63**	USA	AJ705034	AJ716095	AJ745706
B. sinoallii	**OnionBC-23**	China	EU519217	EU514488	EU514479
B. sinoviticola	GBC-5	China	JN692413	JN692399	JN692427
B. sphaerosperma	MUCL21481	UK	AJ705035	AJ716096	AJ745708
B. squamosa	**MUCL1107**	USA	AJ705037	AJ716098	AJ745710
B. tulipae	BT9001	The Netherlands	AJ705040	AJ716101	AJ745712
Monilinia fructigena	9201	The Netherlands	AJ705043	AJ716047	AJ745715

[a] Culture numbers in bold type represent ex-type or other authentic specimens. *Monilinia fructigena* (9201) was added as an outgroup. [b] Sequences in italics were generated in this study. [c] Isolates used for pathogenicity tests.

The phylogenetic relationships among the specimens were thoroughly analyzed utilizing two distinct optimality search methodologies: maximum likelihood (ML) and Bayesian inference (BI). For ML analysis, the GTR+G+I evolutionary model was employed within MEGA v. 7 [13], and the stability of the inferred clades was rigorously assessed through 1000 bootstrap replications, with gaps treated as missing data. In contrast, for BI analysis, MrBayes v. 3.2.6 [14] was utilized, preceded by determining the optimal model for each locus using MrModeltest 2.3 based on the corrected Akaike Information Criterion (AICc). Specifically, TN93+G+I was selected for G3PDH, K2+G for HSP60, and GTR+G+I for RPB2. The BI analysis involved simultaneously running four Markov chains over 1×10^7 generations, with samples collected every 1000 generations. To ensure robustness, the initial 25% of generations were discarded as burn-in, and posterior probabilities (PP) were subsequently calculated for the majority rule consensus tree. Additionally, *Monilinia fructigena* (strain 9201) was designated as the outgroup to provide a comparative framework for the phylogenetic analysis.

2.3. Morphological Identification

For macroscopic and microscopic characterization, fungal isolates were cultured on PDA at 25 °C. Colony appearance (color, texture, and pigment production), radial growth rate, conidiomata presence, and the formation pattern as well as abundance were assessed 5, 10, and 20 days post-inoculation (dpi). The morphological characteristics of conidiophores, conidia, sclerotia, and conidiogenous cells were observed under a Zeiss Axio Imager A2m microscope (Carl Zeiss Ltd.; Baden-Württemberg; Germany). At least 50 measurements were made for each fungal structure, using an ocular micrometer. The colony growth rate was estimated by measuring two perpendicular directions on three replicates per isolate and calculating the average value for each isolate. Fungal isolates and specimens used in the present study were deposited in Jiangsu Vocational College Agriculture and Forestry (JSAFC).

2.4. Pathogenicity Studies

Two representative isolates (JSAFC 2180 and JSAFC 2213) from leaf spots and two (JSAFC 2188 and JSAFC 2296) from fruit rot of pecan were selected to test for pathogenicity.

Spore suspensions of each isolate were prepared by flooding 15-day-old PDA cultures with sterile water and adjusting the collected suspension to 10^6 spores/mL using a hemacytometer. Prior to inoculation, leaves and fruits were harvested from symptomless pecan trees and surface sterilized as described above. One wound was made on each leaf and fruit (0.2 cm in depth) by stabbing with a sterilized insect pin (0.71 mm in diameter). A conidial suspension (20 μL of 10^6 spores/mL) was placed on each wound site, with each isolate applied to ≥5 leaves or fruits. Control leaves and fruits were mock-inoculated with sterile water. The experiment was carried out three times.

The inoculated leaves and fruits were carefully placed in transparent plastic boxes lined with sterile filter papers and incubated at 25 °C under a controlled 12-hour photoperiod. To fulfill Koch's postulates, fungal isolates were reisolated from symptomatic leaves and fruits, and their identities were validated through both phylogenetic and phenotypic comparisons. Disease incidence was meticulously recorded as the percentage of inoculated leaves or fruits that exhibited leaf spot symptoms out of the total inoculated samples. Furthermore, disease severity was assessed one week post-inoculation by precisely measuring lesion length in two perpendicular directions. To identify statistically significant differences in severity among species, Tukey's Honestly Significant Difference (HSD) test was employed at a significance level of $p < 0.05$ using IBM SPSS Statistics 24.0 software (SPSS, Inc., Chicago, IL, USA).

3. Results

3.1. Field Symptoms, Loss Survey, and Fungal Isolation

The disease occurs on leaves and fruits of many pecan varieties, including Bigenyuan No. 1, Bigenyuan No. 3, and Mahan, in Nanjing and Jurong, Jiangsu Province, China. The initial symptoms were water-soaked and brown to dark brown spots scattered on infected leaves, and part of them were surrounded by yellowish halos (Figure 1A). Gradually, several lesions expanded and fused to form large necrosis as symptoms progressed (Figure 1B). When the fruits were infected, symptoms initially appeared as sunken, sub-circular or irregular, tan to black rot, which gradually enlarged and coalesced to form a large necrotic area (Figure 1C). In the later stage, the lesion reached the immature nut shell, resulting in premature drop (Figure 1D,E). Finally, the kernels became inedible. A survey in Jurong, Jiangsu Province, revealed that the disease breaks out annually and the affected area is often over 10 hectares, causing approximately 40% economic loss. In total, eleven single-spored isolates were recovered from diseased samples, and all the isolates showed consistent morphology in culture on PDA.

Figure 1. Photographs showing disease symptoms of pecan gray mold in the field. (**A**,**B**) The symptoms of leaves. (**C**–**E**) The symptoms of fruits.

3.2. Molecular Identification and Phylogenetic Analysis

For the purpose of molecular identification, fragments from three loci—*G3PDH*, *HSP60*, and *RPB2*—were successfully amplified from all eleven isolates under investigation (Table 1). The PCR products exhibited varying lengths, with *G3PDH* ranging from 913 to 959 bp, *HSP60* from 1028 to 1082 bp, and *RPB2* from 1181 to 1217 bp. Upon alignment, the concatenated dataset comprising these three loci (*G3PDH* + *HSP60* + *RPB2*) totaled 2958 nucleotides, of which 2289 were constant sites, 295 were parsimony uninformative sites, and 374 were parsimony informative sites. Both Bayesian Inference (BI) and Maximum Likelihood (ML) phylogenetic analyses revealed a compelling result: the eleven isolates studied, including the type specimen of *B. cinerea*, formed a strongly supported monophyletic group (with a bootstrap value of 94 and posterior probabilities of 1.00) and were clearly distinguished from the other *Botrytis* species (Figure 2).

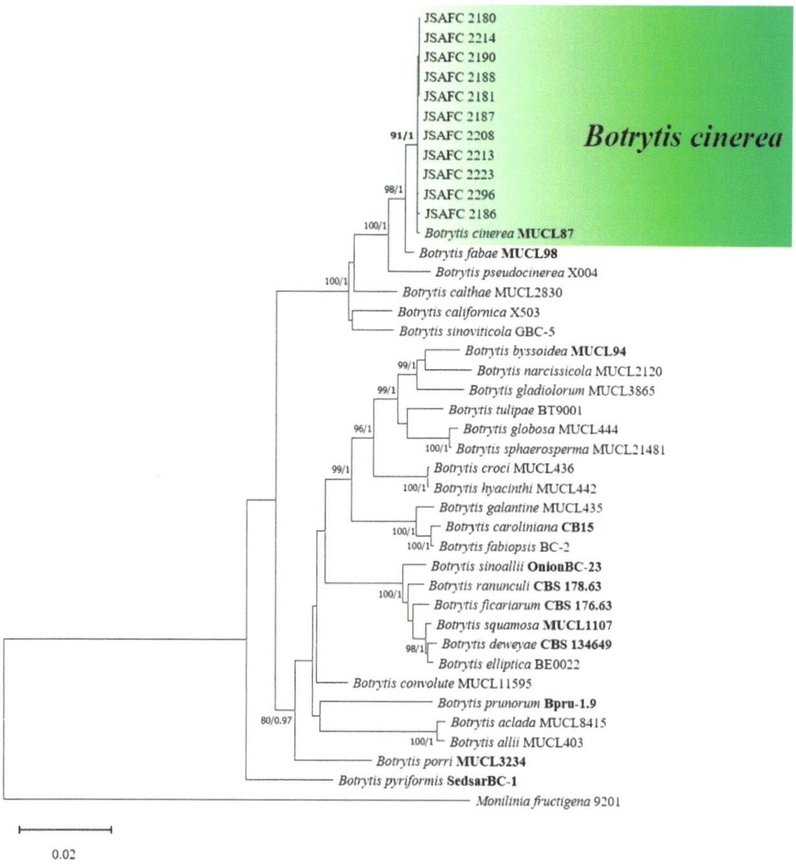

Figure 2. Maximum likelihood (ML) tree of *Botrytis* spp. based on combined *G3PDH*, *HSP60*, and *RPB2* loci. The tree generated by Bayesian inference had a similar topology. Bootstrap values (≥85%) and posterior probabilities (≥0.95) are shown at the nodes. Ex-type or other authoritative cultures are emphasized in bold font. *Monilinia fructigena* (9201) was used as an outgroup. The scale bar indicates the average number of substitutions per site.

3.3. *Morphology*

All eleven isolates examined produced rapidly growing colonies with an average growth rate of 2.55 cm/day at 25 °C. No obvious differences in morphology were observed among these isolates. The colonies initially formed fluffy white hyphae on PDA and then turned gray with scarce to abundant sporulation after 2 weeks of incubation. The conidiophores were erect, pale brown, septate, and branched, with sizes ranging from 483.5 to 1982.6 µm in length by 5.3 to 19.6 µm in width (average = 954.2 × 11.1 µm) (Figure 3). The conidia were unicellular, hyaline to light brown, and ellipsoid to ovoid, with the length and width ranging from 7.2 to 14.1 × 4.0 to 6.2 µm (average = 9.6 µm × 5.8 µm) (Figure 3). No sclerotia was observed among those isolates. On the base of the molecular and morphological characteristics, the fungal isolates recovered from the symptomatic samples of pecan were identified as *B. cinerea* Pers.

Figure 3. Morphological characteristics and pathogenicity (wounded inoculation) of *Botrytis cinerea* (isolate JSAFC 2188) from pecan. (**A**) Photograph showing colony morphology. (**B**) Photograph showing conidiophores with conidia. (**C**) Photograph showing conidia. (**E**) Photograph showing symptom on leaf and (**G**) fruit 7 days after inoculation. (**D**,**F**) Mock control. Scale bars (**B**,**C**) = 10 µm.

3.4. *Pathogenicity Tests*

All representative isolates were pathogenic to leaves and fruits of pecan, regardless of whether the wounded inoculation or non-wounded inoculation method was used. However, wound inoculation resulted in larger lesion sizes compared to non-wound inoculation, suggesting that wounds may be an important pathway for the invasion of *B. cinerea*. The pathogens were successfully re-isolated from symptomatic tissues but not from healthy (controls). Inoculation of pecan leaves with a spore suspension of *B. cinerea* isolates resulted in water-soaked lesions followed by necrotic rot (Figure 3). All representative isolates are pathogenic to the leaves of pecan (Table 2). Differences in the diameter of the necrotic lesion were obtained, being JSAFC 2188 the most virulent isolate on both wounded and non-wounded leaves. Pecan fruit inoculations resulted in dark brown, soft, watery decay around the inoculation point at 7 dpi; consistently, JSAFC 2188 was the most virulent isolate ($p < 0.05$). In contrast, mock-inoculated controls remained asymptomatic, exhibiting only faint discoloration around the wound site. The successful re-isolation of the pathogen from all inoculated tissues and its absence from mock-inoculated controls underscores the fulfillment of Koch's postulates.

Table 2. Pathogenicity of *Botrytis cinerea* isolates on *Carya illinoinensis* 7 days after inoculation.

Isolates	Disease Incidence (%) [a]				Lesion Diameter (mm) [b]			
	Leaves		Fruits		Leaves		Fruits	
	Wounded	Non-Wounded	Wounded	Non-Wounded	Wounded	Non-Wounded	Wounded	Non-Wounded
JSAFC 2180	100.0 ± 0.0	100.0 ± 0.0	86.7 ± 9.4	73.3 ± 9.4	23.8 ± 4.7 b	18.6 ± 4.3 ab	8.3 ± 3.3 b	6.5 ± 2.1 b
JSAFC 2188	100.0 ± 0.0	100.0 ± 0.0	100.0 ± 0.0	86.7 ± 18.9	32.4 ± 7.0 a	21.6 ± 6.2 a	11.4 ± 2.6 a	10.2 ± 3.4 a
JSAFC 2213	100.0 ± 0.0	100.0 ± 0.0	86.7 ± 18.9	66.7 ± 9.4	24.6 ± 4.1 b	17.8 ± 5.1 ab	8.4 ± 2.1 b	5.9 ± 1.9 b
JSAFC 2296	100.0 ± 0.0	86.7 ± 9.4	66.7 ± 9.4	26.7 ± 9.4	21.0 ± 4.6 b	15.4 ± 3.5 b	7.6 ± 1.6 b	3.5 ± 1.2 c
Mock control	0.0 ± 0.0	0.0 ± 0.0	0.0 ± 0.0	0.0 ± 0.0	0.0 ± 0.0 c	0.0 ± 0.0 c	0.0 ± 0.0 c	0.0 ± 0.0 d

[a] Percent disease incidence following inoculation of leaves and fruits of pecan. Data represent the mean of 3 replications, each with ≥5 leaves or fruits, respectively. [b] Lesion diameter following inoculation of leaves and fruits of pecan. Data represent the mean of 3 replications, each with ≥5 leaves or fruits, respectively. Data followed by different letters in each column are significantly different based on Tukey's tests at the $p < 0.05$ level.

4. Discussion

In the present study, a collection of eleven isolates of *B. cinerea* was systematically acquired from instances of gray mold disease affecting pecan trees in Jurong, Jiangsu Province, China. Remarkably, all of these isolates consistently exhibited a morphological phenotype that is characteristic of *B. cinerea*, as described by Mirzaei et al. [15]. While phenotypic and genotypic diversity among *B. cinerea* isolates has been extensively explored in diverse geographical locations worldwide, a comprehensive analysis specifically focusing on *B. cinerea* isolates causing gray mold disease in pecan trees remains elusive in the published literature to date.

Despite the widespread perception of *Botrytis* species as aggressive plant pathogens, the *Botrytis* population displays remarkable diversity, encompassing isolates with various morphotypes and virulence levels that coexist within the same host. Notably, this population includes mycelial-type isolates that lack the ability to initiate infections [16]. This phenotypic variability is potentially further influenced by the presence of transposable elements and/or mycoviruses within the fungal genome, which may modulate fitness attributes, particularly virulence, among *Botrytis* isolates [17]. In our study, all isolated strains of *B. cinerea* proved to be pathogenic, causing necrotic lesions on pecan leaves and manifesting as rot symptoms on the fruits. Among these, JSAFC 2188 emerged as the most virulent isolate. Interestingly, despite no apparent morphological differences among representative isolates, their pathogenicity varies significantly, particularly in the case of JSAFC 2296, which exhibits relatively low pathogenicity when inoculated onto leaves or fruits. This may be related to the mycoviruses it carries. More profoundly, given the importance of *B. cinerea* in agricultural and forestry production, subsequent research could aim to isolate the mycoviruses from hypovirulent strains and attempt to apply them in the control of gray mold disease.

Botrytis cinerea boasts an exceptionally broad host range, encompassing 616 genera and exceeding 1600 plant species, solidifying its position as the second most significant phytopathogen globally, both scientifically and economically [3,18]. This fungus is notorious for causing gray mold, a pervasive and devastating disease that jeopardizes the yields of numerous fruits, vegetables, and medicinal plants [19–21]. Given its potential as a source of inoculum for gray mold in grapes, pears, and cherries in China [22–24], the threat posed by *B. cinerea* is particularly acute. In Jiangsu Province, where cherry orchards, pear orchards, and grape vineyards are often situated in close proximity to commercial pecan orchards, the risk of cross-infection between these hosts escalates significantly. This proximity facilitates the spread of the pathogen, underscoring the need for vigilant monitoring and management strategies to mitigate the devastating impacts of *B. cinerea* on agricultural productivity.

5. Conclusions

In conclusion, our comprehensive analysis, incorporating symptom observation, cultural and microscopic characterization, molecular data analysis, and pathogenicity testing, definitively identified *B. cinerea* as the fungal pathogen responsible for gray mold disease on pecan. Notably, this represents the first documented case of *B. cinerea* causing gray mold in *C. illinoinensis* in China. These findings contribute valuable insights to the knowledge base on gray mold on pecan and highlight the need for targeted control measures to mitigate its impact on agricultural productivity.

Author Contributions: Conceptualization, J.-J.C.; methodology, X.-R.Z.; software, X.-R.Z.; validation, X.-X.H.; formal analysis, X.-X.H.; investigation, X.-X.H.; resources, X.-R.Z.; data curation, J.-F.P.; writing—original draft preparation, X.-R.Z.; writing—review and editing, X.-R.Z. and X.-X.H.; visualization, X.-R.Z. and J.-F.P.; supervision, J.-J.C. and Y.G.; project administration, J.-J.C.; funding acquisition, X.-R.Z. and X.-X.H. All authors have read and agreed to the published version of the manuscript.

Funding: This research was funded by the Science Fund of the Jiangsu Vocational College of Agriculture and Forestry (2023kj25, 2024rc55), Zhenjiang Innovation Capacity Building Project (SS2024010), and the Natural Science Foundation of the Jiangsu Higher Education Institutions of China (20KJB220003).

Institutional Review Board Statement: Not applicable.

Informed Consent Statement: Not applicable.

Data Availability Statement: The original contributions presented in the study are included in the article, further inquiries can be directed to the corresponding author.

Conflicts of Interest: The authors declare no conflicts of interest.

References

1. Zhang, R.; Peng, F.R.; Li, Y.R. Pecan production in China. *Sci. Hortic.* **2015**, *197*, 719–727. [CrossRef]
2. Zhang, C.C.; Ren, H.D.; Yao, X.H.; Wang, K.L.; Chang, J.; Shao, W.Z. Metabolomics and transcriptomics analyses reveal regulatory networks associated with fatty acid accumulation in Pecan kernels. *J. Agric. Food Chem.* **2022**, *70*, 16010–16020. [CrossRef] [PubMed]
3. Singh, R.; Caseys, C.; Kliebenstein, D.J. Genetic and molecular landscapes of the generalist phytopathogen. *Mol. Plant Pathol.* **2024**, *25*, e13404. [CrossRef] [PubMed]
4. Yang, D.; Yao, J.W.; Wang, B.B.; Zheng, J.L.; Cao, C.X.; Huang, D.Y. First report of *Botrytis cinerea* causing gray mold on tea (*Camellia sinensis*) in China. *Plant Dis.* **2024**, *108*, 203. [CrossRef]
5. Xiao, G.L.; Zhang, Q.H.; Zeng, X.G.; Chen, X.Y.; Liu, S.J.; Han, Y.C. Deciphering the molecular signatures associated with resistance to *Botrytis cinerea* in strawberry flower by comparative and dynamic transcriptome analysis. *Front. Plant Sci.* **2022**, *13*, 888939. [CrossRef]
6. Dwivedi, M.; Singh, P.; Pandey, A.K. Botrytis fruit rot management: What have we achieved so far? *Food Microbiol.* **2024**, *122*, 104564. [CrossRef]
7. Ferrada, E.E.; Latorre, B.A.; Zoffoli, J.P.; Castillo, A. Identification and characterization of *Botrytis* Blossom Blight of Japanese Plums caused by *Botrytis cinerea* and *B. prunorum* sp. nov in Chile. *Phytopathology* **2016**, *106*, 155–165. [CrossRef]
8. Moparthi, S.; Parikh, L.P.; Troth, E.G.; Burrows, M.E. Identification and prevalence of seedborne *Botrytis* spp. in Dry Pea, Lentil, and Chickpea in Montana. *Plant Dis.* **2023**, *107*, 382–392. [CrossRef]
9. Wan, Y.; Li, D.W.; Si, Y.Z.; Li, M.; Huang, L.; Zhu, L.H. Three new species of *Diaporthe* causing leaf blight on *Acer palmatum* in China. *Plant Dis.* **2023**, *107*, 849–860. [CrossRef]
10. White, T.; Bruns, T.; Lee, S.; Taylor, J.; Innis, M.; Gelfand, D.; Sninsky, J. Amplification and direct sequencing of fungal ribosomal RNA genes for phylogenetics. In *PCR Protocols: A Guide to Methods and Applications*; Academic Press: Cambridge, MA, USA, 1990; pp. 315–322.
11. Staats, M.; van Baarlen, P.; van Kan, J.A.L. Molecular phylogeny of the plant pathogenic genus *Botrytis* and the evolution of the specificity. *Mol. Biol. Evol.* **2005**, *22*, 333–346. [CrossRef]

12. Zhong, S.; Zhang, J.; Zhang, G.Z. *Botrytis polyphyllae*: A new *Botrytis* species causing gray mold on *Paris polyphylla*. *Plant Dis.* **2019**, *103*, 1721–1727. [CrossRef] [PubMed]
13. Kumar, S.; Stecher, G.; Tamura, K. MEGA7: Molecular Evolutionary Genetics Analysis Version 7.0 for Bigger Datasets. *Mol. Biol. Evol.* **2016**, *33*, 1870–1874. [CrossRef] [PubMed]
14. Ronquist, F.; Teslenko, M.; van der Mark, P.; Ayres, D.L.; Darling, A.; Höhna, S.; Larget, B.; Liu, L.; Suchard, M.A.; Huelsenbeck, J.P. MrBayes 3.2: Efficient Bayesian Phylogenetic Inference and model choice across a large model space. *Syst. Biol.* **2012**, *61*, 539–542. [CrossRef]
15. Mirzaei, S.; Goltapeh, E.M.; Shams-Bakhsh, M.; Safaie, N. Identification of *Botrytis* spp. on plants grown in Iran. *J. Phytopathol.* **2008**, *156*, 21–28. [CrossRef]
16. Morel, W.A.; Marques-Costa, T.M.; Santander-Gordón, D.; Fernández, F.A.; Zabalgogeazcoa, I.; de Aldana, B.R.V.; Sukno, S.A.; Díaz-Mínguez, J.M.; Benito, E.P. Physiological and population genetic analysis of *Botrytis* field isolates from vineyards in Castilla y León, Spain. *Plant Pathol.* **2019**, *68*, 523–536. [CrossRef]
17. Potgieter, C.A.; Castillo, A.; Castro, M.; Cottet, L.; Morales, A. A wild-type *Botrytis cinerea* strain co-infected by double-stranded RNA mycoviruses presents hypovirulence-associated traits. *Virol. J.* **2013**, *10*, 220. [CrossRef]
18. Dean, R.; Van Kan, J.A.L.; Pretorius, Z.A.; Hammond-Kosack, K.E.; Di Pietro, A.; Spanu, P.D.; Rudd, J.J.; Dickman, M.; Kahmann, R.; Ellis, J.; et al. The Top 10 fungal pathogens in molecular plant pathology. *Mol. Plant Pathol.* **2012**, *13*, 414–430. [CrossRef]
19. Bi, K.; Liang, Y.; Mengiste, T.; Sharon, A. Killing softly: A roadmap of *Botrytis cinerea* pathogenicity. *Trends Plant Sci.* **2023**, *28*, 211–222. [CrossRef]
20. Romanazzi, G.; Smilanick, J.L.; Feliziani, E.; Droby, S. Integrated management of postharvest gray mold on fruit crops. *Postharvest Biol. Technol.* **2016**, *113*, 69–76. [CrossRef]
21. Latorre, B.A.; Elfar, K.; Ferrada, E.E. Gray mold caused by *Botrytis cinerea* limits grape production in Chile. *Cienc. Investig. Agrar.* **2015**, *42*, 305–330. [CrossRef]
22. Yin, W.X.; Adnan, M.; Shang, Y.; Lin, Y.; Luo, C.X. Sensitivity of *Botrytis cinerea* from Nectarine/Cherry in China to six fungicides and characterization of resistant isolates. *Plant Dis.* **2018**, *102*, 2578–2585. [CrossRef] [PubMed]
23. Zhang, Y.; Li, X.; Shen, F.; Xu, H.; Li, Y.; Liu, D. Characterization of *Botrytis cinerea* isolates from grape vineyards in China. *Plant Dis.* **2018**, *102*, 40–48. [CrossRef] [PubMed]
24. Zhang, M.; Wu, H.Y.; Wang, X.J.; Sun, B. First report of *Botrytis cinerea* causing fruit rot of *Pyrus sinkiangensis* in China. *Plant Dis.* **2014**, *98*, 281. [CrossRef] [PubMed]

horticulturae

Article

Exiguobacterium acetylicum Strain SI17: A Potential Biocontrol Agent against *Peronophythora litchii* Causing Post-Harvest Litchi Downy Blight

Shilian Huang [1], Xinmin Lv [1], Li Zheng [2,*] and Dongliang Guo [1,*]

1 Institute of Fruit Tree Research, Guangdong Academy of Agricultural Sciences; Key Laboratory of South Subtropical Fruit Biology and Genetic Resource Utilization, Ministry of Agriculture and Rural Affairs; Guangdong Provincial Key Laboratory of Science and Technology Research on Fruit Trees, Guangzhou 510640, China; shiil_huang@163.com (S.H.); lvxinmin@gdaas.cn (X.L.)
2 Key Laboratory of Green Prevention and Control on Fruits and Vegetables in South China of Ministry of Agriculture and Rural Affairs, Innovative Institute for Plant Health, College of Agriculture and Biology, Zhongkai University of Agriculture and Engineering, Guangzhou 510225, China
* Correspondence: bluestar183@163.com (L.Z.); guodongliang@gdaas.cn (D.G.); Tel.: +86-20-38765541 (D.G.)

Abstract: Litchi downy blight (LDB) caused by *Peronophythora litchii* destroys 20–30% of litchi fruit every year and causes significant economic losses. Some *Exiguobacterium* strains exhibit considerable promise in both agricultural and industrial sectors. *E. acetylicum* SI17, isolated from the litchi fruit carposphere, demonstrated significant biocontrol activity against LDB through pre-harvest treatment. To elucidate its underlying regulatory mechanisms, the genome of SI17 was sequenced and analyzed, revealing a circular chromosome spanning 3,157,929 bp and containing 3541 protein-coding genes and 101 RNA genes. Notably, 94 genes were implicated in the production of secondary metabolites. Among the 29 *Exiguobacterium* strains so far sequenced, SI17 possessed the largest genome. In the phylogenomic analysis encompassing the entire genome, SI17 was clustered into Group I. Treating litchi fruit with SI17 before harvesting resulted in a decrease in H_2O_2 content in the fruit peel and an increase in superoxide dismutase activity, thus enhancing resistance to LDB. Interestingly, SI17 did not display plate antagonism against *Peronophythora litchii* SC18. It can be inferred that SI17 generates secondary metabolites, which enhance litchi's resistance to LDB. This study represents the first documentation of an *Exiguobacterium* strain exhibiting a role in litchi plant disease and showcasing significant potential for the biological control of LDB.

Keywords: *Exiguobacterium acetylicum* SI17; litchi downy blight; genome sequencing; defense activation; biocontrol efficacy

1. Introduction

The litchi fruit (*Litchi chinensis*) belongs to the Sapindaceae family and is typically found in subtropical areas of Asia, South Africa, Australia, the United States (especially Hawaii), and Israel [1]. Originating from South China, it is predominantly cultivated in the southern provinces of Fujian, Guangdong, and Hainan. Unfortunately, harvested litchi fruit are highly susceptible to spoilage due to fungal diseases caused by pathogens such as *Peronophythora litchii*, *Geotrichum candidum*, and *Colletotrichum gloeosporiodes*, resulting in a 20–30% loss. Among these, litchi downy blight (LDB), caused by *P. litchii*, is a significant issue [2]. *P. litchii* infects litchi tender leaves, flowers, and mature fruits in the field and post-harvest storage. This hemibiotrophic, homothallic, and polycyclic oomycete produces oospores spherical in shape with smooth walls and a light-yellow color. Oospores can survive in the soil or plant debris for several years and germinate directly or indirectly under favorable conditions. Lemon-shaped deciduous sporangia with papillae at the apex are also produced. Sporangium, wind or water dispersal, germinates directly, producing germ tubes and penetration hyphae able to penetrate the receptive organs. Alternatively, sporangia can form secondary sporangia or release zoospores. Zoospores, kidney shaped, have two lateral flagella, and rapidly germinate in cool, wet conditions. *P. litchii* causes withering and watery brown spots on the infection sites of tender leaves or fruit and produces downy white sporangiophores, which trigger pre- and post-harvest fruit decay [3]. LDB also affects the growth and development of young leaves, although it is most detrimental during the period from bloom to fruit ripening, particularly in extremely humid and rainy conditions [4]. At present, the prevailing methods for managing LDB largely rely on post-harvest applications of chemical fungicides (like dimethomorph and propamocarb). Regrettably, these residues may persist in fruits, posing carcinogenic hazards to consumers and contributing to environmental pollution [5]. Hence, it is crucial and urgent to explore alternative techniques like biological control to minimize post-harvest decay of litchi fruit. This method provides safe, environmentally friendly, and sustainable results, in line with the goals of global strategic initiatives for eco-conscious and/or organic farming practices [6].

In recent years, there has been rapid progress in utilizing biological control agents (BCAs) to manage post-harvest diseases [7]. These applications have exhibited effectiveness in suppressing *P. litchii* and preventing decay in litchi fruit. Among the BCAs employed, there are various antagonistic bacteria like *Bacillus subtilis* [1,8,9], endophytic strains such as *B. amyloliquefaciens* TB2 and LY-1 [10,11], and *Lactobacillus plantarum* LAB [12]. Additionally, certain metabolites have demonstrated antagonistic effects against *P. litchii*, including hypothemycin from *Paecilomyces* sp. SC0924 [2], zeamines from *Dickeya zeae* [13], and 4-ethylphenol from *Streptomyces fimicarius* BWL-H1 [14]. The primary mechanisms through which BCAs inhibit diseases include the synthesis of antibiotics, release of extracellular enzymes such as protease, chitinase, cellulase, and β-1,3-glucanase, production of siderophores, and generation of indole acetic acid (IAA). They also occupy ecological niches, enhance microbial diversity in the phyllosphere and carposphere [15], and prompt plant defense responses such as systemic acquired resistance, induced systemic resistance, and/or priming [16].

Exiguobacterium exhibits significant promise for utilization across various sectors like industry and agriculture, owing to the presence of certain strains that possess diverse functionalities such as enzyme production, bioremediation, degradation of toxic substances, and facilitation of plant growth [17]. For instance, *E.* sp. VSG-1 produces hydrolytic enzymes like cellulase, pectinase, mannanase, xylanase, and tannase, which bolster the conversion of fermented sugar from sugarcane bagasse into bioethanol [18]. Additionally, *E. profundum*, known for its protease production ability, has been employed in the extraction of chitin from shrimp shells [19]. Furthermore, various other hydrolytic enzymes like lipase, amylase, and pullulanase have been isolated from different *Exiguobacterium* strains, further expanding their potential applications [20–23]. *E. alkaliphilum* B-3531D, *E.* sp. AO-11, and *E. mexicanum* M7 exhibit the capacity to metabolize crude oil, suggesting promise for the development of bioremediation solutions for oil-contaminated marine ecosystems [24–26]. Additionally, four *Exiguobacterium* isolates have demonstrated the ability to reduce Cr (VI) levels in polluted environments [27–30]. Moreover, *Exiguobacterium* exhibits significant potential for agricultural usage, with several strains possessing traits that promote plant growth across various crops [31–35]. While disease control is vital in agriculture, there have been no reported applications of *Exiguobacterium* in plant disease management thus far.

The genus *Exiguobacterium* includes species and strains engaged in industry and agriculture, comprising enzyme production, bioremediation, degradation of toxic substances, and plant growth-promoting properties [17]. Cellulase, pectinase, mannanase, xylanase, and tannase produced by *E.* sp. VSG-1 make steam-exploded sugarcane bagasse useful to biofuel fabrication by *Saccharomyces cerevisiae* fermentation [18]. A protease-producing strain of *E. profundum* has been employed in the extraction of chitin from shrimp shells [19]. Lipase, amylase, and pullulanase have been isolated from different *Exiguobacterium* strains, increasing their potential applications [20–23]. *Exiguobacterium* sp. AO-11, *E. alkaliphilum*

B-3531D, and *E. mexicanum* M7 metabolize crude oil, suggesting promise for the bioremediation of oil-contaminated marine ecosystems [24–26]. *Exiguobacterium* strains ZM-2, GS1, PY14, and *E. mexicanum* from chromite mines reduce Cr (VI) levels in polluted environments [27–30]. Moreover, several *Exiguobacterium* strains exhibit significant potential for agricultural applications as plant growth promoting bacteria [31–35]. *Exiguobacterium* has been reported to suppress fungal diseases of cereal crops in Australia [36], and *E. acetylicum* strain 1P MTCC 8707 inhibited the in vitro growth of *Rhizoctonia solani*, *Sclerotium rolfsii*, Pythium sp., and *Fusarium oxysporum* [37]. No applications of *Exiguobacterium* in pre- and post-harvest plant disease management have been reported thus far.

The genomes of several *Exiguobacterium* strains have been sequenced. Based on the concatenated single-copy core genes and 16S rRNA gene sequences, the *Exiguobacterium* strains were clustered into two groups. Group I comprised strains capable of thriving at temperatures of 7 °C or lower, while Group II encompassed the remaining strains [38,39]. The proliferation of transporters, crucial for conveying vital substrates and conferring resilience against environmental stresses, was a key factor propelling the genome expansion in Group I strains, thus broadening their ecological niche [39]. Additionally, genome sequencing enables us to elucidate the agricultural utility of these strains.

E. acetylicum SI17, sourced from the carposphere of litchi fruit, has demonstrated efficacy in combating LDB through biocontrol measures [40]. The volatile organic compounds (VOCs) emitted by SI17, particularly those containing α-Farnesene (AF), exhibited the ability to hinder the growth of *P. litchii*, thus alleviating the severity of LDB [40]. The purpose of the study was to gain deeper insights into the attributes of SI17. We carried out genome sequencing and comparative analysis of its genome.

2. Materials and Methods

2.1. *E. acetylicum SI17 Bioinformatics Analysis*

2.1.1. Bacterial Culture and DNA Extraction

E. acetylicum SI17 was isolated from the interior of litchi fruit sarcocarp. The cells of *E. acetylicum* SI17 were cultured in a 500 mL flask containing 100 mL of Luria–Bertani (LB) medium (tryptone 10 g/L, yeast extract 5 g/L, and NaCl 10 g/L) at 28 °C with shaking at 200 rpm for 24 h. Subsequently, the cells were harvested at 4 °C through centrifugation at 10,000× *g* for 5 min. Genomic DNA extraction was carried out immediately using the method outlined by Hoffman and Winston [41]. The purity and integrity of the DNA were evaluated by agarose gel electrophoresis, and its concentration was measured using a Qubit™ fluorometer (Invitrogen, Carlsbad, CA, USA).

2.1.2. Genome Sequencing, Assembly, Annotation, and Phylogenomic Analyses

The genome of *E. acetylicum* SI17 was sequenced, assembled, and annotated following the methodology outlined by Zheng et al. [42]. The genome was sequenced on the Nanopore PromethION. Further, twenty-eight complete amino acid sequences of various *Exiguobacterium* strains were retrieved from the NCBI database (https://www.ncbi.nlm.nih.gov/genome/ (accessed on 24 May 2021)). A phylogenomic tree, based on whole genomes, was then constructed using CVTree3 [43], employing *Bacillus subtilis* as the out-group.

2.2. *Biocontrol (Antagonistic) Activity on Litchi Fruit*

2.2.1. Inoculation Experiments on Litchi Fruit

The pathogen *P. litchii* SC18 and the biocontrol agent *E. acetylicum* SI17 were cultured using the method described by Zheng et al. [40]. The pathogen was cultured on carrot agar (CA) medium for 7 days. "Feizixiao" litchi fruits, approximately 60% ripe (about 65 days after flowering), were treated by spraying with *E. acetylicum* SI17 suspension at a concentration of 5 × 10^7 CFU/mL or a 1/10 dilution of LB broth (as control) until completely covered. After seven days, the harvested fruits were moved to a greenhouse maintained at a steady temperature of 25 °C, 95% humidity, with 12 h day and night cycles. Twenty-four hours later, the treatments were sprayed with 30 mL of *P. litchii* SC18 at a concentration of 5×10^4 sporangia/mL. Disease severity was measured every 12 h from 60 to 96 hpi. Each treatment contained 3 replicates, with 30 fruits per replicate. The disease index was calculated following the procedure outlined by Zheng et al. [40].

2.2.2. Relative Quantification of *P. litchii* SC18 in Pericarp

Total RNA was extracted from the litchi pericarp by using the RNAprep Pure Plant Kit (Tiangen, Beijing, China). The purity and integrity of the RNA were assessed using agarose gel electrophoresis and the Agilent 2100 system (Agilent Technologies, Santa Clara, CA, USA), while the RNA concentration was measured with the Nanodrop (Thermo Scientific, Waltham, CA, USA) and Qubit 2.0 (Life Technologies, Carlsbad, CA, USA) systems. For cDNA synthesis, one microgram of RNA was used with the PrimeScript^{TM} RT Reagent Kit with gDNA Eraser (Takara, Kusatsu, Japan). The qRT-PCR assay was carried out using TB Green^{TM} Premix Ex Taq^{TM} II (Takara, Kusatsu, Japan), and the data were obtained using the Thermal Cycler^{TM} Dice Real Time System III (Takara, Kusatsu, Japan). The relative expression levels were determined by calculating the fold change based on the $2^{-\triangle\triangle Ct}$ method. The primers were designed with Primer 3.0 (ABI, Carlsbad, CA, USA), and the sequences were as follows: *P. litchii* Actin-F: ACATTGCCCTGGACTTCG, *P. litchii* Actin-R: AGCTCCTTGGTCATACGC, litchi Actin-F: CGGGAAATTGTCCGTGAC, litchi Actin-R: GAGGACTTCTGGGCAACG. *P. litchii Actin* was quantified using litchi *Actin* as a reference gene, and the results indicated the relative amount of *P. litchii* SC18 in the litchi peel.

2.2.3. Enzyme Activity in the Pericarp

In 2021, samples were collected from Litchi cv. "Feizixiao" fruits. Nine fruits were taken for each treatment at each time point (0, 48, 84, 96, 108, and 120 h post-inoculation (hpi)). The pericarp was then collected, promptly enveloped in tin foil, flash-frozen in liquid nitrogen, and stored at −80 °C until required for experimental analyses. The levels of H_2O_2 and the activities of the enzymes catalase (CAT, EC 1.11.1.6), superoxide dismutase (SOD, EC 1.15.1.1), and peroxidase (POD, EC 1.11.1.7) were meticulously measured. The detection kits for these analyses were obtained from Nanjing Jiancheng Biological Engineering Institute, Nanjing, China (http://www.njjcbio.com/ (accessed on 18 November 2021)).

2.3. Data Analysis

All statistical analyses were performed using SPSS 25.0 (IBM, Armonk, NY, USA). Significant differences between more than two groups were determined using one-way ANOVA and Duncan's multiple comparison test. Statistical significance was set at $p < 0.05$. Histograms were constructed using Microsoft Office Excel 2016 (Microsoft, Los Angeles, CA, USA).

3. Results

3.1. Genome and Phylogenomic Analysis of *E. acetylicum* SI17

3.1.1. Genome Sequencing, Assembly, and Functional Annotation

The data gathered from the Nanopore platform were used to construct a circular between chromosomes, with a length of 3,157,929 base pairs (bp) and a GC content of 47.28% (Figure 1, Table 1). Additionally, four plasmids were detected, measuring 119,838 bp, 43790 bp, 6694 bp, and 68,355 bp in length, with GC contents of 40.05%, 43.16%, 37.50%, and 37.92%, respectively (Figure 1, Table 1). The genome comprises 3541 predicted protein-coding sequences (CDS), accounting for 88.98% of the total genomic content. These sequences possess an average length of 846 bp, as shown in Table 1. Analysis of the genome components revealed the presence of 154 interspersed repeats, 105 tandem repeats (TR), 70 tRNA genes, 27 rRNA genes, 1 sRNA gene, 3 other RNA genes, 6 genomic islands (GIs), and 5 prophages. Notably, no clustered regularly interspaced short palindromic repeats (CRISPR) sequences were identified, as specified in Table 2.

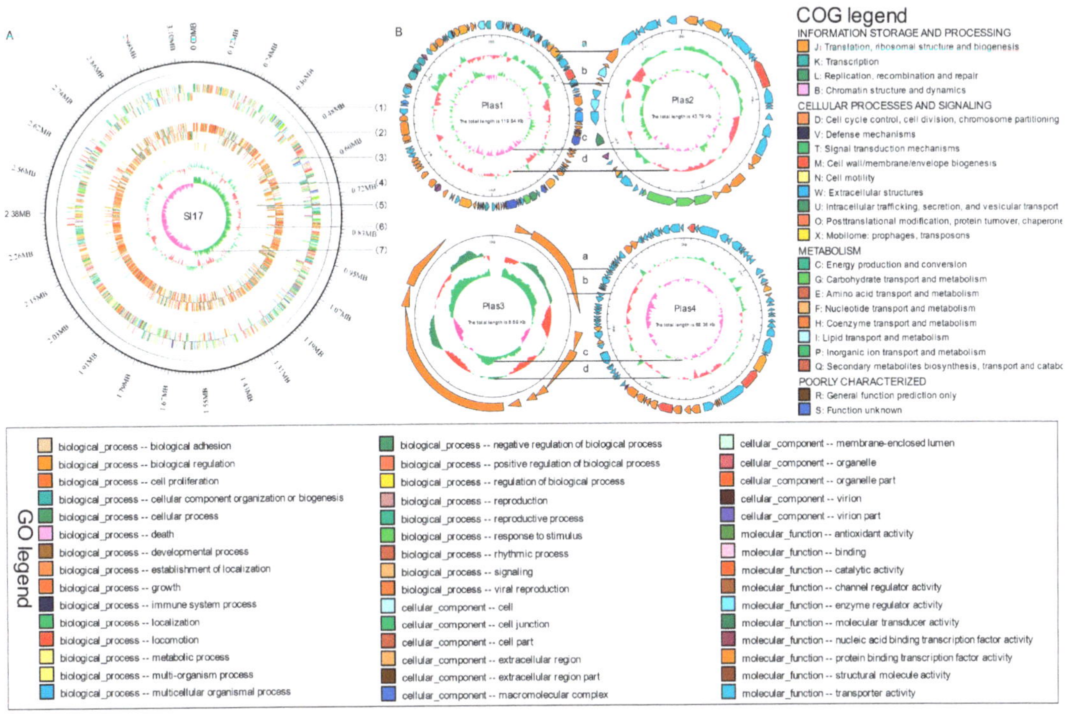

Figure 1. Map of chromosome and four plasmids of *Exiguobacterium acetylicum* SI17. (**A**) chromosome, from outside to inside, the map shows the (1) position of the genome, (2) coding genes on the + and −, strands, (3) COG annotation, (4) GO annotation, (5) non-coding RNA, and (6) GC content. The outer green indicates that the GC content in this region is higher than the genome-wide average, while the inner red indicates the opposite. The height of the peak represents the magnitude of the difference from the average GC content. Higher peaks indicate greater differences from the average GC content. (7) GC skew value (G–C/G + C). When the value is positive, there is a high possibility that CDS is transcribed from the positive chain; Otherwise, there is a high possibility that CDS is transcribed from the negative chain. (**B**) plasmids, from outside to inside, the map shows the (a) COG annotation, (b) position of the genome, (c) GC content, (d) GC skew value.

Table 1. Genomic features of strain SI17.

Features	Value				
Number of reads	716,365				
Number of bases (bp)	8,189,664,288				
Mean read length (bp)	11,432.3				
N50 read length (bp)	14,606				
Mean read quality	10.1				
Gene number	3541				
Gene total length (bp)	3,022,368				
Gene average length (bp)	846				
Gene/genome (%)	88.98				
Features	Chromosome	Plasmid 1	Plasmid 2	Plasmid 3	Plasmid 4
Genome size (bp)	3,157,929	119,838	43,790	6694	68,355
G + C content (%)	47.28	40.05	43.16	37.50	37.92

Table 2. Genome component analyses of the strain SI17.

Non-coding RNA (ncRNA)				
Type	Number	Average_Length	Total_Length	In Genome (%)
tRNA	70	77	5394	0.1588
5s rRNA	9	115	1035	
16s rRNA	9	1550	13,950	1.2128
23s rRNA	9	2912	26,208	
sRNA	1	86	86	0.0025
Genomics Islands (GIs)				
GIs_ID	Start	End	Length	GC%
GIs001 (chr)	1,353,417	1,359,120	5704	41.09
GIs002 (chr)	1,511,852	1,532,374	20,523	39.63
GIs003 (chr)	1,710,372	1,718,135	7764	42.17
GIs004 (plas 1)	57,004	61,841	4838	36.59
GIs005 (plas 1)	96,328	103,839	7512	50.12
GIs006 (plas 2)	12,956	20,897	7942	36.15
Prophage				
Prophage_ID	Start	End	Length	GC%
Prophage_1 (Chr1)	804,985	828,528	23,544	47.52
Prophage_2 (Chr1)	1,237,842	1,295,459	57,618	47.55
Prophage_3 (Chr1)	1,333,287	1,341,796	8,510	47.56
Prophage_4 (Chr1)	1,600,271	1,629,792	29,522	49.76
Prophage_5 (Chr1)	1,845,601	1,882,853	37,253	45.65
Interspersed Repeat				
Type	Number	Total Length (bp)	Average length (bp)	In Genome (%)
LTR	89	7967	90	0.2346
DNA	14	1005	72	0.0296
LINE	35	3408	103	0.1003
SINE	14	1110	79	0.0327
RC	2	101	50	0.003
Total	154	13,043	0.384	90
Tandem Repeat				
Type	Number	Repeat Size (bp)	Total Length (bp)	In Genome (%)
Tandem repeat (TR)	55	8~201	5332	0.157
Minisatellite DNA	50	11~51	4348	0.128
Microsatellite DNA	0	0~0	0	0

Functional annotation was conducted by aligning the predicted gene protein sequences using the diamond alignment tool against several frequently used databases. In the context of Clusters of Orthologous Groups (COG) analysis, a total of 2574 proteins were sorted into 23 distinct functional groups. Apart from the general function prediction category, both translation, ribosomal structure, and biogenesis, as well as the transcription category, exhibited a higher percentage among the 22 categories, encompassing 237 and 230 genes, respectively (Figure S1). In the Gene Ontology (GO) analysis, the 2409 proteins were categorized into three primary groups: biological processes, molecular functions, and cellular components. Interestingly, categories of metabolic processes and cellular processes, catalytic activity and binding, as well as cell part and cell, demonstrated higher gene abundance (Figure S2). Moreover, in the context of Kyoto Encyclopedia of Genes and Genomes (KEGG) pathways, the metabolism pathway emerged as particularly notable (Figure S3). The Carbohydrate-Active enZYmes (CAZy) database, dedicated to carbohydrate enzymes, covers a broad spectrum of enzyme families pivotal in catalyzing the degradation, alteration, and biosynthesis of carbohydrates. It features five main groups: glycoside hydrolase

(GH), glycosyl transferase (GT), polysaccharide lyase (PL), carbohydrate esterase (CE), and auxiliary activity (AA). Notably, there were 45, 36, 1, 5, 0, and 27 genes categorized under GH, GT, PL, CE, AA, and CBM (carbohydrate-binding module), respectively (Figure S4). Additionally, the Transporter Classification Database (TCDB) analysis (Figure S5) showed that genes classified as porters, including uniporters, symporters, and antiporters (94 genes), along with P-P-bond-hydrolysis-driven transporters (129 genes), collectively represented 74.33% of all predicted transporters (300 genes).

3.1.2. Phylogenomic and Comparative Genomic Analysis

The 29 *Exiguobacterium* genomes were divided into two groups. Notably, *E. acetylicum* SI17 was clustered in the first group, showing greater homology to *E. indicum* and *E. enclensis* (Figure 2). Despite differences in GC content, there were no significant variations in genome size, gene count, proteins, rRNAs, or tRNAs between the two groups (Table S1). To the best of our knowledge, this is the first documented instance of using *Exiguobacterium* strains for controlling plant diseases. This suggests that the Group I strains may have potential applications in agriculture.

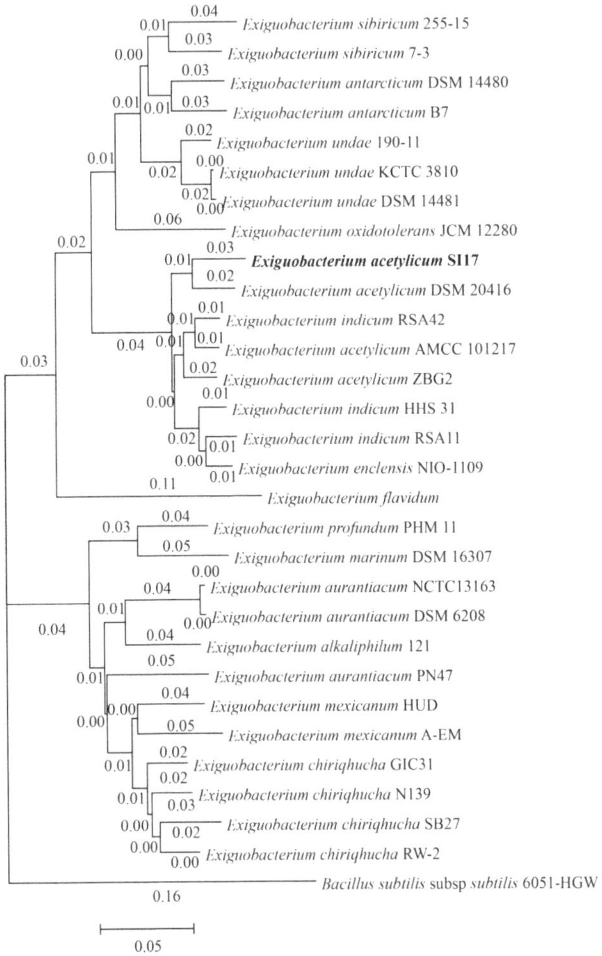

Figure 2. Phylogenomic tree based on whole-genome sequences of 29 *Exiguobacterium* strains. *Bacillus subtilis* was used as an out-group. Strain *E. acetylicum* SI17 is highlighted in bold.

3.2. Biocontrol Activity of *E. acetylicum* SI17 on LDB

3.2.1. Pre-Harvest *E. acetylicum* SI17 Treatment Suppressed LDB

Pre-harvest inoculation experiments on litchi fruit indicated that *E. acetylicum* SI17-treated fruits had a significantly lower disease index at 60, 72, and 84 h, but this difference lessened over time (Figure 3A,B).

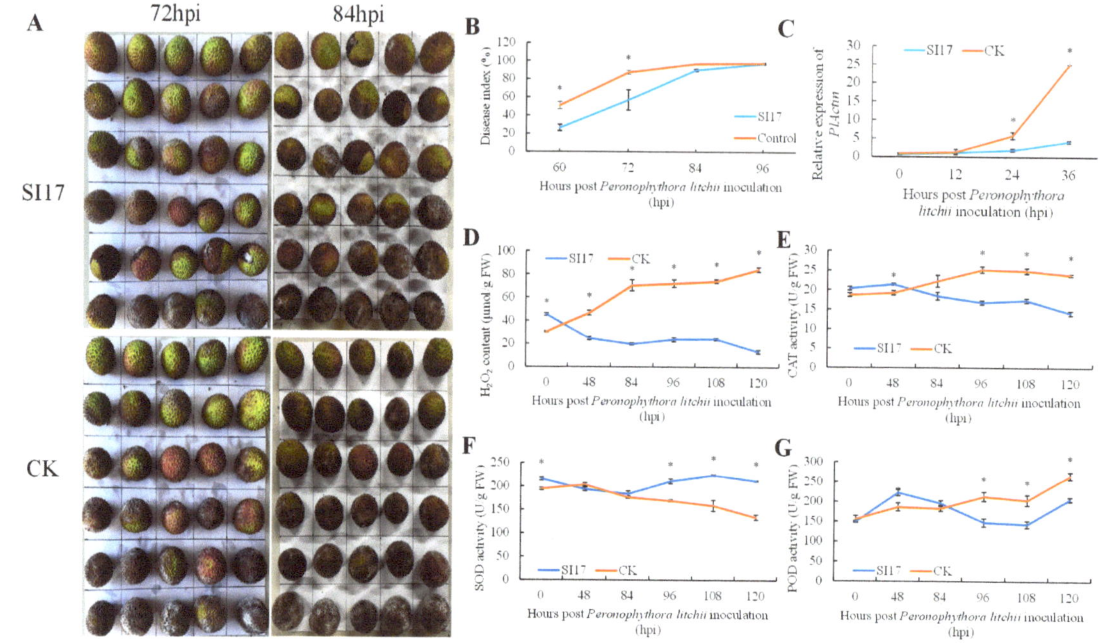

Figure 3. Interaction between *Exiguobacterium acetylicum* strain SI17 pre-harvest treatment and *Peronophythora litchi* on litchi cv. "Feizixiao" fruits: (**A**) litchi downy blight symptoms and (**B**) disease index (B) development, (**C**) *PlActin* expression, (**D**) H_2O_2 content, (**E**) catalase (CAT), (**F**) superoxide dismutase (SOD), (**G**) peroxidase (POD) activities. CK = control. The * above bars indicate significant differences at $p < 0.05$ in Duncan's multiple range test.

3.2.2. Relative Quantification of *P. litchii* SC18 in Pericarp after *E. acetylicum* SI17 Treatment

Following the treatment with *P. litchii* SC18, the relative quantity of pathogens was measured using qRT-PCR. In both the *E. acetylicum* SI17 treatment and control groups, relative pathogen levels were low before 12 hpi (Figure 3C). By 24 hpi, pathogen levels in the *E. acetylicum* SI17 treatment group were three times lower than those in the control group, and by 36 hpi, they were six times lower (Figure 3C). As time progressed, the pathogen growth rate in the control group was much higher than that in the *E. acetylicum* SI17 treatment group.

3.2.3. Pre-Harvest *E. acetylicum* SI17 Treatment Enhanced the Activity of Defense-Related Enzymes

After inoculation with *P. litchii*, H_2O_2 levels in the *E. acetylicum* SI17-treated litchi pericarp steadily declined, unlike the control group, which showed a rapid H_2O_2 buildup before 84 h, followed by a gradual increase. Overall, H_2O_2 levels in the *E. acetylicum* SI17-treated litchi pericarp were lower compared to the control (Figure 3D). In the treatment group, the activity of SOD initially decreased before 84 h and then peaked at 108 h. Conversely, in the control group, SOD activity continued to decline over time (Figure 3F). CAT activity displayed an upward trend in the control group, whereas in the *E. acetylicum*

SI17-treated litchi pericarp, it showed a downward trend (Figure 3E). Prior to 84 h, CAT activity in the *E. acetylicum* SI17 treatment group surpassed that of the control group but subsequently declined (Figure 3E). POD activity in the control group increased steadily over time, whereas in the *E. acetylicum* SI17 treatment group, it initially rose, then declined before rising again, peaking at 48 h and reaching its lowest point at 108 h (Figure 3F).

4. Discussion

E. acetylicum has been observed to enhance wheat seedling growth [32] and break down shrimp shell waste [44]. Yet, its impact on plant diseases remains unexplored until now. Our research reveals that *E. acetylicum* SI17 demonstrates efficacy in managing LDB. By sequencing the whole genome of *E. acetylicum* SI17, we identified the genetic basis for this functional trait. Compared to the other 28 *Exiguobacterium* genomes, *E. acetylicum* SI17 stands out with the largest genome size and the highest counts of genes, proteins, and tRNAs (Table S1). *Exiguobacterium* is clustered into two groups based on its ability to withstand temperatures as low as 7 °C, yet there was no significant difference in functionality between these groups. *E. oxidotolerans* (Group I) and *E. profundum* PHM11 (Group II) demonstrated the capacity to enhance the growth and yield of *Bacopa monnieri* and rice, respectively, particularly under conditions of salt stress [33,45]. Additionally, *E. acetylicum* (Group I) and *E. marinum* a-1 (Group II) exhibited capabilities in degrading harmful substances, such as cyanide-containing effluents and polypropylene, respectively [46,47]. This suggests that not only species within Group I but also those within Group II possess the ability to manage plant diseases.

Existing research has shown that using post-harvest *E. acetylicum* SI17 treatment is highly effective in controlling LDB. Additionally, exposing litchi fruits to SI17 VOCs before harvest significantly decreases the severity of LDB [40]. In this study, we examined the effectiveness of pre-harvest *E. acetylicum* SI17 treatment in controlling LDB to explore its potential for field application. Our findings indicate promising results (Figure 3A,B). H_2O_2, as a representative of reactive oxygen species (ROS), can be excessively generated in response to various biotic and abiotic stresses, resulting in its heightened reactivity and toxic effects on plants [48]. The accumulation of ROS is a widespread defense mechanism for higher plants to resist pathogen attacks. However, excessive ROS can cause plant damage and facilitate pathogen infection [49]. SOD, CAT, and POD are essential enzymes that act as antioxidants, responsible for neutralizing ROS [50]. Many studies have shown that biocontrol agents can effectively clear excess ROS that accumulate due to pathogen induction. For example, *Bacillus velezensis* could enhance the activity of ROS-scavenging enzymes, thereby enhancing the eggplant fruits' disease resistance [51]. A similar situation occurred in *Bacillus amyloliquefaciens* GJ1 against citrus Huanglongbing [52] and *Streptomyces griseorubiginosus* LJS06 against cucumber anthracnose [53]. Litchi pre-treated with SI17 demonstrated an increased ability to scavenge H_2O_2 (Figure 3D), primarily due to heightened SOD activity rather than CAT and POD activities (Figure 3E–G).

Fermentation (C), suspension (F), and supernatant (Q) of *E. acetylicum* SI17 demonstrated no inhibitory effect on the growth of *P. litchii* SC18 (Figure S6). Therefore, it is speculated that *E. acetylicum* SI17 may control disease by enhancing plant resistance rather than directly suppressing *P. litchii* SC18 growth. This is one way for a biocontrol agent to resist disease. Biocontrol agent *Pseudomonas chlororaphis* PA23 could induce systemic acquired resistance against *Sclerotinia sclerotiorum* of *Brassica napus* [54]. Furthermore, various strains of *Exiguobacterium* have been found to produce diverse hydrolytic enzymes [17]. Genome analysis provided additional insights into the key features of *E. acetylicum* SI17 in its ability to combat LDB. Within this strain, four gene clusters, comprising a total of 94 genes, were identified to be linked with the biosynthesis of secondary metabolites (Table S2). These clusters comprised one non-ribosomal peptide synthetase (NRPS), one siderophore, and two terpenes (Table S2). Various metabolites produced by biocontrol agents showed disease resistance in plants. Phenazine-1-carboxylic acid produced by *Pseudomonas chlororaphis* YL-1 showed effectiveness against *Acidovorax citrulli* [55]. Nep1-like proteins generated by

Pythium oligandrum enhanced plant resistance against *Phytophthora* pathogens and *Sclerotinia sclerotiorum* [56]. Hence, it can be deduced that *E. acetylicum* SI17 could enhance plant resistance through the production of diverse secondary metabolites, although the specific functionalities of these metabolites necessitate further verification.

5. Conclusions

The *Exiguobacterium acetylicum* SI17 exhibited a biocontrol effect on LDB caused by *P. litchii*. This study marks the initial documentation of an *Exiguobacterium* strain being employed for managing plant diseases. To deepen our comprehension of its regulatory mechanism, we conducted a full genome sequencing of *E. acetylicum* SI17, revealing a larger genome size with high numbers of genes, proteins, and tRNAs. Additionally, the genome was clustered into Group I. Litchi fruit pre-treated with *E. acetylicum* SI17 showed increased SOD activity and improved capacity to scavenge H_2O_2. Pre-harvest treatment with *E. acetylicum* SI17 suppressed LDB, suggesting promising prospects for its practical application. These findings lead to the hypothesis that *E. acetylicum* SI17 might generate a metabolite that enhances litchi's resistance, highlighting the need for more detailed exploration and identification.

Supplementary Materials: The following supporting information can be downloaded at: https://www.mdpi.com/article/10.3390/horticulturae10080888/s1, Table S1: Characteristic of *Exiguobacterium* genomes used for phylogenomic analysis; Table S2: Secondary metabolism gene clusters in *E. acetylicum* SI17; Figure S1: Clusters of Orthologous Groups (COG) function classifications. B: Chromatin structure and dynamics; C: Energy production and conversion; D: Cell cycle control, cell division, chromosome partitioning; E: Amino acid transport and metabolism; F: Nucleotide transport and metabolism; G: Carbohydrate transport and metabolism; H: Coenzyme transport and metabolism; I: Lipid transport and metabolism; J: Translation, ribosomal structure and biogenesis; K: Transcription; L: Replication, recombination and repair; M: Cell wall/membrane/envelope biogenesis; N: Cell motility; O: Posttranslational modification, protein turnover, chaperones; P: Inorganic ion transport and metabolism; Q: Secondary metabolites biosynthesis, transport and catabolism; R: General function prediction only; S: Function unknown; T: Signal transduction mechanisms; U: Intracellular trafficking, secretion, and vesicular transport; V: Defense mechanisms; W: Extracellular structures; X: Mobilome, prophages, transposons; Figure S2: Gene Ontology (GO) analysis. Biological process (from left to right): biological adhesion, biological regulation, cell proliferation, cellular component organization or biogenesis, cellular process, death, developmental process, establishment of localization, growth, immune system process, localization, locomotion, metabolic process, multi-organism process, multicellular organismal process; negative regulation of biological process, positive regulation of biological process, regulation of biological process, reproduction, reproductive process, response to stimulus; rhythmic process, signaling; viral reproduction; cellular component (from left to right): cell, cell junction, cell part, extracellular region, extracellular region part, macromolecular complex, membrane-enclosed lumen, organelle, organelle part, virion, virion part; molecular function (from left to right): Antioxidant activity, binding, catalytic activity, channel regulator activity, enzyme regulator activity, molecular transducer activity, nucleic acid binding transcription factor activity, protein binding transcription factor activity, structural molecule activity, transporter activity; Figure S3: Kyoto Encyclopedia of Genes and Genomes (KEGG) pathway annotation. Cellular Processes (A): Transport and catabolism (1), Cellular community-prokaryotes (2), cell motility (3), cell growth and death (4); Environmental Information Processing (B): Signal transduction (5), Membrane transport (6); Genetic Information Processing (C): Translation (7), Transcription (8), Replication and repair (9), Folding, sorting and degradation (10); Human Diseases (D): Substance dependence (11), Neurodegenerative diseases (12), Infectious diseases (13), Immune diseases (14), Endocrine and metabolic diseases (15), Drug resistance (16), Cardiovascular diseases (17), Cancers (18); Metabolism (E): Xenobiotics biodegradation and metabolism (19), Nucleotide metabolism (20), Metabolism of terpenoids and polyketides (21), Metabolism of other amino acids (22), Metabolism of cofactors and vitamins (23), Lipid metabolism (24), Glycan biosynthesis and metabolism (25), Energy metabolism (26), Carbohydrate metabolism (27), Biosynthesis of other secondary metabolites (28), Amino acid metabolism (29); Organismal Systems (F): Nervous system (30), Immune system (31), Excretory system (32), Environmental adaptation (33), Endocrine system (34), Digestive system (35),

Aging (36); Figure S4: Carbohydrate-Active enZYmes Database (CAZy) annotation. AA: auxiliary activity; CBM: carbohydrate-binding module; CE: carbohydrate esterase; GH: glycoside hydrolase; GT: glycosyl transferase; PL: polysaccharide lyase; Figure S5: Transporter Classification Database (TCDB) annotation. 1.A: alpha-Type Channels; 1.B: beta-Barrel Porins; 1.C: Pore-Forming Toxins (Proteins and Peptides); 1.E: Holins; 2.A: Porters (uniporters, symporters, antiporters); 3.A: P-P-bond-hydrolysis-driven transporters; 3.D: Oxidoreduction-driven transporters; 3.E: Light absorption-driven transporters; 4.A: Phosphotransfer-driven Group Translocators; 4.C: Acyl CoA ligase-coupled transporters; 5.A: Transmembrane 2-electron transfer carriers; 8.A: Auxiliary transport proteins; 9.A: Recognized transporters of unknown biochemical mechanism; 9.B: Putative transport proteins; Figure S6: Plate antagonism against *Peronophythora litchii* SC18 of fermentation (C), suspension (F), and supernatant (Q) of SI17.

Author Contributions: S.H.: data curation, methodology, investigation, writing—original draft, writing—review and editing; X.L.: data curation, methodology, supervision, writing—original draft; L.Z.: conceptualization, methodology, supervision, validation, funding acquisition, writing—review and editing; D.G.: conceptualization, data curation, formal analysis, funding acquisition, project administration, supervision, validation, writing—original draft, writing—review and editing. All authors have read and agreed to the published version of the manuscript.

Funding: This research was supported by grants from Guangdong Basic and Applied Basic Research Foundation (2022A1515010470), Key Laboratory of Biology and Genetic Resources Utilization of South Tropical Fruit Trees, Ministry of Agriculture and Rural Affairs (202102), the Innovative team program of the Department of Education of Guangdong Province (2023KCXTD018) and Guangdong Province Rural Revitalization Strategic Project "Orchard Agricultural Machinery and Agronomy Integration and Production Management Informatization" (2024-TS-2-4).

Data Availability Statement: The Genbank accession number of *Exiguobacterium acetylicum* SI17 genome and four plastid sequences are CP075897.1-CP075901.1. All other datasets for this study are included in the article/Supplementary Materials.

Conflicts of Interest: The authors declare no conflicts of interest.

References

1. Jiang, Y.M.; Zhu, X.R.; Li, Y.B. Postharvest control of litchi fruit rot by *Bacillus subtilis*. *Food Sci. Tech.* **2001**, *34*, 430–436. [CrossRef]
2. Xu, L.; Xue, J.; Wu, P.; Wang, D.; Lin, L.; Jiang, Y.; Duan, X.; Wei, X. Antifungal activity of hypothemycin against *Peronophythora litchii* in vitro and in vivo. *J. Agric. Food Chem.* **2013**, *61*, 10091–10095. [CrossRef] [PubMed]
3. Situ, J.; Xi, P.; Lin, L.; Huang, W.; Song, Y.; Jiang, Z.; Kong, G. Signal and regulatory mechanisms involved in spore development of *Phytophthora* and *Peronophythora*. *Front. Microbiol.* **2022**, *13*, 984672. [CrossRef] [PubMed]
4. Zhang, Z.; Wang, T.; Liu, G.; Hu, M.; Yun, Z.; Duan, X.; Cai, K.; Jiang, G. Inhibition of downy blight and enhancement of resistance in litchi fruit by postharvest application of melatonin. *Food Chem.* **2021**, *347*, 129009. [CrossRef] [PubMed]
5. Jiang, Y.M.; Wang, Y.; Song, L.; Liu, H.; Lichter, A.; Kerdchoechuen, O.; Joyce, D.C.; Shi, J. Postharvest characteristics and handling of litchi fruit-an overview. *Aust. J. Exp. Agric.* **2006**, *46*, 1541. [CrossRef]
6. Dukare, A.S.; Paul, S.; Nambi, V.E.; Gupta, R.K.; Singh, R.; Sharma, K.; Vishwakarma, R.K. Exploitation of microbial antagonists for the control of postharvest diseases of fruits: A review. *Crit. Rev. Food Sci.* **2019**, *9*, 1498–1513. [CrossRef]
7. Ippolito, A.; Nigro, F. Impact of preharvest application of biological control agents on postharvest diseases of fresh fruits and vegetables. *Crop Prot.* **2000**, *19*, 715–723. [CrossRef]
8. Sivakumar, D.; Zeeman, K.; Korsten, L. Effect of a biocontrol agent (*Bacillus subtilis*) and modified atmosphere packaging on postharvest decay control and quality retention of litchi during storage. *Phytoparasitica* **2007**, *35*, 507–518. [CrossRef]
9. Sivakumar, D.; Arrebola, E.; Korsten, L. Postharvest decay control and quality retention in litchi (cv. McLean's Red) by combined application of modified atmosphere packaging and antimicrobial agents. *Crop Prot.* **2008**, *27*, 1208–1214. [CrossRef]
10. Cai, X.Q.; Lin, N.; Chen, W.; Hu, F.P.; DongLiang, Q.; Mitra, S.K.; Diczbalis, Y. Control effects on litchi downy blight disease by endophytic bacterial strain TB2 and its pathogenesis-related proteins. *Acta Hortic.* **2010**, *863*, 631–636. [CrossRef]
11. Wu, Y.; Lin, H.; Lin, Y.; Shi, J.; Xue, S.; Hung, Y.C.; Chen, Y.; Hui, W. Effects of biocontrol bacteria *Bacillus amyloliquefaciens* LY-1 culture broth on quality attributes and storability of harvested litchi fruit. *Postharvest Biol. Tec.* **2017**, *132*, 81–87. [CrossRef]
12. Martínez-Castellanos, G.; Pelayo-Zaldívar, C.; Pérez-Flores, L.J.; López-Luna, A.; Gimeno, M.; Bárzana, E.; Shirai, K. Postharvest litchi (*Litchi chinensis* Sonn.) quality preservation by *Lactobacillus plantarum*. *Postharvest Biol. Tec.* **2011**, *59*, 172–178. [CrossRef]
13. Liao, L.; Zhou, J.; Wang, H.; Fei, H.; Liu, S.; Jiang, Z.; Chen, S.; Zhang, L.H. Control of litchi downy blight by zeamines produced by *Dickeya zeae*. *Sci. Rep.* **2015**, *5*, 15719. [CrossRef]
14. Xing, M.; Zheng, L.; Deng, Y.; Xu, D.; Xi, P.; Li, M.; Kong, G.; Jiang, Z. Antifungal activity of natural volatile organic compounds against litchi downy blight pathogen *Peronophythora litchii*. *Molecules* **2018**, *23*, 358. [CrossRef]

15. Stéphane, H.; Schadt, C.W. Towards a holistic understanding of the beneficial interactions across the *Populus* microbiome. *New Phytol.* **2015**, *205*, 1424–1430.
16. Conrath, U.; Beckers, G.J.; Flors, V.; Garcã-Agustã, P.; Jakab, G.; Mauch, F.; Newman, M.A.; Pieterse, C.M.J.; Poinssot, B.; Pozo, M.J.; et al. Priming: Getting ready for battle. *Mol. Plant Microbe Interact.* **2006**, *19*, 1062–1071. [CrossRef] [PubMed]
17. Kasana, R.C.; Pandey, C.B. *Exiguobacterium*: An overview of a versatile genus with potential in industry and agriculture. *Crit. Rev. Biotechnol.* **2018**, *38*, 141–156. [CrossRef] [PubMed]
18. Vijayalaxmi, S.; AnuAppaiah, K.A.; Jayalakshmi, S.K.; Mulimani, V.H.; Sreeramuluet, K. Production of bioethanol from fermented sugars of sugarcane bagasse produced by lignocellulolytic enzymes of *Exiguobacterium* sp. VSG-1. *Appl. Biochem. Biotech.* **2013**, *2171*, 246–260. [CrossRef]
19. Xie, J.; Xie, W.; Yu, J.; Xin, R.; Shi, Z.; Song, L.; Yang, X. Extraction of chitin from shrimp shell by successive two-step fermentation of *Exiguobacterium profundum* and *Lactobacillus acidophilus*. *Front. Microbiol.* **2021**, *12*, 677126. [CrossRef]
20. Ali, C.H.; Zhang, J.J.; Mbadinga, S.M.; Mu, B.Z. Screening, isolation and optimization of an extracellular lipase producing *Exiguobacterium* sp. BBXS-7 segregated from waste cooking oil contaminated sites. *Wulfenia J.* **2015**, *2*, 185–201.
21. Mojallali, L.; ShahbaniZahiri, H.; Rajaei, S.; Noghabi, K.A.; Haghbeenet, K. A novel ~34-kDa a-amylase from psychrotroph *Exiguobacterium* sp. SH3: Production, purification, and characterization. *Biotechnol. Appl. Bioc.* **2014**, *61*, 118–125. [CrossRef] [PubMed]
22. Qiao, Y.; Peng, Q.; Yan, J.; Wang, H.; Ding, H.; Shi, B. Gene cloning and enzymatic characterization of alkali-tolerant type I pullulanase from *Exiguobacterium acetylicum*. *Lett. Appl. Microbiol.* **2015**, *60*, 52–59. [CrossRef]
23. Parthasarathy, A.; Miranda, R.R.; Eddingsaas, N.C.; Chu, J.; Freezman, I.M.; Tyler, A.C.; Hudson, A.O. Polystyrene degradation by *Exiguobacterium* sp. RIT 594: Preliminary evidence for a pathway containing an atypical oxygenase. *Microorganisms* **2022**, *10*, 1619. [CrossRef] [PubMed]
24. Sakdapetsiri, C.; Kaokhum, N.; Pinyakong, O. Biodegradation of crude oil by immobilized *Exiguobacterium* sp. AO-11 and shelf life evaluation. *Sci. Rep.* **2021**, *11*, 12990. [CrossRef]
25. Delegan, Y.; Kocharovskaya, Y.; Bogun, A.; Sizova, A.; Solomentsev, V.; Iminova, L.; Lyakhovchenko, N.; Zinovieva, A.; Goyanov, M.; Solyanikova, I. Characterization and genomic analysis of *Exiguobacterium alkaliphilum* B-3531D, an efficient crude oil degrading strain. *Biotechnol. Rep.* **2021**, *32*, e00678. [CrossRef]
26. Barghoth, M.G.; Desouky, S.E.; Radwan, A.A.; Shah, M.P.; Salem, S.S. Characterizations of highly efficient moderately halophilic toluene degrading *Exiguobacterium mexicanum* M7 strain isolated from Egyptian saline sediments. *Biotechnol. Genet. Eng.* **2023**, *2*, 1–19. [CrossRef]
27. Alam, M.Z.; Malik, A. Chromate resistance, transport and bioreduction by *Exiguobacterium* sp. ZM-2 isolated from agricultural soil irrigated with tannery effluent. *J. Basic. Microb.* **2008**, *48*, 416–420. [CrossRef] [PubMed]
28. Okeke, B.C. Bioremoval of hexavalent chromium from water by a salt tolerant bacterium, *Exiguobacterium* sp. GS1. *J. Industrial Microbiol. Biot.* **2008**, *35*, 1571–1579. [CrossRef]
29. Das, S.; Bikash, C.B.; Ranjan, K.M.; Biswaranjan, P.; Mathummal, S.; Anindita, C.; Hrudayanath, T. Reduction of hexavalent chromium by *Exiguobacterium mexicanum* isolated from chromite mines soil. *Chemosphere* **2021**, *282*, 131135. [CrossRef]
30. Huang, Y.; Tang, J.; Zhang, B.; Long, Z.E.; Ni, H.; Fu, X.; Zou, L. Influencing factors and mechanism of Cr(VI) reduction by facultative anaerobic *Exiguobacterium* sp. PY14. *Front. Microbiol.* **2023**, *14*, 1242410. [CrossRef]
31. Dastager, S.G.; Kumaran, D.C.; Pandey, A. Characterization of plant growth-promoting rhizobacterium *Exiguobacterium* NII-0906 for its growth promotion of cowpea (*Vigna unguiculata*). *Biologia* **2010**, *65*, 197–203. [CrossRef]
32. Selvakumar, G.; Kundu, S.; Joshi, P.; Nazim, S.; Gupta, A.D.; Gupta, H.S. Growth promotion of wheat seedlings by *Exiguobacterium acetylicum* 1P (MTCC 8707) a cold tolerant bacterial strain from the Uttarakhand Himalayas. *Indian J. Microbiol.* **2010**, *50*, 50–66. [CrossRef] [PubMed]
33. Bharti, N.; Yadav, D.; Barnawal, D.; Maji, D.; Kalra, A. *Exiguobacterium oxidotolerans*, a halotolerant plant growth promoting rhizobacteria, improves yield and content of secondary metabolites in *Bacopa monnieri* (L.) Pennell under primary and secondary salt stress. *World J. Microb. Biot.* **2013**, *29*, 379–387. [CrossRef]
34. Venkadesaperumal, G.; Amaresan, N.; Kumar, K. Plant growth promoting capability and genetic diversity of bacteria isolated from mud volcano and lime cave of Andaman and Nicobar Islands. *Braz. J. Microbiol.* **2014**, *45*, 1271–1281. [CrossRef]
35. Marfetán, J.A.; Gallo, A.L.; Farias, M.E.; Vélez, M.L.; Pescuma, M.; Ordoñez, O.F. *Exiguobacterium* sp. as a bioinoculant for plant-growth promotion and selenium biofortification strategies in horticultural plants. *World J. Microb. Biot.* **2023**, *39*, 134. [CrossRef]
36. Barnett, S.J.; Anstis, S.T.; Roget, D.K.; Ryder, M.H. Suppression of *Rhizoctonia solani* AG-8 induced disease on wheat by the interaction between *Pantoea*, *Exiguobacterium*, and *Microbacteria*. *Aust. J. Soil. Res.* **2006**, *44*, 331–342. [CrossRef]
37. Selvakumar, G.; Joshi, P.; Nazim, S.; Mishra, P.K.; Kundu, S.; Gupta, H.S. *Exiguobacterium acetylicum* strain 1P MTCC 8707, a novel bacterial antagonist from the North Western Indian Himalayas. *World J. Microbiol. Biotechnol.* **2009**, *25*, 131–137. [CrossRef]
38. Vishnivetskaya, T.A.; Kathariou, S.; Tiedje, J.M. The *Exiguobacterium* genus: Biodiversity and biogeography. *Extremophiles* **2009**, *13*, 541–555. [CrossRef] [PubMed]
39. Zhang, D.; Zhu, Z.; Li, Y.; Li, X.; Guan, Z.; Zheng, J. Comparative genomics of *Exiguobacterium* reveals what makes a cosmopolitan bacterium. *mSystems* **2021**, *6*, e0038321. [CrossRef]

40. Zheng, L.; Situ, J.; Zhu, Q.; Xi, P.; Zheng, Y.; Liu, H.; Zhou, X.; Jiang, Z.D. Identification of volatile organic compounds for the biocontrol of postharvest litchi fruit pathogen *Peronophythora litchii*. *Postharvest Biol. Technol.* **2019**, *155*, 37–46. [CrossRef]
41. Hoffman, C.S.; Winston, F. A ten-minute DNA preparation from yeast efficiently releases autonomous plasmids for transformation of *Escherichia coli*. *Gene* **1987**, *57*, 267–272. [CrossRef]
42. Zheng, L.; Huang, S.; Hsiang, T.; Yu, G.; Guo, D.; Jiang, Z.; Li, J. Biocontrol using *Bacillus amyloliquefaciens* PP19 against litchi downy blight caused by *Peronophythora litchii*. *Front. Microbiol.* **2021**, *11*, 619423. [CrossRef]
43. Zuo, G.; Hao, B. CVTree3 Web server for whole-genome-based and alignment-free prokaryotic phylogeny and taxonomy. *Genom. Proteom. Bioinf.* **2015**, *13*, 321–331. [CrossRef]
44. Sorokulova, I.; Krumnow, A.; Globa, L.; Vodyanoy, V. Efficient decomposition of shrimp shell waste using *Bacillus cereus* and *Exiguobacterium acetylicum*. *J. Ind. Microbiol. Biot.* **2009**, *36*, 1123–1126. [CrossRef]
45. Srivastava, A.K.; Srivastava, R.; Bharati, A.P.; Singh, A.K.; Sharma, A.; Das, S.; Tiwari, P.K.; Srivastava, A.K.; Chakdar, H.; Kashyap, P.L.; et al. Analysis of biosynthetic gene clusters, secretory, and antimicrobial peptides reveals environmental suitability of *Exiguobacterium profundum* PHM11. *Front. Microbiol.* **2022**, *12*, 785458. [CrossRef] [PubMed]
46. Mekuto, L.; Alegbeleye, O.O.; Ntwampe, S.K.; Ngongang, M.M.; Mudumbi, J.B.; Akinpelu, E.A. Co-metabolism of thiocyanate and free cyanide by *Exiguobacterium acetylicum* and *Bacillus marisflavi* under alkaline conditions. *3 Biotech.* **2016**, *6*, 173. [CrossRef] [PubMed]
47. Sun, Y.; Zhang, Y.; Hao, X.; Zhang, X.; Ma, Y.; Niu, Z. A novel marine bacterium *Exiguobacterium marinum* a-1 isolated from in situ plastisphere for degradation of additive-free polypropylene. *Environ. Pollut.* **2023**, *336*, 122390. [CrossRef] [PubMed]
48. Mhamdi, A.; Van Breusegem, F. Reactive oxygen species in plant development. *Development* **2018**, *145*, dev164376. [CrossRef] [PubMed]
49. Tian, L.; Li, J.; Huang, C.; Zhang, D.; Xu, Y.; Yang, X.; Song, J.; Wang, D.; Qiu, N.; Short, D.P.G.; et al. Cu/Zn superoxide dismutase (VdSOD1) mediates reactive oxygen species detoxification and modulates virulence in *Verticillium dahliae*. *Mol. Plant Pathol.* **2021**, *22*, 1092–1108. [CrossRef]
50. Gill, S.S.; Tuteja, N. Reactive oxygen species and antioxidant machinery in abiotic stress tolerance in crop plants. *Plant Physiol. Bioch.* **2010**, *48*, 909–930. [CrossRef]
51. Zhang, X.; Xin, Y.; Wang, J.; Dhanasekaran, S.; Yue, Q.; Feng, F.; Gu, X.; Li, B.; Zhao, L.; Zhang, H. Characterization of a *Bacillus velezensis* strain as a potential biocontrol agent against soft rot of eggplant fruits. *Int. J. Food Microbiol.* **2024**, *410*, 110480. [CrossRef] [PubMed]
52. Nan, J.; Zhang, S.; Jiang, L. Antibacterial potential of *Bacillus amyloliquefaciens* GJ1 against citrus Huanglongbing. *Plants* **2021**, *10*, 261. [CrossRef] [PubMed]
53. Chai, C.H.; Hong, C.F.; Huang, J.W. Identification and characterization of a multifunctional biocontrol agent, *Streptomyces griseorubiginosus* LJS06, against cucumber anthracnose. *Front. Microbiol.* **2022**, *13*, 923276. [CrossRef]
54. Duke, K.A.; Becker, M.G.; Girard, I.J.; Millar, J.L.; Dilantha Fernando, W.G.; Belmonte, M.F.; de Kievit, T.R. The biocontrol agent *Pseudomonas chlororaphis* PA23 primes *Brassica napus* defenses through distinct gene networks. *BMC Genom.* **2017**, *18*, 467. [CrossRef] [PubMed]
55. Liu, Y.; Zhou, Y.; Qiao, J.; Yu, W.; Pan, X.; Zhang, T.; Liu, Y.; Lu, S.E. Phenazine-1-carboxylic acid produced by *Pseudomonas chlororaphis* YL-1 is effective against *Acidovorax citrulli*. *Microorganisms* **2021**, *9*, 2012. [CrossRef]
56. Yang, K.; Chen, C.; Wang, Y.; Li, J.; Dong, X.; Cheng, Y.; Zhang, H.; Zhai, Y.; Ai, G.; Song, Q.; et al. Nep1-like proteins from the biocontrol agent *Pythium oligandrum* enhance plant disease resistance independent of cell death and reactive oxygen species. *Front. Plant Sci.* **2022**, *13*, 830636. [CrossRef]

horticulturae

Article

Epidemiology and Management of Bean Common Mosaic Virus (BCMV) in Traditional *Phaseolus vulgaris* L. Landraces within Protected Geographical Indications

Sonia Expósito-Goás, Lautaro Gabriel Pinacho-Lieti, Fernando Lago-Pena and Cristina Cabaleiro *

Department of Plant Production and Engineering Projects, EPSE, Campus Terra, University of Santiago de Compostela, E-27002 Lugo, Spain
* Correspondence: cristina.cabaleiro@usc.es; Tel.: +34-600-940-167

Abstract: Protected geographical indications (PGIs) share health problems related to plant propagation material. The PGI "Faba de Lourenzá" encompasses a 1660 km^2 area in northern Galicia, Spain, renowned for its "Faba Galaica" (FG) and Faba do marisco" (FM) bean cultivars. The lack of certified virus-free seeds poses a challenge. From 2019 to 2023, seeds from 60 lots were tested for BCMV. Plants from several plots were tested periodically to develop disease progress curves (DPCs). Control methods (plots out PGI zone, virus-free seedlings, rogueing, corn borders, and intercropping) were tested. Yields in five plots were used to assess BCMV's economic impact. Seed lots were 22.3% FG-infected and <5% FM-infected. The transmission rate of BCMV from infected FG plants to their seeds was 25.5 ± 5%, while for FM it was 12 ± 3%. FG yield losses were on average 31.6 ± 4.5%. Combining virus-free seedlings and infected plant removal in plots outside the PGI area proved effective at reducing infection rates; combining with intercropping resulted in the lowest incidence in an FG plot. Farmer training and off-site plot selection to produce healthy sowing beans are key to improving results.

Keywords: *Phaseolus*; PGI; ELISA; virus-free seeds; disease progress curves (DPCs); intercropping

1. Introduction

The germplasm of the common beans, *Phaseolus vulgaris* L. (*Fabaceae* family), brought to Europe after the XV century from different areas of South America, has evolved and adapted to new environments. The occasional crosses and the strong selection for consumer preferences in terms of bean types played an important role in the evolution of new genetic variants of dry beans in Europe. New evolutions that were not present in the American centres of origin appeared in Europe: the new germplasm came from recombination between Mesoamerican ones and Andine genetic improvements, which were better adapted to the conditions of diverse new agroecosystems [1].

In Spain, the North and Northwest of the Iberian Peninsula is the area where greater genetic variability of legumes has been found [1] and where traditional farming systems such as smallholdings and self-consumption agriculture are still maintained. The characteristics of local cultivars grown in special locations are recognized by the EU as Protected Geographical Indications (PGIs). In the North of Galicia (Spain), the area of A Mariña (1660 km^2) (Figure 1) was recognized in 2008 as the PGI "Faba de Lourenzá" [2]. The products protected are dried beans, which are separated from the pod of the local variety known as "Faba Galaica" (FG) that belongs to the International commercial class (ICC) 'Favada' [3]; another traditional cultivar in the same area is "Faba do Marisco" (FM), a different cultivar (also known as "verdina") included in the ICC "flageolet" [4]. Researchers from the "Consejo Superior de Investigaciones Científicas" (CSIC) [3] developed a line of improvement PMB-0382 "Faba Galaica" from the local variety PHA-0917, now in the

Spanish register of commercial varieties. "Faba do Marisco" is also registered by CSIC (2020/0103) [4].

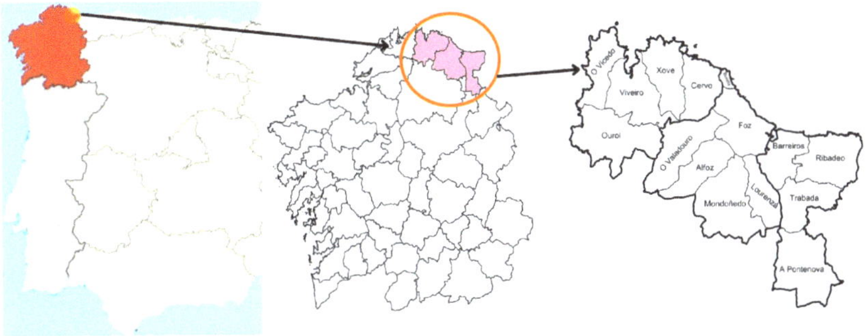

Figure 1. Geographical area included in the PGI "Faba de Lourenzá" in A Mariña, Galicia, Spain.

Cultivars within the PGIs have some qualitative characteristics that differentiate them from other beans and that make them highly demanded by consumers. FG is known for its exceptional culinary quality, due to its low proportion of skin (between 8 and 10%), its high-water absorption capacity (greater than 100%), and its behaviour when cooked, resulting, at the end of the process, in whole and complete grains in which the pastiness of the pulp stands out, free from lumps and barely differentiated from the skin. The plant has a climbing habit and indeterminate growth with long internodes [3]. FM seed is small (0.3–0.4 g), with a pale green colour thanks to the harvest conducted at a certain physiological stage; Its name in Galicia comes from the way it is cooked: a bean stew with shellfish ("marisco"). The plant is a low-stalk variety and has determined growth [4].

Beans can be infected with various pathogens transmitted through seeds, primarily viruses and bacteria. This complicates disease control in the field, especially for potyviruses, which can also be transmitted horizontally by aphids, pollen, or via mechanical means. Viruses transmission via seeds can occur through direct invasion of embryonic tissue, and/or indirectly by infecting pollen grains or ovules, which lead to an infected embryo after fertilization, resulting in an infected seed in a plant in which the virus is not detected [5]. Potyviruses are a major concern for *Phaseolus* spp. [6]. The most common are *bean yellow mosaic virus* (BYMV) and bean common mosaic, caused by two Potyvirus species: *Bean common mosaic virus* (BCMV) and *Bean common mosaic necrosis virus* (BCMNV). The difference between the two viruses is the phenotype generated in resistant cultivars. More than 200 aphid species, mainly in the Aphidinae subfamily (particularly the *Macrosiphum* and *Myzus* genera), can transmit potyviruses. In Spain, most potyviruses affecting beans have been identified [7–9]. Since potyviruses are transmitted non-persistently by aphids, epidemics can develop rapidly from even low seed-borne populations in areas where dry beans are the dominant crop. Utilizing aphicides to control non-persistently transmitted viruses like BCMV is generally ineffective [10].

The average seed transmission rate of BCMV and BCMNV from infected plants usually ranges from 15% to 35% [5,11,12] but can vary widely (3–95%) depending on the viral strain and bean variety [13]. This high seed transmission is a major cause of global BCMV epidemics, as it serves as the primary virus source in the field [14]. BCMV can survive for more than 30 years in seeds that keep their germination potential stored under suitable conditions [14].

In susceptible bean genotypes, BCMV infection induces a wide variety of symptoms, including mosaic, stunting, chlorosis, and leaf curling [15]. Mottling and malformation of the first true leaves indicate that the initial infection occurs through the seed [16]. The severity of these symptoms depends on various factors, including the viral strain, the bean cultivar, and the plant's age [17].

BCMV can seriously affect the common bean's yield or even cause total production loss [18]; however reported yield losses due to BCMV and BCMNV vary between 6 and 98% [6,16], depending on the cultivar, the time of infection [19], and the virus strain [20]. There are references to high losses due to potyvirus infections of aggressive breeds [19,21,22], but even mild or symptomless infections can decrease crop yield by 50% [22]. For many cultivars all over the world, progress has been made in combating BCMV through breeding bean varieties possessing the I gene, a dominant gene conferring resistance to most BCMV strains [6]. Potyvirus-resistant varieties have been the preferred control measure for extensive cultivation of green and dry beans because they are easier and more economical than trying to prevent transmission of potyviruses in the field [14,23,24]; however, for landraces in PGIs, this is not a viable option.

Several strategies can help ensure a healthy crop, including using seeds with low potyvirus levels and cultivating in highlands distant from major legume production areas. Other effective methods for preventing potyvirus transmission include removing infected or symptomatic plants (roughing), oil spraying [25], using straw or reflective mulches to reduce the attraction of vectors [26], using barrier crops, or intercropping [27]. These methods have been combined as an integrated strategy for effective protection against potyvirus in potato, cucurbit, maize, and bean crops [28–30].

Accurate and early diagnosis of BCMV and other potyviruses in *P. vulgaris* tissues is critical to be able to start with as low as possible levels of virus in the seed lots, to check when the epidemic reaches its exponential phase, and to predict the final level in the seeds. The potyvirus can spread easily, thus being able to detect it in asymptomatic plants is crucial. However, the most sensitive molecular methods are out of the question because they are too expensive and complex for routine testing of a high number of seedlings or plants in the field [24]. Some serological methods are very useful for quick and easy virus detection (Lateral flow tests, ImmunoStrips®) but again, their price makes their use impossible for high numbers of samples. Tissue printing assays have many advantages for epidemiological studies and for checking planting material [31]: samples are easy to obtain, deliver, and process; the use of commercial monoclonal and polyclonal antibodies at low concentrations is very efficient; nitrocellulose membranes or different kinds of paper are cheap support for assays; it uses common substances and standard buffers; the period of incubation and development of purple colour in positive samples is short; it is possible to check the tissue where a particular virus is detected; and it is possible to test groups of seedlings/petioles in a single print.

In a preliminary analysis of growers' FG seed lots in 2006 and 2014 in Lourenzá (A Mariña, Lugo, Spain), the prevalence of potyvirus was >95% [32]. Serological analysis indicated that BCMV was the most widespread potyvirus, and the presence of BCMNV was not detected. In addition to viruses, there could also be a high incidence of seeds infected with bacteria that cause rot during germination or plant damage when conditions are favourable, as was found in other local bean cultivars in the North of Spain [33].

Studying the epidemiology of potyvirus in these traditional cultivars, its seed transmission rate and the associated yield losses would undoubtedly help growers—in this and other similar PGIs—make informed decisions about the need for control measures and the advantages of coordination to do it properly. Obtaining seeds with the lowest possible levels of potyvirus could help delay the development of epidemics and reduce economic losses.

2. Materials and Methods

Phaseolus vulgaris seed lots: A total of 43 lots of Faba Galaica (FG) and 14 lots of Faba do Marisco (FM) bean seeds were obtained from various sources: the CSIC collections (Consejo Superior de Investigaciones Científicas), Cooperativa Terras da Mariña S.L., local growers within the PGI, and small shops selling bulk seeds from local farmers outside the PGI (Table 1).

Table 1. Identification of the lots of bean seeds from different groups that were analysed in this study.

Origin	Group	Year	#Lots	Identification [1]
CSIC	I	2014	1	C_G (1)
		2017	1	C_G (2)
		2018	1	C_G (3)
		2019	1	C_G (4)
		2021	1	C_G (5)
TERRAS DA MARIÑA S.COOP.GALEGA (LOURENZÁ)	II	2018	2	CTM_G (1–2)
		2019	6	CTM_G (3–8)
		2021	3	CTM_M (1–2.7)
		2022	4	CTM_M (3–6)
		2022	19	CTM_G (9–27)
ALFOZ & LOURENZÁ (out of cooperative)	III	2019	3	AL_G (1–3)
		2019	3	L_G (1–3)
		2019	1	LuC_G (1)
		2020	1	L_G (4)
		2021	1	L_G (5)
		2021	1	L_G (6)
		2021	1	LuC_G (2)
		<2021	2	LuC_M (1–2)
		2022	5	LuC_M (3–7)
Total # lots			57	

[1] C, CSIC (Consejo Superior de Investigacioes Científicas); G, faba Galaica; CTM, Cooperativa Terras da Mariña; M, faba do Marisco; AL, Alfóz; L, Lourenzá; LuC, Lugo, bulk sale in stores.

A total of 20–40 seeds were randomly selected from each seed lot and germinated in 14 cm diameter petri dishes at 25 °C in the dark. Seed development was monitored daily for signs of bacterial or fungal presence. Seed-borne viruses and bacteria were analysed using either root or leaf tissue from the same plants.

Fields: Tables 2 and 3 show details of the commercial and experimental plots sampled. While most plots were planted with cv. FG, several FM plots were included in the 2023 study. The plot code is as follows: first letter(s) for location (A, Alfoz; L, Lourenzá; M, Mondoñedo; SF, San Fiz; EI, plots in the USC (Campus Lugo), followed by the cultivar code (G, faba Galaica and M faba do Marisco). Next, letters indicate virus transmission control techniques (Mz, maize; B, border; I, Greenhouse). The last two numbers represent the year. Seeds were sown in trays, and all plants were tested before planting to ensure they were BCMV-free. In most cases, the experimental fields were planted with plantlets tested for BCMV originating from the same batch of seeds (CG_5).

Some plots were analysed only once, at the end of the growing season (LG_20, LG21, LM_23) while others were analysed on several dates to check the progress of the disease incidence and draw Disease Progress Curves (DPCs).

Harvest: in plots AG1_20, AG2_20, LG_22, LGBMz_23, and RGI_23, the plants that were sampled during the season were classified according to the date of virus detection. A total of 10–30 plants per class were harvested and the number of pods, the weight of pods before threshing, and the final dry weight of seeds were registered.

Virus transmission to seeds. Samples of seeds from the 7 fields where the incidence of BCMV had been estimated at harvest (Table 2) were germinated in seedbed trays and analysed (root, leaf, or both) to check the percentage of seeds with BCMV.

2020: From the AG1_20 and AG2_20 harvest, sixty seeds were analysed for BCMV and Ps·Ph.

2021: From the LG_21 harvest, the field where all plants tested positive for BCMV, two samples of 200 and 156 seeds were analysed.

2022: From the LG_22 harvest, four replicates of 10 seeds taken randomly from lots on each date of detection of infection were germinated in Petri dishes and analysed for BCMV,

Ps·Ph, and Xa·Ph in roots. Leaves of potted plantlets that were negative for BCMV in roots were tested again for BCMV.

2023: from the LGBMz_23 and RGI-23 harvest, more than 50 seeds taken randomly from the plants on each date of detection of the infection were germinated in pots and leaves analysed for BCMV. For FM, 10–40 seeds from infected plants from EIM_23 and SFM_23 were germinated separately and analysed for BCMV.

Table 2. Plots studied between 2020 and 2023 in several locations.

Plot	Year	cv	Location	Study			
				DPC [1]	Harvest	TTS [2]	One Test [3]
AG1_20	2020	FG	Alfoz	+	+	+	
AG2_20	2020	FG	Alfoz	+	+	+	
LG_20	2020	FG	Lourenzá	-	-	-	+
LG_21	2021	FG	Lourenzá	-	-	+	+
LG_22	2022	FG	Lourenzá	+	+	+	
LGBMz_23	2023	FG	Lourenzá	+	+	+	
MGI_22	2022	FG	Mondoñedo	-	-	-	
MGBMz_22	2022	FG	Mondoñedo	-	-	-	
RGI_23	2023	FG	Ribadeo	+	+	+	
EIG_22	2022	FG	Lugo	+	-	-	
EIM_23	2023	FM	Lugo	+	-	-	
SFM_23	2023	FM	Lugo	+	-	-	
SFGMz_23	2023	FG	Lugo	+	-	-	
LGMz1_23	2023	FG	Lourenzá	-	-	-	+
LGMz2_23	2023	FG	Lourenzá	-	-	-	+
LM_23	2023	FM	Lourenzá	-	-	-	+

[1] DPC: disease progress curve; [2] TTS, Transmission to seeds; [3] One test at the end of the season.

Table 3. Details on initial bean common mosaic virus incidence, sampling frequency, and number of plants analysed on each sampling date in the plots described in Table 2.

Plot	% BCMV Seedlings	Sampling	Frequency	Samples/Date
AG1_20	10%	Same plants [1]	monthly	107
AG2_20	10%	Same plants [1]	monthly	104
LG_20	nt [2] (>20%)	Random	at harvest	>500
LG_21	nt (>20%)	All plants in 2 lines	at harvest	200
LG_22	nt	Same plants [1]	10 days	200 (300)
LGBMz_23	0	Random & same	monthly	335
MGI_22	0	All plants	10 days	360
MGBMz_22	0	All plants	10 days	450
RGI_23	0	All plants	10 days	220
EIG_22	0	All plants	15 days	50
EIM_23	0	All plants	15–30 days	50
SFM_23	0	All plants	15–30 days	50
SFGMz_23	0	All plants	15–30 days	50
LGMz1_23	nt	Random	At harvest	75
LGMz2_23	nt	Random	At harvest	75
LM_23	nt	Random	At harvest	75

[1] after most plants were infected, a new batch of plants was tested; [2] nt: not tested (% based on average seed infection).

Virus and bacteria detection. All the analyses were conducted using direct immunoprinting DIP-ELISA (Enzyme-linked Immunosorbent Assay), following the basic protocol in [34,35], with some modifications/simplifications in sample printing, antibody dilutions (Table 4), and buffers.

Table 4. Optimized dilutions of commercial antibodies used in tissue printing ELISA.

Reactive	Company	Procedure	Dilution	Buffers
BCMV + AP	Loewe® Biochemical GmbH, Sauerlach, Germany	Direct	1:800	BCB
BCMV − Ab		Indirect	1:200	Carbonate buffer
Anty-rabbit + AP	Bio-Rad Laboratories, S.A., Madrid, Spain	Indirect	1:3000	BCB
Ps·ph + AP	Agdia Inc. Elkhart, IN, USA	Direct	1:500	BCB
Xa·ph + AP		Direct	1:100	BCB
Ps·ph − Ab		Indirect	1:100	Carbonate buffer
Xa·ph − Ab		Indirect	1:100	Carbonate buffer

Xa·ph: *Xanthomonas axonopodis* pv phaseoli; Ps·ph *Pseudomonas syringae/savastanoi* pv phaseolicola.

Several plant tissues were used for analysis: roots (collected a few days after germination), leaf petioles, or leaf blades. Roots and petioles were printed, and leaf blades were gently squashed with skewers onto nitrocellulose membranes (0.45 µm pore size) from Sartorius (Göttingen, Germany). The target pathogens were BCMV and the bacteria *Xanthomonas axonopodis* pv. phaseoli (Xa·ph) and *Pseudomonas syringae/savastanoi* pv. phaseolicola (Ps·ph).

After blocking the membranes with a 1.5–2% skim milk solution in distilled water for one hour, two distinct detection procedures were employed:

Direct Procedure: Membranes were incubated with alkaline phosphatase (AP)-conjugated antibodies specific to each pathogen (dilutions in Table 4) in conjugate buffer (BCB, Bioreba AG, Reinach, Swizerland; ref. 110140/-42). Following washes with saline buffer (8.5% NaCl + 0.05% Tween 20) three times for 3–5 min each, the membranes were incubated with a ready-to-use BCIP-NBT liquid substrate (Sigma-Aldrich, St. Louis, MO, USA, ref. B-1911) for colour development.

Indirect Procedure: This method added an extra incubation step. Membranes were first incubated with coating antibodies specific to each pathogen in carbonate buffer for 1.5 h or overnight at 4 °C. After washing, they were incubated with an anti-rabbit antibody conjugated with AP (BIO-RAD, ref. 1706518) for 30 min. Washes and colour development were performed as in the direct procedure.

A blue colour indicating a positive reaction typically developed within 15–30 min, but sometimes later. The reaction was stopped under running tap water once positive controls showed clear distinction.

Disease progress curves for BCMV epidemics: DPC were drawn as "days after first analysis" (DAFA) and BCMV incidence for the 9 plots indicated in Table 2. Analysis of the DPCs was conducted using the EPIMODEL V3.4 software program, which fits temporal disease progress data to five temporal population growth models that are commonly used in the analysis of plant disease epidemics: monomolecular, exponential, logistic, Gompertz, and linear models. This process involves examining disease progress and rate curves and then, based on the shapes of these curves, choosing one or more models that are likely to provide the best fit for the raw data. Transformed datasets are then graphed and evaluated based on coefficients of determination (R^2), mean square errors (MSE), and standard deviations of parameter estimates [36,37].

Statistical analysis: Comparison of BCMV levels in the seed lots (dependent variable) from each zone/origin/plot/date of infection was conducted using analysis of variance (ANOVA) according to a general linear model (GLM); data in percentage (in the range 0–30%) were transformed [$\sqrt{(X + 0.5)}$]. Comparisons of the data for yield/plant in each plot were done using one-way ANOVA with the date of virus detection as the independent variable and yield/plant as the dependent variable. Tukey's b test with $p \leq 0.05$ significance level was used to separate the means of different treatments after ANOVA. To compare data (in percentage) from different seed lots and transmission to seeds, χ^2 tests were performed.

3. Results

3.1. BCMV and Bacteria in Seed Lots

The prevalence of BCMV was >95% in the sampled lots of FG, with only 2 out of 43 seed lots being negative. The percentage of positive seeds in each lot was quite variable (0–70%) and on average (±se), the incidence was 22.3 ± 2.6%, which was not significantly different for lots in the three groups (F: 0.155; *p*: 0.86 > 0.05) (Figure 2). The germination of lots of FG <5 years old was almost as good as fresh (more than 90%) and bacterial rot was a rare event; older seeds kept quite good germination levels if stored at 4–5 °C.

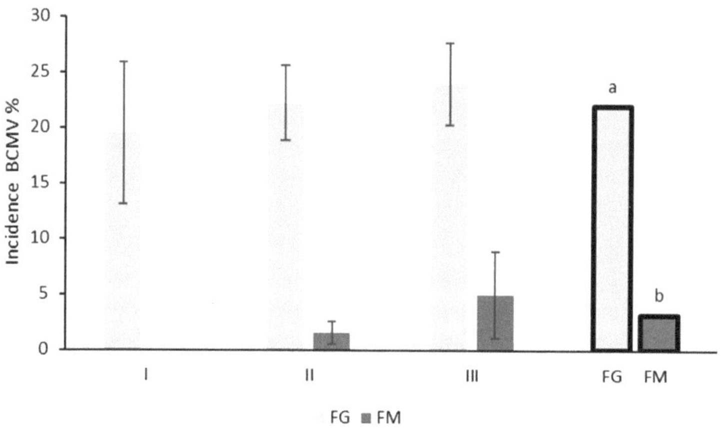

Figure 2. Average incidence of bean common mosaic virus (percentage ± se) in the lots with the three groups of Faba Galaica (FG) and two groups of Faba do Marisco (FM); (I, CSIC; II, Cooperative Terras da Mariña; III: PGI but out of Cooperative). Columns with black borders are the mean values of incidence for the two bean varieties, with different letters for significant differences after one-way ANOVA (F = 11.57; *p* = 0.01).

FM seed lots showed very poor or no germination when they were not from the last harvest. Most ungerminated seeds exhibited typical bacterial soft rot, with oozing observed. Additionally, both Ps·ph and Xa·ph were detected by tissue printing ELISA in these older seeds. BCMV was not detected in either the roots or leaves of six of the ten fresh FM seed lots. The average BCMV incidence in the remaining lots was less than 5% (Figure 2).

BCMV incidence in FG (22%) was significantly higher than in FM (3.2%) (Figure 2).

The prevalence of Ps·ph and Xa·ph in seed lots was 61.1% and 78.9%. However, the proportion of infected seeds within each lot with at least one positive seed was generally low. The average contamination rates were similar for both bacteria, with 9.2% for Xa·ph and 8.8% for Ps·ph. It is important to note that both these values had high standard deviations. Many seeds that tested positive for bacteria failed to complete germination, rotting shortly after radicle emergence. Interestingly, seeds infected with Ps·ph were frequently co-infected with Xa·ph.

3.2. Transmission of BCMV to Seeds

In plots AG1_20 and AG2_20, the seeds from harvest classified according to the date of detection of infection had BCMV levels of between 23 and 45%. In AG_1, there were significant differences between the only two dates of detection of BCMV in mother plants ($\chi^2 = 5.2 > 3.84$, $p < 0.05$, df = 1). No significant differences ($\chi^2 = 4.3 < 7.81$, $p < 0.05$, df = 3) or trends were observed in BCMV levels in AG_2 based on the date of detection of infection (Figure 3).

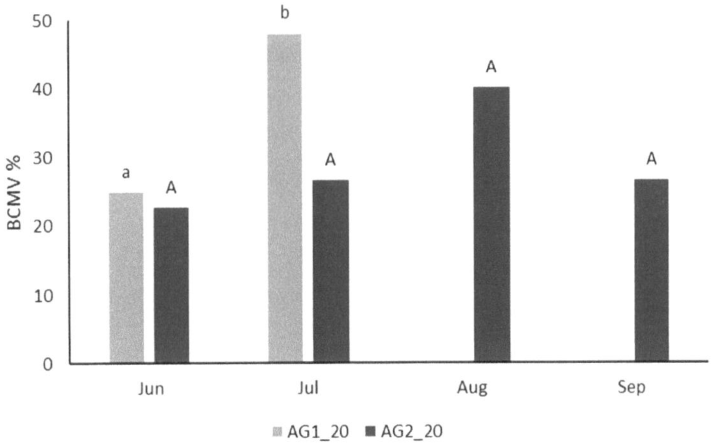

Figure 3. Percentage of transmission of bean common mosaic virus to bean seeds based on the date of detection of infection in mother bean plants in plots AG1_20 and AG2_20. Different letters indicate significant differences based on χ^2 tests for each plot.

Seeds from BCMV from the plot of FG tested at harvest in 2021 (LG_21) were $25.5 \pm 5\%$ infected. The transmission of BCMV to seeds in the harvest from LG_22 showed a slight trend towards a higher percentage of infection when transmission occurred at later dates (Figure 4), but the differences between dates were not significant based on χ^2 test ($\chi^2 = 12.54 < 14.06$; $p = 0.05$; df: 7). The average transmission was $28.4 \pm 3.2\%$.

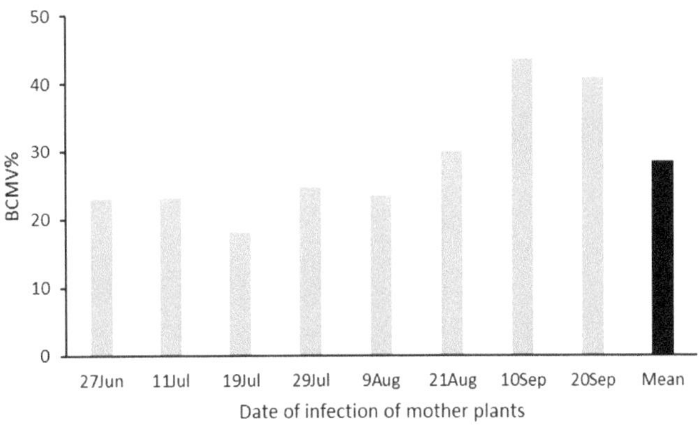

Figure 4. Percentage of transmission of bean common mosaic virus to seeds based on the date of detection of infection of bean plants in plot LG_22. There were no significant differences between dates of infection based on the χ^2 test.

In 2023, significant differences in seed transmission were observed between the two plots: LGBMz_23, surrounded by corn, and RGI_23, grown under protected cultivation (Figure 5) (F: 41.97, $p = 0.003$, df: 1). Seeds from LGBMz_23 had a higher proportion of infected seeds compared to those harvested from plants infected early in the season, but the chi-square test revealed no significant differences in infection rates among dates of detection of infection ($\chi^2 = 4.44 < 5.99$, $p = 0.05$, df = 2). The average BCMV transmission rate in LGBMz_23 seeds was 26.1%. BCMV levels in seeds from greenhouse plants were very low (2.5%), and no significant differences were detected across infection dates ($\chi^2 = 0.73 < 5.99$, $p = 0.05$, df = 2). Since all BCMV-positive plants were removed within the first month, there are no data from plants infected for longer periods in the greenhouse.

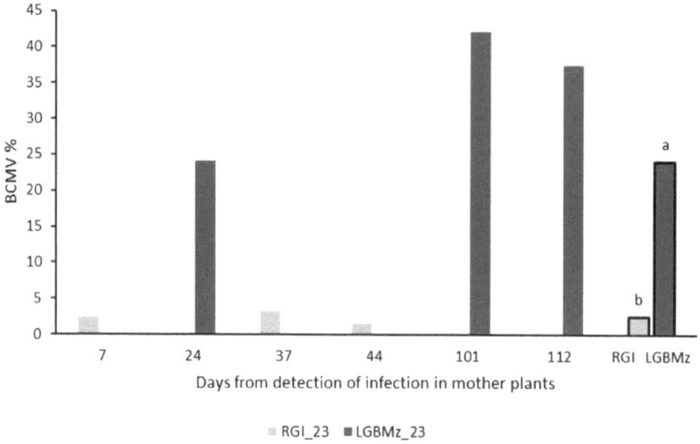

Figure 5. Percentage of transmission of bean common mosaic virus to bean seeds based on the elapsed time between the date of detection of infection of bean plants by BCMV and the harvest in LGBMz_23 and RGI_23. Columns with black borders are the mean incidence for the two 2023 plots; different letters indicate significant differences.

The prevalence of BCMV in seeds harvested from plants of FM in EIM_23 and SFM_23 at harvest was 76% (Figure 6). However, only seeds from two plants (one from each plot) had high levels (>40%) of BCMV detected in the radicle sample. The overall transmission of BCMV to seeds was 12.2 ± 3%. There were no significant differences between the two plots.

Figure 6. Percentage of transmission of bean common mosaic virus to bean seeds harvested from FM plants in EIM_23 and SFM_23; mean percentage of transmission in the two plots.

3.3. *Harvest*

Three out of five plots did not show significant differences in harvest weight based on the date of detection of infection. However, a trend towards higher yields was observed in plants infected at the end of the season and in plants testing negative for BCMV at harvest (Figure 7).

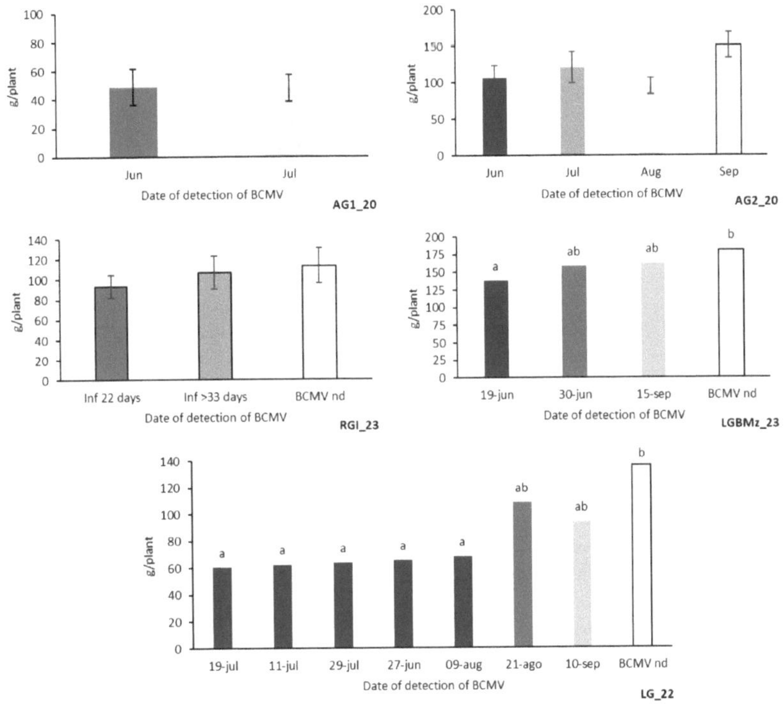

Figure 7. Harvest based on the date of detection of infection by bean common mosaic virus in bean plants in AG1_20, AG2_20, RGI_23, LGBmZ_23, and LG_22. The dry weights of seeds are shown in g/plant ±se when $p > 0.05$; the different letters represent significant differences based on the Tukey' b test, with $p < 0.05$.

In plots where significant differences were observed, they were typically between plants infected earliest and those infected later or not infected at all. The largest yield differences were observed in LG_22, reaching up to 50%, followed by LGBMz_23, with a maximum difference of 24%.

Similar trends were observed for fresh weight, number of pods per plant, and the percentage of seed weight relative to pod weight mirroring the pattern seen for dry seed weight.

3.4. Epidemiology and Control Strategies

As shown in Table 5 and Figure 8, the DPCs for BCMV incidence varied significantly across locations, years, and cultivars. FG consistently exhibited higher infection rates and final BCMV incidence compared to FM. In the absence of control measures, all FG plots (AG1_20, AG2_20, LG_21, LG_22) reached 100% BCMV incidence, often early in the season.

In 2020, two plots (AG1 and AG2) were planted with two lots of seeds with 10% BCMV. The initial infection rate was similar for both fields. However, by July, all sampled plants in AG1 were positive for BCMV, while those in AG2 reached 100% infection only at the final sampling (September). The rate of increase (slope of the linear regression line) was highest for AG1 throughout the study (Table 5). This rapid spread seems to be prevalent within the PGIs, as exemplified by LG_20; by harvest, nearly all plants in LG_20 were infected. To identify uninfected plants for harvest comparison, we had to analyse over 500 additional plants free of potyvirus symptoms, and most were positive for BCMV. Similarly, in 2021, LG_21 had 100% of plants infected at harvest and LG_22 displayed an intermediate spread rate in 2022 (Table 4). However, similar to other plots, nearly all plants tested positive for BCMV at harvest.

Despite starting with virus-free seedlings, removing infected plants initially, and having no neighbouring bean crops, MGI_22 and MGBMz_22 experienced rapid disease spread (40% infection in MGI within a month and 75% infection in MGBMz within two months). This was likely due to poor weed management practices and late or no proper vine training. As a result, the grower abandoned these plots, and no further data collection was possible after July.

During the 2023 season, the DPCs showed promising results in managing BCMV infection using various approaches. In the two plots chosen randomly among those with traditional intercropping corn/bean as the only difference with respect to the standard monoculture, the final level of BCMV infection was under 60%. Despite initially removing infected plants and minimal aphid catches on the yellow sticky traps, the virus became established within the protected greenhouse (RGI_23) (Figure 8). However, by harvest time, only 45% of the plants tested positive for BCMV, indicating a reduction in infection rates over time. In the field plot LGBMz_23, the corn border strategy was initially ineffective in preventing transmission, likely due to delayed corn growth and poor weed management, but the rate of spread of the virus decreased later on and at harvest, it was 75%. As expected, all plots situated outside the PGI area yielded improved results. In SFGMz_23, the combination of techniques successfully delayed the epidemic, resulting in a low infection rate and 25.7% of infected plants at the end of a long season.

The three plots with FM had significantly lower rates of infection inside and outside the IGP area. Despite being sown with untested seeds and surrounded by FG fields that reached 100% infection by August, the 2023 Lourenzá plot (FM) showed a BCMV infection rate of 45% at that time. The two plots in Lugo planted with tested FM seeds had 25 and 35% infection at the end of the season, but considerably lower infection levels (9 and 20%) at the time this cultivar would be typically harvested (before maturity).

Table 5. Results of the epidemiological study. For bean common mosaic virus epidemics with enough data, values of R^2, slope, and SEEy are from the model that best fits the DPC.

Plot	LA [1]	Model	R^2	Slope	SEEy	% BCMV Expected	% BCMV Observed
AG1_20	bh	Linear	0.89	0.009	0.153	>100	100
AG2_20	bh	Linear	0.99	0.02	0.007	100	100
LG_20	bh					-	97
LG_21	bh					-	100
LG_22	bh	Linear	0.95	0.017	0.077	92	98
LGBMz_23	h	Monomolecular	0.82	0.012	0.28	80	75
MGI_22	35 d [2]			-		-	40
MGBMz_22	56 d [2]			-		-	75
RGI_23	h	Gompertz	0.74	0.02	0.750	40.5	45
EIG_22	h	Gompertz	0.80	0.04	0.62	55	60
EIM_23	h	Linear	0.92	0.003	0.037	24	25
SFGMz_23	bh	Gompertz	0.55	0.01	0.49	22	25.7
SFM_23	h	Gompertz	0.84	0.019	0.418	31	34.8
LGMz1_23	bh					-	52.6
LGMz2_23	bh					-	59.7
LM_23	h					-	45

[1] LA: last analysis; bh. before harvest; h, at harvest; ah, after harvest. [2] Monitoring stopped due to wrong management of the plot. SEE standard error of the estimate.

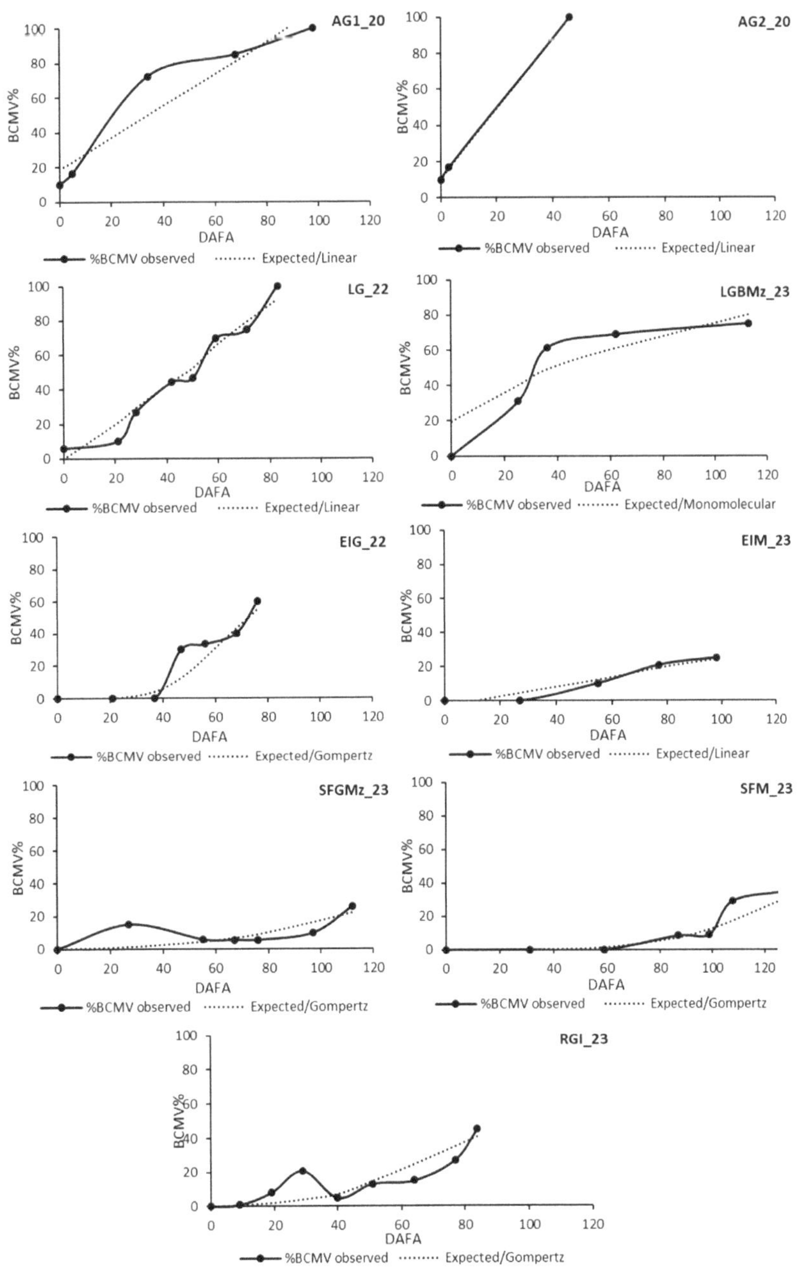

Figure 8. Disease progress curves for 9 of the bean common mosaic virus epidemics studied between 2020 and 2023. DAFA: days after first analysis of bean plants. The regression lines correspond to the expected values based on the epidemiological model that best fits the DPC values (Table 5).

4. Discussion

Dry beans are one of the most important crops in the world and BCMV is considered economically important throughout Africa, Europe, North America, and Latin America, with infection levels that often reach 100% and estimated yield losses of 35–98% [16]. Our findings with Faba Galaica (FG) align with these general trends in BCMV prevalence and damage. For large-scale cultivation, various control strategies exist to minimize losses from seed-borne pathogens, ranging from certified seeds to resistant cultivars (CABI, 2023). In Protected Geographical Indications (PGIs), managing plant and seed health is critical for the long-term viability of traditional cultivars and farming practices. These elements define the unique characteristics (typicity) required for PGI certification. That makes PGIs interesting crop systems to study plant disease epidemics: they are limited to a certain area, use one or few cultivars in small fields with traditional growing systems, and usually get high prices in specific markets. For small PGIs like 'Faba de Lourenzá', the total seed requirement is low, and like in many other PGIs in the EU and bean crops in developing countries [38], farmers often save or exchange seeds from the previous year. In this context, we observed a very high prevalence of seed-borne pathogens, particularly viruses. Despite these findings and potential yield losses (Figure 7), growers often express minimal concern, likely because, similar to potato varieties infected with Potato Virus Y (PVY), plants infected during the season may not exhibit clear disease symptoms.

This study analysed thousands of samples between 2020 and 2024, a throughput impossible with other virus detection methods due to cost constraints. Tissue printing proved to be the most efficient technique for field screening and evaluating propagation material in our previous studies with grapevine and potatoes [29]. Tissue printing has been used since the 1990s for the detection of viruses, bacteria, and MLO in cereals, citrus, plums, and potatoes, and for polyphagous viruses such as cucumber mosaic virus (CMV) or tomato spotted wilt virus (TSWV) among many others [39]. Legume viruses were among the first viruses detected using this technique [34,39]. ELISA tests for bacteria, including tissue printing ELISA, have been approved as one of the European Plant Protection Organization (EPPO) Standards (2010–09) [40] but to our knowledge, it has not been reported for Xa·ph or Ps·ph in bean germinated seeds or leaves.

Analysis of FG seed lots from PGI fields and seeds from known BCMV-infected mother plants yielded similar average infection rates in the range of 20–30%. That seed infection rate translates to near-complete field infection by harvest, as observed in our samplings from Lourenzá (2020–2022) where no control measures were implemented. Cifuentes et al. [23] reported that sowing seeds with only 4% infection resulted in 100% plant infection at harvest. In our study, with a much higher seed infection rate (>20%), this level of virus prevalence will likely be reached well before the end of the season, as exemplified by plot AG2_20. The only exception to the high seed transmission rate observed in FG is RGI_23, which showed a very low value of 2.5%. This could be due to greenhouse conditions, but further confirmation is needed. There are a number of factors that influence seed transmission rates, including environmental conditions [41]. Greenhouse environments differ significantly from open fields in the region, potentially explaining the lower transmission rate observed in RGI_23. FM lower incidence in seed lots (<5%) corresponds to lower levels of transmission to seed (2.5%).

The BCMV levels observed in FG within the PGI Faba de Lourenzá are comparable to those reported for other Spanish PGIs [23,42]. However, Asturias reported a significantly lower incidence of potyviruses [43]. FM, cultivated alongside FG, exhibits a contrasting pattern. FM seed lots have lower infection rates, field plants typically test negative until later in the season, and seed transmission from infected plants is also lower compared to FG. This lower level of BCMV infection in FM might be attributed to two factors: its determinate growth habit leading to a shorter growing season, and the harvesting of green pods before full maturity and drying. The lower BCMV seed transmission in FM compared to FG translates to a reduced initial inoculum for aphids within FM plots. Additionally, late-season transmission from surrounding infected FG fields, via aphids, would likely occur after FM seed harvest. The two FM individual plants with over 40% infected seeds (Figure 6) were likely exceptions, possibly due to harvesting at a later stage when seeds were more mature.

Our findings revealed significant harvest weight differences, primarily between plants infected early in the season and those testing negative at harvest. This highlights the importance of planting seeds with minimal BCMV infection. Yield reductions in fields sown with 20–40% infected seeds might be more substantial than growers acknowledge. Studies in other Spanish PGIs report bean yield losses of 40–50% due to BCMV [42,43]. Though traditional varieties cultivated for decades may seem unaffected, the true impact of potyvirus infections is likely underestimated due to the absence of data on potential yields achievable with disease-free seeds. Local minor cultivars grown in small, often non-irrigated fields, exhibit high yield variability due to various factors. This makes it challenging for farmers to attribute yield losses specifically to viral diseases, especially when symptoms are mild or absent in plants infected later in the season [42]. Tolerance is assumed when symptoms are very mild or not shown at all in plants infected after the first month [42]. Current seed certification programs, effective in reducing seed-borne diseases, are often impractical for small regions cultivating local varieties due to limited economic viability. In these cases where seeds are reused and exchanged and vector control in the field is absent, the risk of reaching 100% infection at harvest and progressive plant degeneration is very high [20]. Moreover, although these traditional beans fetch a higher price in gourmet shops than common varieties (15–25 €/kg in 2023), their production costs are also higher due to being less intensively cultivated and requiring more manual labour compared to fully mechanized improved varieties. Therefore, minimizing virus-related losses becomes crucial for maintaining their profitability.

Our efforts to reduce virus transmission were sometimes hampered by grower decisions related to weed control, planting density, or vine training. In some cases (MGI_22, MGBMz_22, and early stages of LGBMz_23), this was due to inadequate weed and vine management just after planting. This is because weeds, including leguminous species (*Trifolium* spp.) that can harbour BCMV, attract aphids at the beginning of summer. Carazo and Romero [44] identified 15 additional reservoir species for BCMV beyond legumes. In Nigeria, virus incidence in weeds around the legume plots was 2.5% [45]. Studies have shown that mulching reduces early-season infection in seed potatoes [26].

Figure 8 showcases the diverse effects of location, cultivar, and control strategies on Disease Progress Curves (DPCs). While some models provide superior fits to the epidemic data, the slopes of these best-fit DPCs are also demonstrably influenced. Interestingly, natural epidemics were best modelled by linear functions with high R-squared values and low errors (Table 4). Conversely, controlled epidemics were primarily described by Gompertz models, albeit with a consistently lower goodness-of-fit. In most cases, the predicted final infection level aligned well with observed values (Table 5). This information could be valuable for pre-emptively determining crop destination based on anticipated infection severity. Early removal of infected plants most significantly impacts DPC trajectory and model fit.

The parallels between BCMV and Potato Virus Y (PVY) in seed potato production are evident. Consequently, established management strategies for potatoes can be effectively applied to control seed-borne viruses in beans and other crops [26,28,29,46,47]. This strategy combines using low-infection seeds, removing symptomatic young plants, and implementing intercropping. Maize as an intercrop offers distinct advantages over border crops, particularly in small fields [26].

For indeterminate growth cultivars like FG, intercropping with maize is a simple practice. It provides support for climbing vines and leads to significant cost reductions. Interestingly, the traditional cultivation system for FG has been associated with corn. However, this practice is declining in favour of monoculture. While there is no such tradition for FM, intercropping with sweetcorn has been proposed for bush bean cultivars in other regions of Spain as well [48].

Similar to potato intercropping, interplanted crops surrounding the bean plants act as a mechanical barrier for aphids and a virus sink. These companion crops attract aphids away from the main bean crop [26], which likely explains the positive results observed in

SFGMz_23 and the lower level of BCMV at the end of the season in both LGMz_23 plots. An integrated approach that combines multiple techniques like these is most effective. Additionally, utilizing virus detection methods like tissue printing ELISA for pre-planting seed lot evaluation is crucial.

5. Conclusions

The high prevalence of Bean common mosaic virus (BCMV) in FG fields is primarily caused by infected seeds. This is due to the virus's high seed transmissibility and the occurrence of early horizontal transmission by aphids.

The significant differences in BCMV incidence and seed transmission between the two evaluated bean cultivars seem to be related to their different growing seasons and durations in the field. Harvesting before full maturity in FM results in poor germination of seeds older than the previous season ones but also in lower seed transmission.

The best way to reduce the virus incidence and minimize yield losses is through the commitment of farmers in the PGIs to:

(a) locate plots for seed production far from the growing area where a combination of techniques for preventing virus transmission will be applied by trained growers using an IPM approach,
(b) test representative samples of the seeds produced in those plots and use exclusively lots with the lowest possible virus level, and
(c) train growers to pay special attention to soil and plant management from the beginning of crop planting to delay virus transmission.

Author Contributions: Conceptualization, S.E.-G. and C.C.; Methodology, S.E.-G. and C.C.; Software, C.C.; Validation, S.E.-G. and C.C.; Formal Analysis, C.C. and F.L.-P.; Investigation, S.E.-G., C.C. and L.G.P.-L.; Resources, C.C.; Data Curation, S.E.-G., C.C. and L.G.P.-L.; Writing—Original Draft Preparation, C.C.; Writing—Review and Editing, C.C., S.E.-G., F.L.-P. and L.P; Visualization, F.L.-P.; Supervision, C.C.; Project Administration, C.C.; Funding Acquisition, C.C. All authors have read and agreed to the published version of the manuscript.

Funding: This research was partially funded by a contract with Unión de Cooperativas AGACA through the "Axencia Galega da Calidade Alimentaria", Proxectos Piloto CALL (DOG 33 18/022021Exp: FEADER2021/066A).

Data Availability Statement: The raw data supporting the conclusions of this article will be made available by the authors upon request.

Acknowledgments: Thanks to Antonio de Ron (CSIC), the Cooperative "Terras da Mariña" and other growers in and outside the PGI "Faba de Lourenzá" for technical support, seed lots, and field access.

Conflicts of Interest: The authors declare no conflicts of interest.

References

1. Santalla, M.; Rodiño, A.P.; De Ron, A.M. Allozyme evidence supporting southwestern Europe as a secondary center of genetic diversity for common bean. *Theor. Appl. Genet.* **2002**, *104*, 934–944. [CrossRef] [PubMed]
2. De Ron, A.M.; Rodiño, A.P.; Santalla, M.; Martínez-Sierra, V. Denominación de Origen Protegida Faba de Lourenzá. In *Nuevos Retos y Oportunidades de las Leguminosas en el Sector Agroalimentario Español*; De los Mozos, M., Giménez, M.J., Rodríguez Conde, M.F., Sánchez, R., Eds.; Consejería de Agricultura de Castilla—La Mancha: Toledo, Spain, 2006; pp. 387–392.
3. Boletín Oficial del estado-A-2018-6283. Orden APM/477/2018, de 26 de abril, por la que se dispone la inclusión de diversas variedades de distintas especies en el Registro de Variedades Comerciales: 20170135 faba galaica: 49361-49364. Available online: https://www.boe.es/diario_boe/txt.php?id=BOE-A-2018-6283 (accessed on 31 May 2024).
4. Ministerio de Ciencia e Innovación/Consejo Superior de Investigaciones Científicas. Ficha "Faba do Marisco". Available online: https://www.mbg.csic.es/wp-content/uploads/2021/03/Ficha-Faba-do-Marisco.pdf (accessed on 31 May 2024).
5. Pagán, I. Transmission through seeds: The unknown life of plant viruses. *PLoS Pathog.* **2022**, *18*, e1010707. [CrossRef] [PubMed]
6. Dell Olmo, E.; Tiberini, A.; Sigillo, L. Leguminous seedborne pathogens: Seed health and sustainable crop management. *Plants* **2023**, *19*, 2040. [CrossRef] [PubMed]

7. Saiz, M.; de Blas, C.; Carazo, G.; Fresno, J.; Romero, J.; Castro, S. Incidence and characterization of bean common mosaic virus isolates in Spanish bean fields. *Plant Dis.* **1993**, *79*, 79–81. [CrossRef]
8. Melgarejo, P.; García-Jiménez, J.; Jordá, M.C.; López, M.M.; Andrés, M.F.; Duran-Vila, N. *Patógenos de plantas descritos en España;* Ministerio de Medio Ambiente y Medio Rural y Marino: Madrid, Spain, 2010; p. 49.
9. Romero, J. Mosaicos de la judía. In *Enfermedades de las Plantas Causadas por Virus y Viroides;* Ayllón, M.A., Cambra, M., Llave, C., Moriones, E., Eds.; Bubok Publishing S.L. y Sociedad Española de Fitopatología: Madrid, Spain, 2016; pp. 455–466.
10. Walkey, D.G.A. Ecology and Epidemiology of Plant Viruses 2nd ed. In *Applied Plant Virology;* Springer: Dordrecht, The Netherlands, 2012; Chapter 8, pp. 210–225.
11. Morales, F.J.; Castaño, M. Seed transmission characteristics of selected bean common mosaic virus strains in differential bean cultivars. *Plant Dis.* **1987**, *71*, 51–53. [CrossRef]
12. Shukla, D.D.; Ward, C.W.; Brunt, A.A. *The Potyviridae;* CAB International: Wallingford, UK, 1994; 516p.
13. Aishwarya, P.; Rangaswamy, K.T.; Basavaraju, S.; Achari, R.; Prameela, H.A. Evaluation of the seed-borne nature of bean common mosaic virus (BCMV) in cowpea. *Int. J. Curr. Microbiol. Appl. Sci.* **2020**, *9*, 239–245. [CrossRef]
14. Tang, M.; Feng, X. Bean Common Mosaic Disease: Etiology, Resistance Resource, and Future Prospects. *Agronomy* **2023**, *13*, 58. [CrossRef]
15. Gonzalez-Cruces, A.; Arista-Carmona, E.; Díaz-Arias, K.V.; Ramírez-Razo, K.; Hernández-Livera, A.; Acevedo-Sánchez, G.; Mendoza-Ramos, C.; Mora-Aguilera, G. Epidemiology of bean common mosaic virus and *Alternaria alternata* in 12 *Phaseolus vulgaris* genotypes. *Mex. J. Phytopathol.* **2022**, *40*, 188–220. [CrossRef]
16. Galvez, E. Aphid-transmitted viruses. In *Bean Production Problems: Disease, Insect, Soil and Climatic Constraints of Phaseolus vulgaris;* Schwartz, H.F., Gálvez, E., Guillermo, E., Eds.; Centro Internacional de Agricultura Tropical (CIAT): Cali, Colombia, 1980; pp. 211–238.
17. Meziadi, C.; Blanchet, S.; Geffroy, V.; Pflieger, S. Genetic resistance against viruses in *Phaseolus vulgaris* L.: State of the art and future prospects. *Plant Sci.* **2017**, *265*, 39–50. [CrossRef]
18. Mangeni, B.H.; Were, K.; Ndong'a, M.; Mukoye, B. Incidence and severity of bean common mosaic disease and resistance of popular bean cultivars to the disease in western Kenya. *J. Phytopathol.* **2020**, *168*, 501–515. [CrossRef]
19. Worrall, E.A.; Wamonje, F.O.; Mukeshimana, G.; Harvey, J.J.W.; Carr, J.P.; Mitter, N. Bean common mosaic virus and bean common mosaic necrosis virus: Relationships, biology, and prospects for control. *Adv. Virus Res.* **2015**, *93*, 1–46.
20. Sastry, K.S. *Seed-Borne Plant Virus Diseases;* Springer: New Delhi, India, 2013; 353p.
21. Hampton, R.O. The nature of bean yield reduction by bean yellow and bean common mosaic virus. *Phytopathology* **1975**, *65*, 1342–1346. [CrossRef]
22. Morales, F.J. Common Bean. In *Virus and Virus-like Diseases of Major Crops in Developing Countries;* Loebenstein, G., Thottappilly, G., Eds.; Springer: Dordrecht, The Netherlands, 2003; pp. 425–445.
23. Cifuentes, G.; Castro, S.; Romero, J. Los Potyvirus de judía transmitidos por semilla: Incidencia de la enfermedad y su efecto en los rendimientos. In Proceedings of the II Seminario de la Judía de la Península Ibérica, Villaviciosa, Spain, 5–7 September 2000.
24. Çelik, A.; Emiralioğlu, O.; Yeken, M.Z.; Çiftçi, V.; Özer, G.; Kim, Y.; Baloch, F.S.; Chung, Y.S. A novel study on bean common mosaic virus accumulation shows disease resistance at the initial stage of infection in Phaseolus vulgaris. *Front. Genet.* **2023**, *14*, 1136794. [CrossRef] [PubMed]
25. Walkley, A.D.G.; Dance, C.M. The effect of oil sprays on aphid transmission of Turnip Mosaic, Beet Yellows, Bean Common and Bean Yellow Mosaic Viruses. *Plant Dis.* **1979**, *63*, 877–881.
26. Dupuis, B.; Cadby, J.; Goy, G.; Tallant, M.; Derron, J.; Schwaerzel, R.; Steinger, T. Control of potato virus Y (PVY) in seed potatoes by oil spraying, straw mulching and intercropping. *Plant Pathol.* **2017**, *66*, 960–969. [CrossRef]
27. Dupuis, B.; Schwaerzel, R.; Goy, G.; Tallant, M.; Derron, J. Stepwise development of an efficient method to control Potato virus Y spread in seed potato fields. In Proceedings of the European Association for Potato Research, Virology Section, Hamar, Norway, 4–9 July 2010; Spetz, C., Blystad, D.R., Eds.; Bioforsk Fokus: Hamar, Norway, 2010; Volume 5, p. 22.
28. Hooks, C.R.R.; Fereres, A. Protecting crops from non-persistently aphid-transmitted viruses: A review on the use of barrier plants as a management tool. *Virus Res.* **2006**, *120*, 1–16. [CrossRef]
29. Martin-Lopez, B.; Varela, I.; Marnotes, S.; Cabaleiro, C. Use of oils combined with low doses of insecticide for the control of *Myzus persicae* and PVY epidemics. *Pest. Manag. Sci.* **2006**, *62*, 372–378. [CrossRef]
30. Gadhave, K.R.; Gautam, S.; Rasmussen, D.A.; Srinivasan, R. Aphid transmission of Potyvirus: The largest plant-Infecting RNA virus genus. *Viruses* **2020**, *12*, 773. [CrossRef] [PubMed]
31. Abd El-Aziz, M.H. Three modern serological methods to detect plant viruses. *J. Plant Sci. Phytopathol.* **2019**, *3*, 101–106. [CrossRef]
32. Cabaleiro, C.; Marnotes, S.; Couceiro, C.; Martínez, M.A.; Lastra, B.; Martín, B.; García, L.; Alvarez, S. Uso de cultivos borde o asociados para el control de virus no persistentes en patata (PVY) y alubias (BCMV). In Proceedings of the XIV Congreso de la SEF, Lugo, Spain, 15–19 September 2008.
33. Legorburu, F.J.; Ruiz de Galarreta, J.I.; Abad, J.M.; Pérez de San Román, C. Bean Common (BCMV) and Bean common mosaic necrosis (BCMNV) potyvirus in relation to bean landraces in the Basque Country. *Investig. Agr. Prod. Prot. Veg.* **1998**, *13*, 153–158.
34. Lin, N.S.; Hsu, Y.H.; Hsu, H.T. Immunological detection of plant viruses and a mycoplasmalike organism by direct tissue blotting on nitrocellulose membranes. *Phytopathol.* **1990**, *80*, 824–828. [CrossRef]

35. Cambra, M.; Gorris, M.T.; Roman, M.P.; Terrada, E.; Garnsey, S.M.; Camarasa, E.; Olmos, A.; Colomer, M. Routine detection of Citrus tristeza virus by direct Immunoprinting-ELISA method using specific monoclonal and recombinant antibodies. In Proceedings of the 14th Conference IOCV, Campinas, Brazil, 13–18 September 1998; da Graça, J.V., Lee, R.F., Yokomi, R.K., Eds.; International Organization of Citrus Virologists c/o Department of Plant Pathology: Riverside, CA, USA, 2000.
36. Madden, L.V.; Hughes, G.; van den Bosch, F. *The Study of Plant Disease Epidemics*; American Phytopathological Society: St. Paul, MN, USA, 2007; 421p.
37. Nutter, F.W., Jr.; Eggenberger, S.K.; Littlejohn, K.J. Visualizing, describing, and modeling Disease Progress Curves using EPIMODEL. In *Exercises in Plant Disease Epidemiology*, 2nd ed.; Stevenson, K.L., Jeger, M.J., Eds.; American Phytopathological Society: St. Paul, MN, USA, 2015; pp. 21–30.
38. Mwaipopo, B.; Nchimbi-Msolla, S.; Njau, P.; Tairo, F.; William, M.; Binagwa, P.; Kweka, E.; Kilango, M.; Mbanzibwa, D. Viruses infecting common bean (*Phaseolus vulgaris* L.) in Tanzania: A review on molecular characterization, detection and disease management options. *Afr. J. Agric. Res.* **2017**, *12*, 1486–1500.
39. Makkouk, K.M.; Kumari, S.G. Detection of ten viruses by the tissue-blot immunoassay (TBIA). *Arab. J. Plant Protection* **1996**, *14*, 3–9.
40. European Plant Protection Organization. ELISA tests for plant pathogenic bacteria. *Bull. OEPP/EPPO* **2010**, *40*, 369–372. [CrossRef]
41. Simmons, H.; Munkvold, G. Seed Transmission in the Potyviridae. In *Global Perspectives on the Health of Seeds and Plant Propagation Material*; Gullino, M.L., Munkvold, G., Eds.; Plant Pathology in the 21st Century 6; Springer Science+Business Media: Dordrecht, The Netherlands, 2014; pp. 3–15.
42. Campelo, M.P.; Reinoso, B.; González, A.J. Incidencia y transmisión de Potyvirus en semillas de judía de la I.G.P. "Alubia de la Bañeza-León". In Proceedings of the XIII Congreso Nacional de la Sociedad Española de Fitopatología, Murcia, Spain, 18–22 September 2006.
43. González, A.J. Virus fitopatógenos transmisibles por semilla en judía tipo "granja asturiana". *Bol. San. Veg. Plagas* **2004**, *30*, 595–603.
44. Carazo, G.; Romero, J. Malas hierbas reservorios de los Potyvirus que infectan judías. In Proceedings of the IV Seminario de Judías de la Península Ibérica, Valladolid, Spain, 2–4 September 2008.
45. Odedara, O.O.; Kumar, P.L. Incidence and diversity of viruses in cowpeas and weeds in the unmanaged farming systems of savanna zones in Nigeria. *Arch. Phytopathol. Plant Prot.* **2017**, *50*, 1–12. [CrossRef]
46. Boiteau, G.; Singh, M.; Lavoie, J. Crop border and mineral oil sprays used in combination as physical control methods of the aphid-transmitted potato virus Y in potato. *Pest. Manag. Sci.* **2009**, *65*, 255–259. [CrossRef]
47. Lacomme, C.; Pickup, J.; Fox, A.; Glais, L.; Dupuis, B.; Steinger, T.; Rolot, J.L.; Valkonen, J.P.T.; Kruger, K.; Nie, X.; et al. Transmission and Epidemiology of Potato virus Y. In *Potato virus Y: Biodiversity, Pathogenicity, Epidemiology and Management*; Lacomme, C., Glais, L., Bellstedt, D., Dupuis, B., Karasev, A., Jacquot, E., Eds.; Springer: Cham, Germany, 2017; pp. 141–176.
48. Santalla, M.; Rodiño, A.P.; Casquero, P.; de Ron, A.M. Interactions of bush bean intercropped with field and sweet maize. *Eur. J. Agron.* **2001**, *15*, 185–196. [CrossRef]

horticulturae

Communication

Molecular Characteristics and Biological Properties of Bean Yellow Mosaic Virus Isolates from Slovakia

Michaela Mrkvová [1,*], Jana Kemenczeiová [1,2], Adam Achs [2], Peter Alaxin [1,2], Lukáš Predajňa [2], Katarína Šoltys [3], Zdeno Šubr [2] and Miroslav Glasa [1,2,*]

[1] Department of Biology, Faculty of Natural Sciences, Institute of Biology and Biotechnology, University of Ss. Cyril and Methodius in Trnava, Námestie J. Herdu 2, 917 01 Trnava, Slovakia; kemenczeiova1@ucm.sk (J.K.); peter.alaxin@savba.sk (P.A.)

[2] Institute of Virology, Biomedical Research Center of Slovak Academy of Sciences, Dúbravská cesta 9, 845 05 Bratislava, Slovakia; adam.achs@savba.sk (A.A.); lukas.predajna@savba.sk (L.P.); zdeno.subr@savba.sk (Z.Š.)

[3] Department of Microbiology and Virology, Faculty of Natural Sciences, Comenius University in Bratislava, Ilkovičova 6, 842 15 Bratislava, Slovakia; katarina.soltys@uniba.sk

* Correspondence: michaela.mrkvova@ucm.sk (M.M.); miroslav.glasa@savba.sk (M.G.)

Abstract: Analysis of the viromes of three symptomatic *Fabaceae* plants, i.e., red clover (*Trifolium pratense* L.), pea (*Pisum sativum* L.), and common bean (*Phaseolus vulgaris* L.), using high-throughput sequencing revealed complex infections and enabled the acquisition of complete genomes of a potyvirus, bean yellow mosaic virus (BYMV). Based on phylogenetic analysis, the Slovak BYMV isolates belong to two distinct molecular groups, i.e., VI (isolate FA40) and XI (isolates DAT, PS2). Five commercial pea genotypes (Alderman, Ambrosia, Gloriosa, Herkules, Senator) were successfully infected with the BYMV-PS2 inoculum and displayed similar systemic chlorotic mottling symptoms. Relative comparison of optical density values using semi-quantitative DAS-ELISA revealed significant differences among virus titers in one of the infected pea genotypes (Ambrosia) when upper fully developed leaves were tested. Immunoblot analysis of systemically infected Alderman plants showed rather uneven virus accumulation in different plant parts. The lowest virus accumulation was repeatedly detected in the roots, while the highest was in the upper part of the plant stem.

Keywords: *Fabaceae*; high-throughput sequencing; pea; potyvirus; virome; virus accumulation

1. Introduction

Cultivated or wildly grown legumes, i.e., plants belonging to the *Fabaceae* family, are known to naturally host a wide range of viruses [1,2]. Some of these pathogens cause severe symptoms, decreasing yields and the quality of legumes [3]. This results in subsequent economic losses, as legumes represent an important source of protein in human and livestock nutrition [4].

Potyviruses belong to the largest genus in the *Potyviridae* family, comprising more than 200 species (https://ictv.global/taxonomy, accessed on 22 January 2024). Potyviruses are characterized by a positive-sense single-stranded RNA genome of approximately 10 kb encapsidated in flexuous rod-shaped filaments [5,6] (Supplementary Figure S1). Bean yellow mosaic virus (BYMV) is one of the most important legume-infecting potyviruses, although its wide host range also includes various monocots and dicots within both domesticated and wild plant species [7,8]. Based on studies of BYMV molecular variability, at least nine molecular groups have been identified [9–11].

Over the past decade, high-throughput sequencing (HTS) technologies have revolutionized plant virome research by providing an unbiased approach to the identification and characterization of viruses present [12,13]. The application of HTS has proven to be very successful for virus discovery, resolving disease etiology in many agricultural crops, and has enabled the description of new viral pathogens also in various legumes [14–16]. In addition to the discovery of new virus species, the use of HTS has enabled the characterization of divergent forms of known and well-established viruses, thus expanding knowledge of their global molecular variability [17,18].

As for other RNA viruses, mutations and recombination contribute to host adaptation, host-dependent pathogenicity, vector transmissibility, and/or viral accumulation in different hosts [19–21]. Therefore, understanding intra-species diversity may provide essential information to adopt tailored control and preventive measures, e.g., through polyvalent detection or planting of less susceptible plant genotypes [22,23].

Molecular and phylogenetic analyses based on partial genomic sequencing or single genes can provide biased results and misleading interpretations due to unequal evolutionary rates of individual genes or possible recombination event(s) in untargeted parts of the genome. In this work, we obtained three complete genomes of BYMV isolates from Slovakia by HTS-based virome analysis of *Fabaceae* environmental samples and performed their molecular and phylogenetic analyses. To contribute to a better understanding of host/virus interactions, we further evaluated the susceptibility of commercial pea genotypes to BYMV infection and compared the relative accumulation of viral proteins in different plant parts.

2. Materials and Methods

2.1. Determination of Complete BYMV Genomes Using HTS

HTS analysis of the *Fabaceae* samples was performed, as described previously [24]. Briefly, total RNAs were extracted from the leaves of original host plants using a Spectrum^{TM} Plant Total RNA Kit (Sigma-Aldrich, St. Louis, MO, USA). To enrich the viral fraction, ribosomal RNA was depleted using the Zymo-Seq RiboFree Universal cDNA Kit (Zymo Research, Irvine, CA, USA). Ribosome-depleted RNA preparations were used for double-stranded cDNA synthesis using the SuperScript II kit (Thermo Fisher Scientific, Waltham, MA, USA), and the samples were processed with the transposon-based chemistry library preparation kit (Nextera XT, Illumina, San Diego, CA, USA), followed by HTS on an Illumina MiSeq platform (2 × 150 bp paired reads paired-end sequencing, Illumina, San Diego, CA, USA).

High-quality trimmed reads were used for de novo assembly and contigs were aligned to the viral genomes database (ftp://ftp.ncbi.nih.gov/genomes/Viruses/all.fna.tar.gz, accessed on 4 November 2023) using CLC Genomic Workbench and Geneious v.8.1.9 software. Subsequently, the reads were remapped against the BYMV full-length sequence NC_003492 retrieved from GenBank (https://www.ncbi.nlm.nih.gov, accessed on 4 November 2023) and against the genomes of additional viruses identified in the previous step to complement and validate the obtained genomic sequences.

The presence of BYMV in the samples analyzed using HTS was confirmed by Sanger sequencing of RT-PCR products encompassing the *CP* gene, amplified using specific primers designed from the HTS-based sequence (BY_8548F 5′-AGAGAAGCTCAATGCTGGTG-3′, forward/BY_9322R 5′-GACATCTCCTGCTGTGTGTC-3′, reverse). Moreover, double-antibody sandwich (DAS)-ELISA, using commercial BYMV-specific antibodies (DSMZ No. RT-0717, [25]), or western blot analysis, using BYMV-specific polyclonal antibodies [26], were used for initial immunological detection.

2.2. Biological Experiments

The PS2 isolate was mechanically transmitted to *Nicotiana benthamiana* from the original pea plant (Figure 1). Fourteen days p.i., the systemically infected leaves were harvested, sliced, and frozen at −80 °C in order to be used as a homogenous source of BYMV for subsequent biological experiments.

Figure 1. Leaf symptoms observed in *Fabaceae* plants at the time of sampling for HTS analysis (DAT—red clover (*T. pratense*), PS2—pea (*P. sativum*), FA40 common bean (*P. vulgaris*).

Five commercial pea varieties, i.e., Alderman, Ambrosia, Gloriosa, Senator (all supplied by Osiva Moravia Ltd., Olomouc, Czech Republic), and Herkules (supplied by Zelseed Ltd., Horná Potôň, Slovakia) were sown in a sterilized garden soil in classic plastic rooting containers and cultivated under controlled conditions in a growth chamber with a 16 h photoperiod (16 h of light/8 h of darkness), a light intensity of 152 µmol m^{-2} s^{-1} FAR, and a temperature of 22 ± 2 °C.

Pea plants were then mechanically inoculated at the stage of one to two fully developed leaves. The infectious juice was obtained by grinding PS2-infected *N. benthamiana* leaves from the previous step. The development of symptoms was observed visually, and the presence of BYMV was tested by DAS-ELISA 14 days p.i.

2.3. Estimation of Virus Accumulation in Plants

Two-week-old plants of *P. sativum* from five BYMV-susceptible genotypes (Alderman, Ambrosia, Gloriosa, Herkules, Senator) were mechanically inoculated with BYMV-PS2-containing sap (1/15 dilution in Norit buffer, https://www.dsmz.de/fileadmin/_migrated/content_uploads/Inoculation_01.pdf, accessed on 22 April 2020). Overall, homogeneous groups of five to eight plants of each genotype, grown in individual pots, were successfully inoculated. The plants were tested individually by DAS-ELISA [25] 21 days p.i. using the polyclonal antibody (DSMZ set No. RT-0717) to compare relatively the virus antigen accumulation in systemically infected plants. Five leaf discs of ca. 0.15 g were taken from the top fully developed leaf of each inoculated plant and uninfected control pea plants using the bottom of a pipette tip to standardize the amount of tested sample for each plant. The discs were homogenized in PBS (1/25 *w*/*v*) containing 0.05% Tween-20 and 2% polyvinylpyrrolidone 40. Each plant sample (5–8 per genotype) was applied in duplicate to a single ELISA plate, so that absorbance values measured at 405 nm for four genotypes could be compared. Statistical analysis was conducted in R version 4.3.2 [27], using the "stats" package.

Whole proteins were extracted from six different parts of the Alderman pea plants (pod, upper stem, upper leaf, bottom stem, bottom leaf, root) using the Pierce^{TM} Plant Total Protein Extraction Kit (Thermo Fisher Scientific, Waltham, MA, USA). A standard curve prepared using bovine serum albumin was applied for the calculation of the protein concentrations, determined using the Pierce TM BCA Protein Assay Kit (Thermo Fisher Scientific, Waltham, MA, USA). For each sample, the same amount of proteins (50 µg) was mixed 1:1 with Laemmli sample buffer, subjected to SDS-PAGE in 12% gel, transferred to a PVDF membrane by a semidry blotting apparatus, and detected using a polyclonal anti-BYMV antibody (DSMZ No. AS-0717)/goat anti-rabbit IgG antibody conjugated with alkaline phosphatase (Sigma, St. Louis, MO, USA) and NBT/BCIP as substrate (described in Nováková et al. [28]). The membrane was recorded using a camera, and densitometry analysis of specific bands was performed using the Image Studio Lite Quantification Software v. 5.2 (LI-COR Biosciences, Lincoln, NE, USA).

3. Results and Discussion

In an effort to unravel the virome of three symptomatic *Fabaceae* plants, i.e., red clover (*T. pratense*), pea (*P. sativum*), and common bean (*P. vulgaris*), leaf samples were subjected to HTS analysis.

For two samples (DAT, FA40), HTS analysis revealed multiple viral infections. While the DAT sample was infected with BYMV, soybean dwarf virus (SDV, genus *Luteovirus*), and blackgrass cryptic virus-2 (unclassified), the infection of FA40 involved BYMV and cucumber mosaic virus (CMV, genus *Cucumovirus*). On the contrary, only a single BYMV infection was found in the PS2 sample (Table 1). Currently, it is largely proven through virome analyses that mixed infections of plants are frequently found in nature, complicating the elucidation of disease etiology [24,29,30]. The presence of multiple viruses in both DAT and FA40 plants made it difficult to decipher the role of any single virus in causing the observed symptoms (Figure 1).

Table 1. List of *Fabaceae* samples used for HTS analysis and their characteristics.

Sample	Natural Host	Location	Year of Sampling	Leaf Symptoms	Viruses Identified in the Sample [1]
DAT	*Trifolium pratense*	Bratislava	2020	Mosaic	BYMV, SbDV, BGCV-2
PS2	*Pisum sativum*	Pezinok	2021	Mottling, mild leaf distortions	BYMV
FA40	*Phaseolus vulgaris*	Vrbová nad Váhom	2022	Severe mosaic, yellowing	BYMV, CMV

[1] SbDV: soybean dwarf virus (genus *Luteovirus*), BGCV-2: black grass cryptic virus-2 (unassigned), CMV: cucumber mosaic virus (genus *Cucumovirus*).

3.1. Genome Characterization of Slovak BYMV Isolates

Although BYMV is a worldwide pathogen infecting a huge range of hosts, information on virus molecular variability from Slovakia was lacking. Three complete BYMV genomes have been assembled from HTS data (Table 2), further confirming the robustness and applicability of new sequencing technologies for plant virome studies [12,13]. Although the 5′ and 3′ extremities of the Slovak BYMV genomes were not properly confirmed by RACE, the high coverage of reads in both untranslated regions enabled their proper determination in silico. The complete genomic BYMV sequences obtained have been deposited in the GenBank database under the accession numbers OR791742 (DAT), OR791741 (PS2) and OR791743 (FA40).

Table 2. Analysis of HTS data from the three *Fabaceae* samples related to BYMV genome reconstruction.

Sample	Total Number of Reads/Average Length (bp)	Reads Mapped against BYMV Reference NC_003492	Percentage of the Full-Length Genome Covered	Coverage Depth	The Closest BLAST Relative
DAT	4,449,806/137.1	662,105	99.9%	9764.5×	AB373203, pea, Japan
PS2	4,025,394/103.7	414,925	99.9%	4794.5×	AB373203, pea, Japan
FA40	16,948,908/117.5	1,075,143	99.9%	14,139.7×	HG970866, lupine, Australia

For all three isolates, a single variant BYMV genome was identified, and no indication of intra-isolate heterogeneity was obvious in the individual datasets. However, mixed infection of a plant with different genetic variants belonging to the same virus species has been previously reported for some viruses [31–33].

Previous works have revealed the broad intra-species molecular diversity of BYMV [20], resulting in their classification into nine molecular groups [9]. Phylogenetic analysis generated from complete BYMV genomes revealed the similar grouping of DAT and PS2 isolates into molecular group IX, together with pea and broad bean isolates from different continents (Figure 2). The closest sequence found by the BLAST search belonged to a Japanese pea isolate (AB373203). On the contrary, the FA40 isolate clustered in group VI contains geographically distant isolates from lupine and broad bean. Interestingly, the closest relative of FA40 is the HG970866 isolate from Australia [9]. Therefore, the phylogenetic analysis showed the absence of geographical- or host-based clustering of BYMV, indicating the long evolutionary history of the virus and its effective dissemination within regions facilitated by hardly controlled seed transmission [34,35].

The complete coding sequences of the three Slovak BYMV isolates were collinear and determined to be 9168 nucleotides (nt) in length, putatively encoding a polyprotein of 3056 aa (347.3–347.7 kDa). Computer analysis showed that the potyviral motifs were conserved in the BYMV DAT, PS2, and FA40 polyproteins, confirming the functionality of the obtained genomes. Specifically, three separate motifs associated with aphid transmission were conserved in the HC-Pro, i.e., RITC (aa position 335–339, appearing generally as KITC in potyviruses) [36], CCC (aa 575–577), and PTK (aa 593–595). The metal-binding motif FRNK-X12-CDNQLD, affecting symptom expression [37], was found at aa position 464–485.

The CI protein, thought to function as an RNA helicase in genome replication, contained the expected NTP-binding motif GAVGSGKST (aa 1227–1235) as well as other motifs characteristic of helicase proteins: DECH (aa 1316–1319), KVSAT (aa 1343–1347), LVYV (aa 1394–1397), VATNIIENGVTL (aa 1445–1456), and GERIQRLGRVGR (aa 1489–1500), while the VLLLEPTRPL motif appeared as VLM(I/V)ESTRPL (aa 1247–1256) [38–40].

Typical conserved motifs of potyviral polymerases QPSTVVDN and GDD, found in the NIb protein at aa positions 2577–2584 and 2615–2617, respectively, correspond to a broader motif SG-(X)3-T-(X)3-NT-(X)30-GDD (aa 2575–2617), proposed as the active site of RNA-dependent RNA polymerases [41]. Other conserved motifs, including SLKAEL (aa 2435–2440, RNA polymerase activity), CVDDFN (aa 2468–2473), CHADGS (as CDADGS, aa 2510–2515, and RNA-dependent polymerase activity [42], were present.

Similarly, CP of all Slovak isolates contained several highly conserved potyvirus motifs. The DAG motif, associated with aphid transmission, appeared as NAG (aa 2790–2792), similar to other BYMV [20,43]. Other conserved motifs found included MVWCIEN (2906–2912), AFDF (2989–2992), QMKAAA (3009–3014), and ENTERH 3034–3039 [44,45].

3.2. Experimental Infection of Pea Genotypes and Analysis of Virus Accumulation in Plants

In recent years, mixed viral infections and complex viromes have been frequently detected within a single plant [47,48], which can complicate the biological characterization of respective viruses or plant/virus interaction studies. In our work, HTS analysis revealed a single BYMV infection in only one of the three analyzed samples (pea PS2, Table 1). Therefore, this sample was used as a source of inoculum to establish systemic infection in experimental *N. benthamiana* plants by mechanical inoculation. Twenty days p.i., infected tobacco plants displayed mottling and mosaic symptoms (Figure 3) and tested positive in DAS-ELISA and RT-PCR. Verification of the CP sequence from *N. benthamiana* by Sanger sequencing (primed byBY_8548F/BY_9322R) did not reveal any nucleotide changes compared to the original HTS-derived sequence.

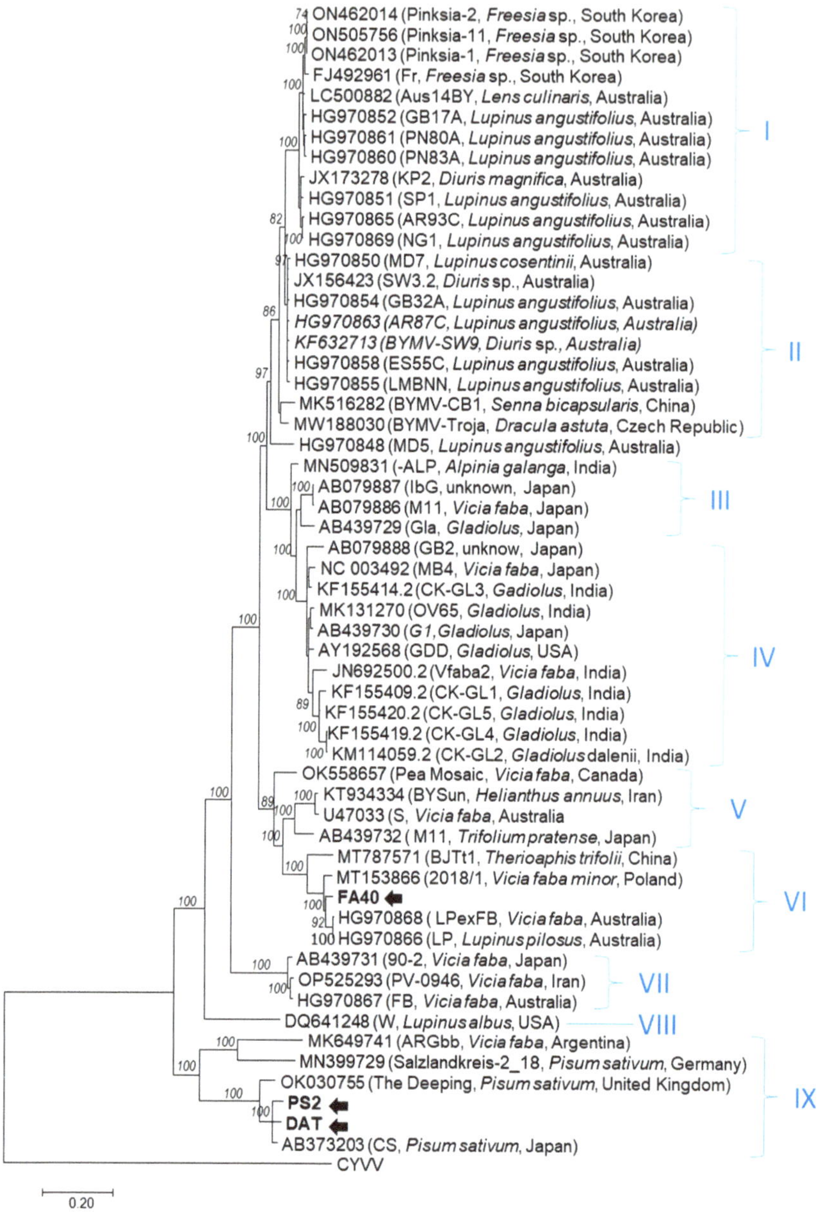

Figure 2. Phylogenetic tree generated from the complete genome nucleotide sequences of bean yellow mosaic (BYMV) isolates. Phylogenetic analysis was inferred using maximum likelihood (ML) based on the General Time Reversible (GTR) model, selected as the best fitting nucleotide substitution model based on the Bayesian Information Criterion (BIC) implemented in MEGA X [46]. Isolates are identified by their name, GenBank accession number, and country of origin. Slovak BYMV isolates sequenced in this study are highlighted in bold and marked with an arrow. Bootstrap values higher than 70% (500 bootstrap resamplings) are indicated. A closely related potyvirus, clover yellow vein virus (CYVV, Genbank accession number NC_003536), was used as an outgroup. Scale bars indicate genetic distances of 0.2.

Figure 3. Systemic symptoms developed on *N. benthamiana* and five commercial pea genotypes after experimental mechanical inoculation with BYMV-PS2 isolate.

As an important part of virus disease management, attention has been given to breeding for resistance in order to provide growers with plant genotypes that are resistant or suffer less from infection [3,49]. In order to evaluate host susceptibility to BYMV-PS2 infection, homogenous lots of five commercially available pea genotypes were mechanically inoculated and tested (cvs. Alderman, Ambrosia, Gloriosa, Herkules, Senator). Based on repeated experiments, all five genotypes were repeatedly infected by BYMV, as determined by symptom evaluation (Figure 3), ELISA, and western blot analysis. The infection rate was 62.5% (Gloriosa) or 100% (Alderman, Ambrosia, Herkules, Senator). Previous reports have revealed different degrees of susceptibility of pea to experimental BYMV infection, including resistant genotypes [50–52]. It has been reported that the recessive resistance gene *mo* controls resistance to BYMV and *wlv* confers specific resistance to BYMV-W [50,53]. As these reports may be biased, continued biological experiments with molecularly different isolates are needed to confirm potential resistance of pea genotypes to BYMV due to the possible emergence of resistance-breaking BYMV variants.

In our experiments, all five BYMV-PS2-infected pea genotypes displayed similar symptoms, consisting of chlorotic mottling and mosaics on leaves, while the growth of plants was not affected compared to non-inoculated controls. In order to comparatively assess the accumulation of virus in the top leaves of different pea genotypes, a semi-quantitative ELISA was performed. One-way ANOVA revealed significant differences between OD values of different plants [$F(6, 69) = 65.49$, $p < 0.001$)]. Differences between individual pea genotypes were further addressed by a post hoc TukeyHSD test (Figure 4), which revealed no significant difference between Herkules, Senator, and Alderman cultivars but a significant difference between Ambrosia and all other genotypes (Figure 4).

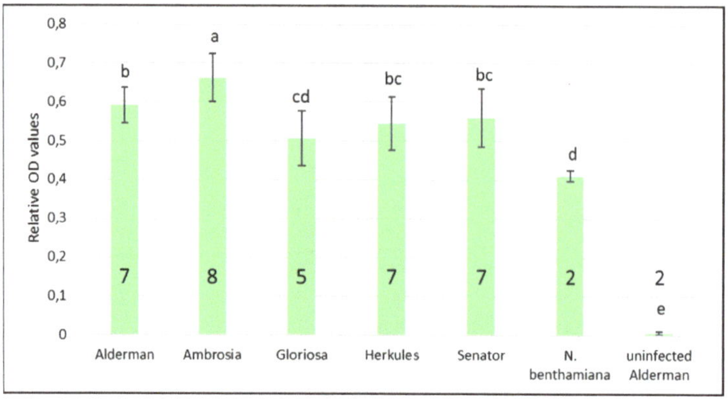

Figure 4. Semiquantitative comparison of virus accumulation, measured as the absorbance at 405 nm in DAS-ELISA, in five different pea genotypes 21 days p.i. Infected *N. benthamiana* and mock-inoculated pea cv. Alderman were used as controls. The graph shows the mean OD values and standard deviation error bars calculated only from BYMV-positive plants, the number of which is indicated. Deviations from the initial number of eight inoculated plants per genotype are caused by plant death, not due to virus infection (Alderman, Hercules, Senator) or uncomplete infection (Gloriosa). Compact letter display method was used to interpret results of post-hoc TukeyHSD test, showing differences between individual plant groups OD values. Groups sharing the same superscript letter are not significantly different ($p > 0.05$).

The Alderman genotype was selected for subsequent experiments aimed at evaluating the accumulation of virus in different parts of the pea plant, as tested by immunoblot analysis. Separately, for each of the four Alderman plants tested, a relative comparison was made between different standardized samples consisting of equal amounts of protein extracted from the root, stems, and leaves from the bottom and upper parts of the plant and from the immature pod (Figure 5A). Immunoblot analyses showed that the level of virus accumulation was specific for each tissue. The results from the four replicates, although with broad variability and no statistical significance, suggested a higher virus accumulation in the actively growing parts of the plant. The relative virus accumulation (compared between different samples as the optical density of the corresponding specific band) was indeed 2.2–13.1-fold higher in the upper stem compared to the root (Figure 5B).

Although it is well-documented that potyviruses systemically infect a wide range of plant species, information on the distribution of the virus within a single plant is still limited. Uneven distribution of potyviruses in different hosts has been reported [54–56]. Rajamäki and Valkonen [57] noted differences in the accumulation of potato virus A in the roots and systemic leaves of *Solanum commersonii* between different isolates of the virus.

Analysis of the correlation between virus titer and symptom severity yielded conflicting results depending on the particular plant/potyvirus model. Furthermore, the results suggest that viral load may or may not correlate (TuMV, [58]; PVY, [59]) with symptom severity. Overall, knowledge of virus distribution in the plant is critical for disease management (although it appears to be virus- and host-specific) and may also have practical implications, e.g., for efficient sampling before testing.

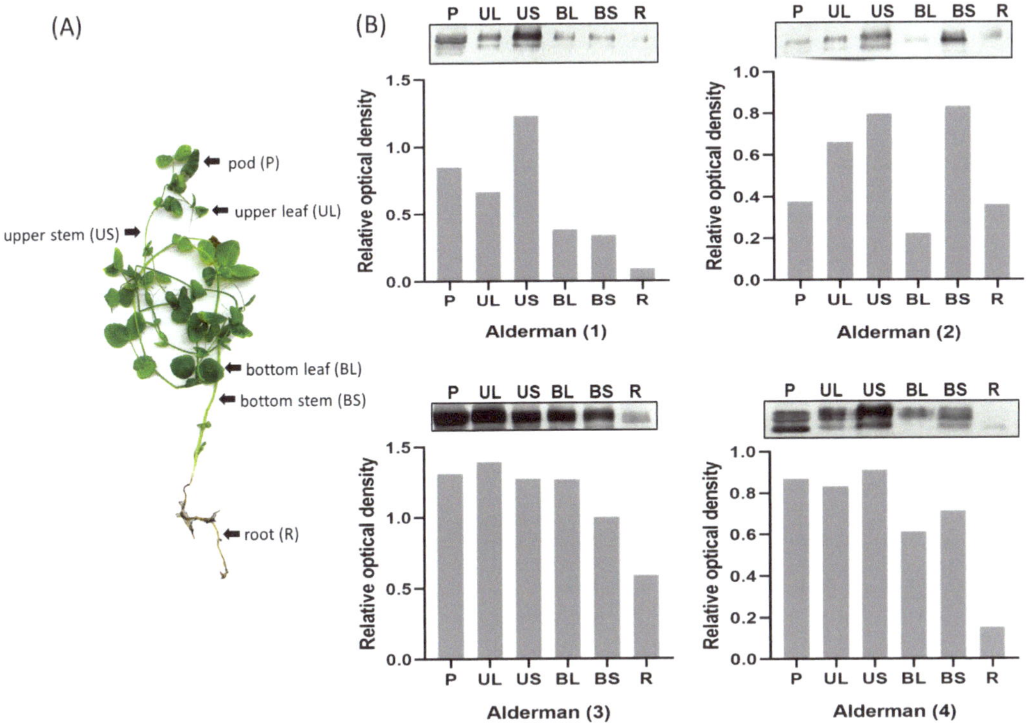

Figure 5. Schematic representation of the collection of different samples for the purpose of conducting relative comparison of virus accumulation within a single pea plant by immunoblot analysis (**A**). Four independent tests of infected cv. Alderman showing relative densitometric analysis of specific bands, performed using Image Studio Lite quantification software v. 5.2. (**B**).

Supplementary Materials: The following supporting information can be downloaded at: https://www.mdpi.com/article/10.3390/horticulturae10030262/s1, Figure S1: Schematic representation of the BYMV genome.

Author Contributions: Conceptualization, M.M. and M.G.; methodology, M.M., K.Š. and M.G.; validation, J.K., A.A. and L.P.; formal analysis, M.M., L.P., K.Š., Z.Š. and M.G.; investigation, M.M., J.K., A.A., P.A., L.P., K.Š., Z.Š. and M.G.; resources, M.M., Z.Š. and M.G., data curation, M.M., L.P., K.Š. and M.G.; writing—original draft preparation, M.M. and M.G.; writing—review and editing, M.M., A.A., K.Š., Z.Š. and M.G.; supervision, M.G.; funding acquisition, M.G. All authors have read and agreed to the published version of the manuscript.

Funding: This research was supported by grant VEGA 2/0036/24 from the Scientific Grant Agency of the Ministry of Education and the Slovak Academy of Sciences and partially by APVV-20-0015 from the Slovak Research and Development Agency.

Data Availability Statement: The nucleotide sequences reported in this paper are deposited in the GenBank database (www.ncbi.nlm.nih.gov) under the accession numbers listed in the text.

Acknowledgments: J.K. acknowledges the support from the Research Support Fund at the University of Ss. Cyril and Methodius in Trnava (FPPV-07-2024). The authors thank Dr Richard Hančinský for his help with the statistical analyses.

Conflicts of Interest: The authors declare no conflicts of interest. The funders had no role in the design of the study; in the collection, analyses, or interpretation of data; in the writing of the manuscript; or in the decision to publish the results.

References

1. Rashed, A.; Feng, X.; Prager, S.M.; Porter, L.D.; Knodel, J.J.; Karasev, A.; Eigenbrode, S.D. Vector-Borne Viruses of Pulse Crops, With a Particular Emphasis on North American Cropping System. *Ann. Entomol. Soc. Am.* **2018**, *111*, 205–227. [CrossRef]
2. Chatzivassiliou, E.K. An Annotated List of Legume-Infecting Viruses in the Light of Metagenomics. *Plants* **2021**, *10*, 1413. [CrossRef] [PubMed]
3. Jha, U.C.; Nayyar, H.; Chattopadhyay, A.; Beena, R.; Lone, A.A.; Naik, Y.D.; Thudi, M.; Prasad, P.V.V.; Gupta, S.; Dixit, G.P.; et al. Major viral diseases in grain legumes: Designing disease resistant legumes from plant breeding and OMICS integration. *Front. Plant Sci.* **2023**, *14*, 1183505. [CrossRef]
4. Affrifah, N.S.; Uebersax, M.A.; Amin, S. Nutritional significance, value-added applications, and consumer perceptions of food legumes: A review. *Legume Sci.* **2023**, *5*, e192. [CrossRef]
5. Wylie, S.J.; Adams, M.; Chalam, C.; Kreuze, J.; López-Moya, J.J.; Ohshima, K.; Praveen, S.; Rabenstein, F.; Stenger, D.; Wang, A.; et al. ICTV Virus Taxonomy Profile: *Potyviridae*. *J. Gen. Virol.* **2017**, *98*, 352–354. [CrossRef]
6. Yang, X.; Li, Y.; Wang, A. Research Advances in Potyviruses: From the Laboratory Bench to the Field. *Annu. Rev. Phytopathol.* **2021**, *59*, 1–29. [CrossRef] [PubMed]
7. Wylie, S.J.; Coutts, B.A.; Jones, M.G.K.; Jones, R.A.C. Phylogenetic analysis of Bean yellow mosaic virus isolates from four continents: Relationship between the seven groups found and their hosts and origins. *Plant Dis.* **2008**, *92*, 1596–1603. [CrossRef]
8. Kehoe, M.A.; Coutts, B.A.; Buirchell, B.J.; Jones, R.A. Split personality of a Potyvirus: To specialize or not to specialize? *PLoS ONE* **2014**, *9*, e105770. [CrossRef]
9. Kehoe, M.A.; Coutts, B.A.; Buirchell, B.J.; Jones, R.A. Plant virology and next generation sequencing: Experiences with a Potyvirus. *PLoS ONE* **2014**, *9*, e104580. [CrossRef]
10. Maina, S.; Zheng, L.; King, S.; Aftab, M.; Nancarrow, N.; Trębicki, P.; Rodoni, B. Genome Sequence and Phylogeny of a Bean Yellow Mosaic Virus Isolate Obtained from a 14-Year-Old Australian Lentil Sample. *Microbiol. Resour. Announc.* **2020**, *9*, e01437-19. [CrossRef]
11. Baradar, A.; Hosseini, A.; Ratti, C.; Hosseini, S. Phylogenetic analysis of a Bean yellow mosaic virus isolate from Iran and selecting the phylogenetic marker by comparing the individual genes and complete genome trees of BYMV isolates. *Physiol. Mol. Plant Pathol.* **2021**, *114*, 101632. [CrossRef]
12. Villamor, D.E.V.; Ho, T.; Al Rwahnih, M.; Martin, R.R.; Tzanetakis, I.E. High Throughput Sequencing For Plant Virus Detection and Discovery. *Phytopathology* **2019**, *109*, 716–725. [CrossRef]
13. Maclot, F.; Candresse, T.; Filloux, D.; Malmstrom, C.M.; Roumagnac, P.; van der Vlugt, R.; Massart, S. Illuminating an Ecological Blackbox: Using High Throughput Sequencing to Characterize the Plant Virome Across Scales. *Front. Microbiol.* **2020**, *11*, 578064. [CrossRef]
14. Bejerman, N.; Giolitti, F.; Trucco, V.; de Breuil, S.; Dietzgen, R.G.; Lenardon, S. Complete genome sequence of a new enamovirus from Argentina infecting alfalfa plants showing dwarfism symptoms. *Arch. Virol.* **2016**, *161*, 2029–2032. [CrossRef]
15. Filardo, F.F.; Thomas, J.E.; Webb, M.; Sharman, M. Faba bean polerovirus 1 (FBPV-1); a new polerovirus infecting legume crops in Australia. *Arch. Virol.* **2019**, *164*, 1915–1921. [CrossRef]
16. Naito, F.Y.; Melo, F.L.; Fonseca, M.E.; Santos, C.A.; Chanes, C.R.; Ribeiro, B.M.; Gilbertson, R.L.; Boiteux, L.S.; de Cássia Pereira-Carvalho, R. Nanopore sequencing of a novel bipartite new world begomovirus infecting cowpea. *Arch. Virol.* **2019**, *164*, 1907–1910. [CrossRef] [PubMed]
17. Glasa, M.; Šoltys, K.; Predajňa, L.; Sihelská, N.; Budiš, J.; Mrkvová, M.; Kraic, J.; Mihálik, D.; Ruiz- Garcia, A.B. High-throughput sequencing of Potato virus M from tomato in Slovakia reveals a divergent variant of the virus. *Plant Protect. Sci.* **2019**, *55*, 159–166. [CrossRef]
18. Elmore, M.G.; Groves, C.L.; Hajimorad, M.R.; Stewart, T.P.; Gaskill, M.A.; Wise, K.A.; Sikora, E.; Kleczewski, N.M.; Smith, D.L.; Mueller, D.S.; et al. Detection and discovery of plant viruses in soybean by metagenomic sequencing. *Virol. J.* **2022**, *19*, 149. [CrossRef]
19. Moury, B.; Simon, V. dN/dS-Based methods detect positive selection linked to trade-offs between different fitness traits in the coat protein of potato virus Y. *Mol. Biol. Evol.* **2011**, *28*, 2707–2717. [CrossRef]
20. Nigam, D.; LaTourrette, K.; Souza, P.F.N.; Garcia-Ruiz, H. Genome-Wide Variation in Potyviruses. *Front. Plant Sci.* **2019**, *10*, 1439. [CrossRef]
21. Stobbe, A.; Roossinck, M.J. Plant Virus Diversity and Evolution. *Curr. Res. Top. Plant Virol.* **2016**, *22*, 197–215. [CrossRef]
22. Rubio, L.; Galipienso, L.; Ferriol, I. Detection of Plant Viruses and Disease Management: Relevance of Genetic Diversity and Evolution. *Front Plant Sci.* **2020**, *11*, 1092. [CrossRef] [PubMed]
23. Majumdar, A.; Sharma, A.; Belludi, R. Natural and Engineered Resistance Mechanisms in Plants against Phytoviruses. *Pathogens* **2023**, *12*, 619. [CrossRef] [PubMed]
24. Mrkvová, M.; Hančinský, R.; Predajňa, L.; Alaxin, P.; Achs, A.; Tomašechová, J.; Šoltys, K.; Mihálik, D.; Olmos, A.; Ruiz-García, A.B.; et al. High-Throughput Sequencing Discloses the Cucumber Mosaic Virus (CMV) Diversity in Slovakia and Reveals New Hosts of CMV from the Papaveraceae Family. *Plants* **2022**, *11*, 1665. [CrossRef]
25. Clark, M.F.; Adams, A.N. Characteristics of the microplate method of enzyme-linked immunosorbent assay for the detection of plant viruses. *J. Gen. Virol.* **1977**, *34*, 475–483. [CrossRef]
26. Šubr, Z.; Matisová, J. Preparation of diagnostic monoclonal antibodies against two potyviruses. *Acta Virol.* **1999**, *43*, 255–257.

27. R Core Team. *R: A Language and Environment for Statistical Computing*; R Foundation for Statistical Computing: Vienna, Austria, 2018. Available online: https://www.R-project.org/ (accessed on 18 February 2023).
28. Nováková, S.; Klaudiny, J.; Kollerová, E.; Šubr, Z.W. Expression of a part of the Potato virus A non-structural protein P3 in Escherichia coli for the purpose of antibody preparation and P3 immunodetection in plant material. *J. Virol. Methods* **2006**, *137*, 229–235. [CrossRef] [PubMed]
29. Moreno, A.B.; López-Moya, J.J. When Viruses Play Team Sports: Mixed Infections in Plants. *Phytopathology* **2020**, *110*, 29–48. [CrossRef]
30. Tomašechová, J.; Hančinský, R.; Predajňa, L.; Kraic, J.; Mihálik, D.; Šoltys, K.; Vávrová, S.; Böhmer, M.; Sabanadzovic, S.; Glasa, M. High-Throughput Sequencing Reveals Bell Pepper Endornavirus Infection in Pepper (*Capsicum annum*) in Slovakia and Enables Its Further Molecular Characterization. *Plants* **2020**, *9*, 41. [CrossRef]
31. Maliogka, V.I.; Salvador, B.; Carbonell, A.; Sáenz, P.; León, D.S.; Oliveros, J.C.; Delgadillo, M.O.; García, J.A.; Simón-Mateo, C. Virus variants with differences in the P1 protein coexist in a Plum pox virus population and display particular host-dependent pathogenicity features. *Mol. Plant Pathol.* **2012**, *13*, 877–886. [CrossRef]
32. Della Bartola, M.; Byrne, S.; Mullins, E. Characterization of Potato virus Y Isolates and Assessment of Nanopore Sequencing to Detect and Genotype Potato Viruses. *Viruses* **2020**, *12*, 478. [CrossRef] [PubMed]
33. Glasa, M.; Hančinský, R.; Šoltys, K.; Predajňa, L.; Tomašechová, J.; Hauptvogel, P.; Mrkvová, M.; Mihálik, D.; Candresse, T. Molecular Characterization of Potato Virus Y (PVY) Using High-Throughput Sequencing: Constraints on Full Genome Reconstructions Imposed by Mixed Infection Involving Recombinant PVY Strains. *Plants* **2021**, *10*, 753. [CrossRef] [PubMed]
34. Sasaya, T.; Iwasaki, M.; Yamamoto, T. Seed Transmission of Bean Yellow Mosaic Virus in Broad Bean (*Vicia faba*). *Ann. Phytopath. Soc. Jpn.* **1993**, *59*, 559–562. [CrossRef]
35. Mali, V.R.; Šubr, Z.; Kúdela, O. Seed Transmission of Como and Potyviruses in Fababean and Vetch Genotypes Introduced into Slovakia. *Acta Phytopathol. Entomol. Hung.* **2003**, *38*, 87–97. [CrossRef]
36. Atreya, C.D.; Pirone, T.P. Mutational analysis of the helper component-proteinase gene of a potyvirus: Effects of amino acid substitutions, deletions, and gene replacement on virulence and aphid transmissibility. *Proc. Natl. Acad. Sci. USA* **1993**, *90*, 11919–11923. [CrossRef] [PubMed]
37. Shiboleth, Y.M.; Haronsky, E.; Leibman, D.; Arazi, T.; Wassenegger, M.; Whitham, S.A.; Gaba, V.; Gal-On, A. The conserved FRNK box in HC-Pro, a plant viral suppressor of gene silencing, is required for small RNA binding and mediates symptom development. *J. Virol.* **2007**, *81*, 13135–13148. [CrossRef]
38. Kadaré, G.; Haenni, A. Virus-encoded RNA helicases. *J. Virol.* **1997**, *71*, 2583–2590. [CrossRef]
39. Deng, P.; Wu, Z.; Wang, A. The multifunctional protein CI of potyviruses plays interlinked and distinct roles in viral genome replication and intercellular movement. *Virol. J.* **2015**, *12*, 141. [CrossRef]
40. Li, Y.; Xia, F.; Wang, Y.; Yan, C.; Jia, A.; Zhang, Y. Characterization of a highly divergent Sugarcane mosaic virus from Canna indica L. by deep sequencing. *BMC Microbiol.* **2019**, *19*, 260. [CrossRef]
41. Hong, Y.; Hunt, A.G. RNA polymerase activity catalyzed by a potyvirus-encoded RNA-dependent RNA polymerase. *Virology* **1996**, *226*, 146–151. [CrossRef]
42. Shen, W.; Shi, Y.; Dai, Z.; Wang, A. The RNA-Dependent RNA Polymerase NIb of Potyviruses Plays Multifunctional, Contrasting Roles during Viral Infection. *Viruses* **2020**, *12*, 77. [CrossRef]
43. Wylie, S.J.; Kueh, J.; Welsh, B.; Smith, L.J.; Jones, M.G.; Jones, R.A. A non-aphid-transmissible isolate of bean yellow mosaic potyvirus has an altered NAG motif in its coat protein. *Arch. Virol.* **2002**, *147*, 1813–1820. [CrossRef]
44. Worrall, E.A.; Hayward, A.C.; Fletcher, S.J.; Mitter, N. Molecular characterization and analysis of conserved potyviral motifs in bean common mosaic virus (BCMV) for RNAi-mediated protection. *Arch. Virol.* **2019**, *164*, 181–194. [CrossRef]
45. Zheng, L.; Wayper, P.J.; Gibbs, A.J.; Fourment, M.; Rodoni, B.C.; Gibbs, M.J. Accumulating variation at conserved sites in potyvirus genomes is driven by species discovery and affects degenerate primer design. *PLoS ONE* **2008**, *3*, e1586. [CrossRef]
46. Kumar, S.; Stecher, G.; Li, M.; Knyaz, C.; Tamura, K. MEGA X: Molecular Evolutionary Genetics Analysis across computing platforms. *Mol. Biol. Evol.* **2018**, *35*, 1547–1549. [CrossRef]
47. Fowkes, A.R.; McGreig, S.; Pufal, H.; Duffy, S.; Howard, B.; Adams, I.P.; Macarthur, R.; Weekes, R.; Fox, A. Integrating High throughput Sequencing into Survey Design Reveals Turnip Yellows Virus and Soybean Dwarf Virus in Pea (*Pisum sativum*) in the United Kingdom. *Viruses* **2021**, *13*, 2530. [CrossRef] [PubMed]
48. Hasiów-Jaroszewska, B.; Boezen, D.; Zwart, M.P. Metagenomic Studies of Viruses in Weeds and Wild Plants: A Powerful Approach to Characterise Variable Virus Communities. *Viruses* **2021**, *13*, 1939. [CrossRef] [PubMed]
49. Meziadi, C.; Blanchet, S.; Geffroy, V.; Pflieger, S. Genetic resistance against viruses in phaseolus vulgaris l.: State of the art and future prospects. *Plant Sci.* **2017**, *265*, 39–50. [CrossRef] [PubMed]
50. Schroeder, W.T.; Provvidenti, R. A Common Gene for Resistance to Bean Yellow Mosaic Virus and Watermelon Mosaic Virus 2 in *Pisum sativum*. *Phytopathology* **1971**, *61*, 846–848. [CrossRef]
51. Jurík, M.; Lebeda, A.; Gallo, J. Resistance of green peas to legume viruses. *Acta Virol.* **1994**, *38*, 97–99. [PubMed]
52. van Leur, J.A.G.; Kumari, S.G.; Aftab, M.; Leonforte, A.; Moore, S. Virus resistance of Australian pea (*Pisum sativum*) varieties. *N. Z. J. Crop Hortic. Sci.* **2013**, *41*, 86–101. [CrossRef]
53. Provvidenti, R.; Hampton, R.O. Inheritance of resistance to White lupin mosaic virus in common pea. *HortScience* **1993**, *28*, 836–837. [CrossRef]

54. Green, S.K.; Kuo, Y.J.; Lee, D.R. Uneven distribution of two potyviruses (feathery mottle virus and sweet potato latent virus) in sweet potato plants and its implication on virus indexing of meristem derived plants. *Trop. Pest Manag.* **1988**, *34*, 298–302. [CrossRef]
55. Dovas, C.I.; Mamolos, A.P.; Katis, N.I. Fluctuations in concentration of two potyviruses in garlic during the growing period and sampling conditions for reliable detection by ELISA. *Ann. Appl. Biol.* **2002**, *140*, 21–28. [CrossRef]
56. Kogovšek, P.; Kladnik, A.; Mlakar, J.; Znidarič, M.T.; Dermastia, M.; Ravnikar, M.; Pompe-Novak, M. Distribution of Potato virus Y in potato plant organs, tissues, and cells. *Phytopathology* **2011**, *101*, 1292–1300. [CrossRef] [PubMed]
57. Rajamäki, M.L.; Valkonen, J.P. Viral genome-linked protein (VPg) controls accumulation and phloem-loading of a potyvirus in inoculated potato leaves. *Mol. Plant Microbe Interact.* **2002**, *15*, 138–149. [CrossRef]
58. Glasa, M.; Šoltys, K.; Predajňa, L.; Sihelská, N.; Nováková, S.; Šubr, Z.; Kraic, J.; Mihálik, D. Molecular and Biological Characterisation of *Turnip mosaic virus* Isolates Infecting Poppy (*Papaver somniferum* and *P. rhoeas*) in Slovakia. *Viruses* **2018**, *10*, 430. [CrossRef] [PubMed]
59. Mehle, N.; Kovač, M.; Petrovič, N.; Pompe Novak, M.; Baebler, Š.; Krečič Stres, H.; Gruden, K.; Ravnikar, M. Spread of potato virus Y NTN in potato cultivars (*Solanum tuberosum* L.) with different levels of sensitivity. *Physiol. Mol. Plant Pathol.* **2004**, *64*, 293–300. [CrossRef]

horticulturae

MDPI

Article

Biocontrol Potential of *Trichoderma asperellum* CMT10 against Strawberry Root Rot Disease

Ping Liu [1], Ruixian Yang [1,*], Zuhua Wang [1], Yinhao Ma [1], Weiguang Ren [1], Daowei Wei [1] and Wenyu Ye [2]

[1] School of Environmental Engineering and Chemistry, Luoyang Institute of Science and Technology, Luoyang 471002, China; lylp76lp@163.com (P.L.); zuowu_zhang@163.com (Z.W.); guoqing_m@163.com (Y.M.); renweiguang02@163.com (W.R.); weidaowei05@163.com (D.W.)

[2] College of JunCao Science and Ecology (College of Carbon Neutrality), Fujian Agriculture and Forestry University, Fuzhou 350002, China; wenyuye08@163.com

* Correspondence: fairy19790805@163.com

Abstract: Strawberry root rot caused by *Neopestalotiopsis clavispora* is one of the main diseases of strawberries and significantly impacts the yield and quality of strawberry fruit. Currently, the only accessible control methods are fungicide sprays, which could have an adverse effect on the consumers of the strawberries. Biological control is becoming an alternative method for the control of plant diseases to replace or decrease the application of traditional synthetic chemical fungicides. *Trichoderma* spp. are frequently used as biological agents to prevent root rot in strawberries. In order to provide highly effective biocontrol resources for controlling strawberry root rot caused by *Neopestalotiopsis clavispora*, the biocontrol mechanism, the control effects of *T. asperellum* CMT10 against strawberry root rot, and the growth-promoting effects on strawberry seedlings were investigated using plate culture, microscopy observation, and root drenching methods. The results showed that CMT10 had obvious competitive, antimycotic, and hyperparasitic effects on *N. clavispora* CMGF3. The CMT10 could quickly occupy nutritional space, and the inhibition rate of CMT10 against CMGF3 was 65.49% 7 d after co-culture. The inhibition rates of volatile metabolites and fermentation metabolites produced by CMT10 were 79.67% and 69.84% against CMGF3, respectively. The mycelium of CMT10 can act as a hyperparasite by contacting, winding, and penetrating the hyphae of CMGF3. Pot experiment showed that the biocontrol efficiency of CMT10 on strawberry root rot caused by *Neopestalotiopsis clavispora* was 63.09%. CMT10 promoted strawberry growth, plant height, root length, total fresh weight, root fresh weight, stem fresh weight, and root dry weight by 20.09%, 22.39%, 87.11%, 101.58%, 79.82%, and 72.33%, respectively. Overall, this study showed the ability of *T. asperellum* CMT10 to control strawberry root rot and its potential to be developed as a novel biocontrol agent to replace chemical fungicides for eco-friendly and sustainable agriculture.

Keywords: strawberry root rot; *Trichoderma asperellum*; biocontrol mechanism; biocontrol efficacy; growth-promoting effect

1. Introduction

Strawberries (*Fragaria ananassa*), perennial herbaceous plants belonging to the genus *Fragaria* in the Rosaceae family, are renowned for their short cultivation cycle and high economic yield. This fruit is popular among consumers because of its exceptional taste and nutritional value. Strawberries are an important economic crop both in China and globally [1]. According to data from the Food and Agriculture Organization (FAO) of the United Nations, as of 2020, China boasted a strawberry cultivation area of over 127,000 hm^2, with a production surpassing 3.336 million tons, ranking it as the world's leading producer of strawberries [2]. The predominant method of cultivation in China is greenhouse cultivation, which involves enclosed spaces, elevated temperatures, and high humidity. Continuous cultivation practices have led to the accumulation of pathogens, resulting in frequent outbreaks of strawberry diseases and economic losses, which hinder the sustainable development of the strawberry industry [3]. One of the major diseases affecting strawberries is root rot, particularly in continuously cultivated fields [4]. The complex array of pathogens that contribute to strawberry root rot includes *Neopestalotiopsis clavispora*, *Phytophthora fragariae*, *Fusarium solani*, *Fusarium oxysporum*, *Rhizoctonia solani*, *Colletotrichum acutatum*, and *Armillaria mellea* [5,6]. The primary method of controlling strawberry root rot in current production practices is to use chemicals, because of the diversity of pathogens and the lack of strawberry varieties with a high resistance to root rot [7,8]. However, the use of fungicides on edible strawberry fruits poses a potential risk to human health. Hence, there is an urgent need to explore novel control strategies for strawberry root rot. Biological control measures are particularly effective to reduce soil-borne pathogens. The screening and application of biocontrol microorganisms to control root rot is very important for the sustainable development of the strawberry industry.

Trichoderma species have been used as biological control agents (BCAs) and as an alternative to synthetic fungicides to control a variety of plant diseases [9,10]. The biocontrol mechanisms of *Trichoderma* are based on the activation of multiple mechanisms, either indirectly, by competing for space and nutrients, promoting plant growth, plant defensive mechanisms, and antibiosis, or directly, by mycoparasitism [11,12]. Studies have indicated that *Trichoderma* spp. can increase the resistance of strawberries to root pathogens. Zhang et al. [13] found that *T. harzianum* M10-3-2 could significantly inhibit *F. solani*, which was an agent of strawberry root rot. *T. asperellum* D7-3 had remarkable growth-promoting effects on strawberries, whereas *T. koningiopsis* M0-3-3 enhanced the biocontrol efficiency of other strains against strawberry root rot. They also proved that the combination of the three *Trichoderma* strains (M10-3-2, D7-3, and M0-3-3) was more effective than individual treatments. Mercado et al. [14] discovered that *T. harzianum* could effectively control the strawberry root rot caused by *C. acutatum*. Rees et al. [15] found that *T. atrobrunneum* significantly reduces the incidence of strawberry root rot caused by *A. mellea*. Mirzaeipour et al. [16] obtained three *Trichoderma* strains with effective control against the strawberry root rot caused by *R. solani*. Despite substantial research on the use of *Trichoderma* strains to control strawberry root rot, focus has mainly been on the control of pathogens including *R. solani*, *C. acutatum*, *F. solani*, and *A. mellea*. Studies are still relatively lacking on the screening of *Trichoderma* strains against the strawberry root rot caused by *N. clavispora*.

In this study, the potential role of *T. asperellum* CMT10, isolated from healthy strawberry rhizosphere soil, was investigated as a biological control agent of the strawberry root rot caused by *N. clavispora*. To achieve this goal, plate culture, microscopy observation, and root drenching methods were employed to investigate the biocontrol mechanism and control efficiency of *T. asperellum* CMT10 on strawberry root rot, as well as its growth-promoting effect on strawberry seedlings. This study reveals that *T. asperellum* CMT10 can effectively control the occurrence of the strawberry root rot caused by *N. clavispora* and had an obvious promotion effect on strawberry seedling growth. These results indicate that *T. asperellum* CMT10 is a promising biocontrol microorganism for controlling strawberry root rot.

2. Materials and Methods

2.1. Plant Pathogen and Plant Materials

Neopestalotiopsis clavispora CMGF3 was isolated from strawberries with symptoms of root rot through the tissue isolation method, and the fungus was cultured on potato dextrose agar (PDA; 20% potato, 2% dextrose, 1.5% agar) for 7 d at 28 °C. *N. clavispora* CMGF3 was identified based on morphological characteristics and molecular identification. One-year-old strawberry seedlings of the commercial cultivar "Hongyan" were provided by the "Shilixiang" strawberry seedling cultivation facility.

2.2. Isolation and Screening of Trichoderma Strains

Soil samples were collected from the healthy strawberry rhizosphere soil of the "Shilixiang" strawberry planting field (112°57′14.51″ E, 34°79′42.23″ N) in Luoyang, Henan Province. One gram of soil was taken in a Falcon tube (50 mL) containing sterile distilled water (SDW) and shaken (180 rpm) for 1 h. The samples were diluted from 10^{-1}- to 10^{-5}-fold with sterile distilled water, and a 100 μL dilution (10^{-5}-fold) was spread onto potato dextrose agar (PDA) plates [17]. The *Trichoderma* colonies were transferred to a new PDA medium 7 d after incubation at 28 °C for purification.

A total of 10 *Trichoderma* strains were screened for antagonistic activity against the mycelial growth of *N. clavispora* CMGF3 using a dual-culture plate assay, as described by Pimentel et al. [18]. One mycelial disc (4 mm diameter) of each *Trichoderma* sp. and *N. clavispora* was excised from the growing edges of 7-day-old cultures and placed 2 cm apart on opposite sides of PDA plates (90 mm). The plates were incubated for 7 days at 28 °C. A control of *N. clavispora* alone on PDA plates was used. The experimental design was completely randomized, with 20 treatments and three replicates. The growth rate of *N. clavispora* was determined by measuring the colony diameter. The percent inhibition was calculated as follows: percent inhibition (%) = [(pathogen colony diameter in the control treatment − pathogen colony diameter in the challenge treatment)/pathogen colony diameter in the control treatment] × 100.

2.3. Morphological and Molecular Identification of Trichoderma CMT10

Purified *Trichoderma* CMT10 was inoculated on a PDA plate medium and cultured in the dark for 7 days at 28 °C. Macroscopic morphology was observed, including the color and texture of the colony surface verse and reverse, the presence or absence of pigmentation, and the pattern of growth and sporulation, and images of the colonies were obtained. Microscopic morphologies such as conidia and conidiophores were observed using an optical microscope (ZEISS Axio Scope5, Oberkochen, Germany). Morphological identification relied on the descriptions found in previous research [19,20].

Trichoderma CMT10 was cultured in PDA medium at 28 °C for 7 days. Mycelia were harvested from the cultures, and genomic DNA (gDNA) was extracted using a DNA extraction kit (TIANGEN Biotech, Beijing, China). The extracted DNA was used as a template to amplify the internal transcribed spacer (ITS) region and the translation elongation factor-1α (*tef1-α*) region. The primers were designed with reference to previous studies [21,22]. All amplified loci, primers, and PCR conditions are presented in Table 1. PCR was performed using the TIANGEN Golden Easy PCR kit (TIANGEN Biotech, Beijing, China). The PCR products were subjected to direct automated sequencing using fluorescent terminators on an ABI 377 Prism Sequencer (Sangon Biotech, Shanghai, China). The sequences were confirmed with a BLAST (Basic Local Alignment Search Tool) search of the NCBI (National Center for Biotechnology Information) database (https://www.ncbi.nlm.nih.gov/, accessed on 23 September 2023), and a phylogenetic tree was constructed using the neighbor-joining (NJ) method, with 1000 bootstrap replications in the MEGA 10.0 package. Phylogenetic analysis with ITS-*tef1-α* gene sequences was performed to determine the position of *Trichoderma* CMT10. After identification, the sequences were submitted to Genbank. The strains used in this study and their corresponding GenBank accession numbers are listed in Table 2.

Table 1. Amplification sites, primer sequences, and PCR conditions used in this study.

Gene[a]	Primer	Primer Sequence (5′-3′)	PCR Conditions	Reference
ITS	ITS1 ITS4	TCCGTAGGTGAACCTGCGG TCCTCCGCTTATTGATATGC	94 °C for 5 min (94 °C for 30 s, 55 °C for 30 s and 72 °C for 40 s) × 35 cycles, 72 °C for 7 min	[21]
tef-1α	TEF-F TEF-R	TGGGCCATCAACTGAGAAAGA TCTCCCTACACTTCAACTGCACA	94 °C for 5 min (94 °C for 30 s, 53 °C for 30 s, and 72 °C for 1 min) × 35 cycles, 72 °C for 7 min	[22]

Genes[a]: ITS, internal transcribed spacer; *tef-1α*, translation elongation factor.

Table 2. The ITS and *tef-1α* gene sequences of *Trichoderma* strains from the NCBI database, used for the construction of the phylogenetic tree used in this study.

Code	Culture Accession Number(s)	Original Name	Accession no. ITS	Accession no. *tef-1α*
1	CEN1463	*T. asperellum*	MK714888	MK696646
2	T34	*T. asperellum*	LC123614	EU077228
3	ZJSX5002	*T. asperellum*	JQ040324	JQ040480
4	KUFA0403	*T. asperellum*	OM169354	OP132635
5	RM-28	*T. asperellum*	MK092975	MK095221
6	TR5	*T. longibrachiatum*	KC859426	KC572116
7	Tr5	*T. harzianum*	OP938774	OP948262
8	DUCC001	*T. citrinoviride*	JF700484	JF700485
9	S206	*T. caerulescens*	JN715590	JN715624
10	TW20050	*T. gamsii*	KU523894	KU523895
11	YMF1.02659	*T. kunmingense*	KJ742800	KJ742802
12	CBS 121219	*T. yunnanense*	GU198302	GU198243

2.4. *In Vitro Biocontrol of Trichoderma CMT10 against N. clavispora*

2.4.1. Inhibitory Effects of Volatile Compounds

To determine the effect of the volatile compounds secreted by *Trichoderma* CMT10 against the growth of *N. clavispora* CMGF3, exposure of Trichoderma's volatile compounds was performed using the confrontation culture method [23]. Mycelial discs of Trichoderma were cut using a sterile cork borer (5 mm diameter) and were placed at the center of a freshly prepared PDA plate and cultured for 3 days at 28 °C in the dark. A mycelial disc (5 mm diameter) of the fungal pathogen *N. clavispora* CMGF3 was placed onto another freshly prepared PDA plate in the same manner. PDA plates inoculated with *N. clavispora* mycelial plugs were placed on top of the PDA plates inoculated with *Trichoderma* CMT10 and the plates were then sealed with parafilm. A control without *Trichoderma* inoculation was used and the inhibition of the mycelial growth of *N. clavispora* was observed at 28 °C for 7 d. The experiment was performed in triplicate.

2.4.2. Inhibitory Effects of Soluble Compounds

The effect of the soluble compounds of *Trichoderma* CMT10 against the growth of the fungal pathogen *N. clavispora* CMGF3 under in vitro conditions was determined as follows: *Trichoderma* CMT10 was diluted with sterile water to obtain a conidial suspension containing 1×10^8 spores/mL, and 100 µL conidial suspension was inoculated into 100 mL of potato dextrose broth (PDB) medium at 28 °C for 4 d under shaking conditions (180 rpm). The fermented liquid was centrifuged at 8000× *g* rpm for 2 min and the supernatant was filtered through a 0.22 µm filter to obtain the sterile filtrate. Therefore, the sterile filtrate was spread onto PDA plates at a ratio of 1:9 and a 7-day-old cultured *N. clavispora* CMGF3 mycelium plug was placed onto a PDA plate. A mixture of sterile water was used as the control. After 7 days of incubation at 28 °C, the diameter of the pathogen was measured and the inhibition rate (IR) was calculated. The experiment was performed in triplicate.

2.4.3. Hyperparasitism of *Trichoderma* CMT10

The hyperparasitism of *Trichoderma* CMT10 on *N. clavispora* CMGF3 was observed using a dual culture method [24]. Under sterile conditions, 1 mL of melted PDA medium was pipetted onto a sterilized glass slide to make a PDA membrane. After the solidification of the medium, *Trichoderma* CMT10 and *N. clavispora* CMGF3 mycelial discs were separately inoculated onto the membrane (with a 6 cm distance between them) at 28 °C for an incubation period of 24–72 h. Growth was recorded at 12 h intervals. After successful fungal superparasitism on the pathogen, the dual culture areas were observed using an optical microscope (ZEISS Axio Scope5, Oberkochen, Germany).

2.5. Biochemical Properties of *Trichoderma* CMT10

The precipitated $Ca_3(PO_4)_2$ on Pikovskaya's agar media (glucose, 10 g; $(NH_4)_2SO_4$, 0.5 g; NaCl, 0.3 g; $MgSO_4$, 0.3 g; $MnSO_4$, 0.03 g; K_2SO_4, 0.3 g; $FeSO_4$, 0.03 g; $Ca_3(PO_4)_2$, 5.0 g; agar, 15.0 g; pH 7.0–7.5) was used for the qualitative detection of the phosphate solubilizing of *Trichoderma* CMT10 [25]. Briefly, *Trichoderma* CMT10 was inoculated in Pikovskaya's agar media. The cultures were incubated for 5 days at 28 °C and the fungal growth was evaluated. Siderophore production was carried out using chrome azure S (CAS) agar media (CAS, 0.06 g; HDTMA, 0.07 g; $FeCl_3 \cdot 6H_2O$, 0.003 g; $NaH_2PO_4 \cdot 2H_2O$, 0.30 g; $Na_2HPO_4 \cdot 12H_2O$, 1.21 g; NH_4Cl, 0.125 g; KH_2PO4, 0.038 g; NaCl, 0.06 g; agar, 9.0 g; pH 6.7–6.9) [26]. *Trichoderma* CMT10 was inoculated in the CAS agar media. The cultures were incubated for 5 days at 28 °C and the fungal growth was evaluated. Nitrogen fixation was determined using nitrogen-free agar medium (KH_2PO_4, 0.20 g; $MgSO_4$, 0.20 g; NaCl, 0.20 g; $CaCO_3$, 5.0 g; mannitol, 10.0 g; agar, 15.0 g; pH 6.9–7.91) [27,28]. Specifically, nitrogen-free agar plates were streaked with a 5 mm disc from a PDA culture. The cultures were incubated for 5 days at 28 °C and the fungal growth was evaluated. IAA production by *Trichoderma* CMT10 was quantitatively tested according to Brick et al. [29]. A 5 mm disc was inoculated to a 100 mL Erlenmeyer flask containing 50 mL DF salts minimal medium. The medium was supplemented with 1.02 g/L tryptophan and incubated at 28 °C with continuous shaking at 150 rpm. After a 5 days incubation, 1.5 mL of culture medium was centrifuged at 10,000 rpm for 10 min. Then, a 50 µL aliquot of the supernatant was mixed vigorously with an equal volume of Salkowski's reagent in a 1.5 mL tube and incubated in the dark at 25 °C for 30 min. Uninoculated culture solution mixed with Salkowski's reagent served as a negative control and 50 µL IAA (50 mg/L) mixed with Salkowski's reagent served as a positive control. IAA production was observed as the development of a pink–red color.

2.6. Control Effects of *Trichoderma* CMT10 on Strawberry Root Rot

Mycelial discs of *Trichoderma* CMT10 and *N. clavispora* CMGF3 were inoculated at the center of PDA plates at 28 °C for 7 d. The conidial suspensions (1×10^8 spores/mL) of *Trichoderma* and pathogen were prepared using sterile water, then stored at 4 °C for later use. One-year-old strawberry seedlings of the commercial cultivar "Hongyan" were used. The seedlings were carefully selected from the nursery with one plant per pot. Each plant was transplanted into a plastic pot (diameter, 28 cm; bottom diameter, 20 cm; height, 30 cm). Plants were grown in soil in a growth chamber at 22 °C and 75% humidity with a 16 h light/8 h dark photo period. After acclimation for 15 d, plants were used for pathogen infection and to assess the control efficiency of *Trichoderma* CMT10 on strawberry root rot. The potting root irrigation method was used for inoculation. The experiment included four treatments: (1) inoculation with *N. clavispora* CMGF3 only; (2) inoculation with *Trichoderma* CMT10 only; (3) inoculation with *N. clavispora* CMGF3 after 3 d followed by *Trichoderma* CMT10; and (4) water inoculation as a control. Each treatment consisted of 5 pots, with 3 replicates. Plants were inoculated with 5 mL of the conidial suspension of CMGF3 and CMT10 through the soil around each plant. All the treatments were followed by 60 days of incubation at 28 °C and 80% relative humidity. The disease severity of the seedlings was assessed using a scoring system of 0–5, modified from the report of Vestberg et al. [30]. Level 0 signifies an entire plant in a healthy state; Level 1 indicates a root disease incidence of $\leq$30%, with normal leaves; Level 2 is characterized by a root disease incidence greater than 30% and equal to or less than 60%, with normal leaves; Level 3 represents a root disease incidence greater than 60% and equal to or smaller than 80%, accompanied by yellowing leaves; Level 4 indicates a root disease incidence exceeding 80%, leading to leaf wilting; and Level 5 signifies complete plant mortality. The disease index and control efficiency were calculated based on the grading results. Disease Index = $\sum$ (disease level $\times$ number of plants at that level)/(total number of plants $\times$ highest disease level) $\times$ 100 and control efficiency (%) = (control disease index $-$ treatment disease index)/control disease index $\times$ 100.

2.7. Growth-Promoting Effects of Trichoderma CMT10 on Strawberry Seedlings

The same method used in Section 2.6 was used in this experiment, which consisted of two treatments: (1) inoculation with *Trichoderma* CMT10 and (2) water inoculation as a control. Each treatment consisted of 5 pots, with 3 replicates. Plants were inoculated with 5 mL of the conidial suspension (1×10^8 spores/mL) of CMT10 through the soil around each plant, and the plants were incubated for 60 days at 28 °C. Afterward, the strawberry seedlings were carefully excavated and their height, root length, and fresh weight (stem and leaf fresh weight, root fresh weight, and total fresh weight) were measured. The roots were dried at 45 °C in an oven, and their dry weight (g) was measured. The growth-promoting rate was calculated as follows: growth promotion rate (%) = (treatment biomass − control biomass)/control biomass × 100.

2.8. Data Statistics and Analysis

Data obtained from the experiments were processed using Excel 2010 and a one-way analysis of variance (ANOVA) was performed using DPS 7.05 statistical software. Duncan's new multiple range test was used to assess the significant differences, and the significance level was set at $p \leq 0.05$.

3. Results

3.1. Screening of Trichoderma Strains with Inhibitory Effects on N. clavispora

Ten *Trichoderma* strains were isolated using the dilution culture method. Two *Trichoderma* isolates, CMT10 and CMT4, were found to inhibit the mycelial growths of *N. clavispora* CMGF3, with inhibitory rates of 65.49% and 51.37%, respectively. CMT10 displayed significant inhibition activity against *N. clavispora* (Table 3). Further observations indicated that the mycelial growth of CMT10 was faster than that of CMGF3 and could, thus, quickly occupy the nutrient space. After 3 d of the dual culture, the mycelia of pathogen CMGF3 only reached one-third of the culture dish and an inhibition zone appeared between CMT10 and CMGF3. Moreover, the mycelia of CMGF3 near the inhibition zone were sparse, indicating a weakened growth. By day seven of the dual culture, the mycelia of CMT10 completely covered the CMGF3 colony and completely inhibited the growth and reproduction of CMGF3 (Figure 1). The results indicate that Trichoderma CMT10 could strongly inhibit the mycelial growth and reproduction of CMGF3, demonstrating a robust competitive advantage against the strawberry root rot pathogen.

N. clavispora | 3 d | 7 d

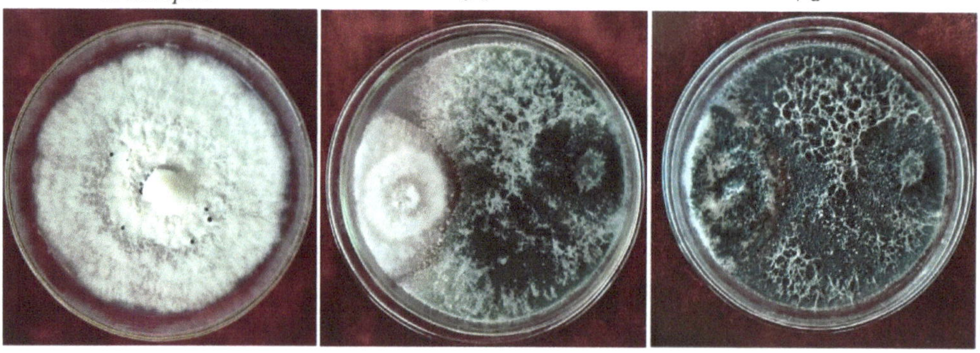

Figure 1. Dual cultures of *Trichoderma* CMT10 against *N. clavispora* on PDA plates.

Table 3. Antagonism test of *Trichoderma* strains against *N. clavispora* on PDA plates.

Treatments	Colony Diameter (cm)	Inhibition Rate (%)
CMT10	2.93 ± 0.153	65.49 a
CMT4	4.13 ± 0.058	51.37 b
CMGF3	8.50 ± 0.000	-

Note: Colony diameter (cm) represented the colony diameter of *N. clavispora* CMGF3 for each treatment. Data were mean ± SD. Different letters in the same column indicated significant difference at the 0.05 level using Duncan's new multiple range test.

3.2. Identification of Trichoderma CMT10

Trichoderma CMT10 displayed a fast growth on PDA medium, with aerial mycelia completely covering the entire culture dish within three days. Initially, the colony appeared white, but it changed to yellow–green and green later. The green conidia were produced and completely covered the plate after five days (Figure 2A). Microscopically, it was observed the branches were pyramidal in type with verticillate, frequently pared lateral branches that arose from main axis with 2–5 phialides clustered at the top. The angle with the main axis was 90°, and the lateral branches re-branched. The phialides were ampulliform, somewhat thicker in the middle, and terminated with conidia (Figure 2B). The conidia were spherical to ellipsoidal, 2 (−3.7) × 3.2 (−4.5) μm, single-celled, and light green (Figure 2C). Based on these cultural and morphological characteristics, the strain CMT10 was identified as *T. asperellum*.

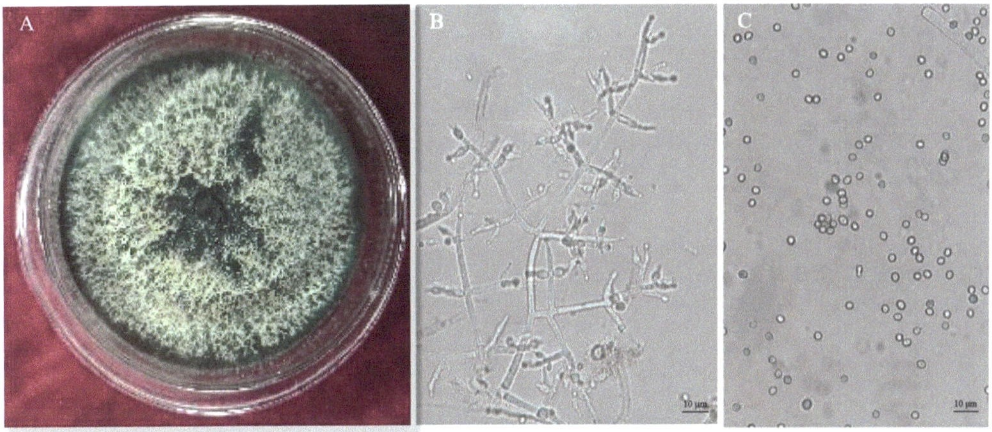

Figure 2. Cultural and morphological characteristics of *Trichoderma* CMT10. (**A**) The colony morphologies on PDA medium incubated at 28 °C for 7 days; (**B**) conidiophores; and (**C**) conidia. Scale bar = 10 μm.

The ITS regions and *tef1-α* regions of *Trichoderma* CMT10 were amplified and sequenced. The GenBank accession numbers were PP126513 and PP171486, respectively. The phylogenetic tree based on the ITS-*tef1-α* gene sequences showed that *Trichoderma* CMT10 was closely related to *T. asperellum* strains CEN1463, T34, ZJSX5002, KUFA0403, and RM-28 (Figure 3). The details of the strain names, origins, and accession numbers are listed in Table 2. Therefore, the CMT10 strain was identified as *T. asperellum*, according to morphological characterization and molecular analysis.

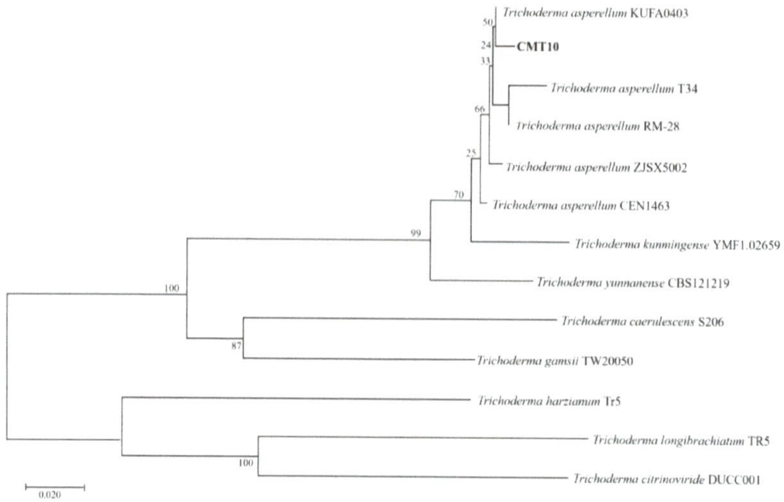

Figure 3. The phylogenetic tree of *Trichoderma* CMT10, based on the ITS-*tef1-α* gene sequences and their homologous sequences. Phylogenetic trees were constructed using the neighbor-joining method of MEGA 10.0 with bootstrap values based on 1000 replications. The accession numbers of the sequences are provided in Table 2. Bootstrap values are shown at branch points. The scale bar indicates 0.020 substitutions per nucleotide position.

3.3. *In Vitro Biocontrol of* Trichoderma *CMT10 against N. clavispora*

3.3.1. Inhibition Rates of Volatile Metabolites from *Trichoderma* CMT10 on *N. clavispora*

The effect of volatile metabolites emitted by *T. asperellum* CMT10 was tested against the growth of *N. clavispora* using the confrontation culture method. The mycelia of pathogenic CMGF3 were inhibited significantly by the volatile metabolites of CMT10, compared to the control. The IR was 69.84% at 7 d after confrontation culture (Figure 4). On the tenth day, the mycelia of the pathogenic CMGF3 had ceased to grow, while the mycelia of *T. asperellum* CMT10 continued to spread and encroach upon the colony of the pathogenic CMGF3.

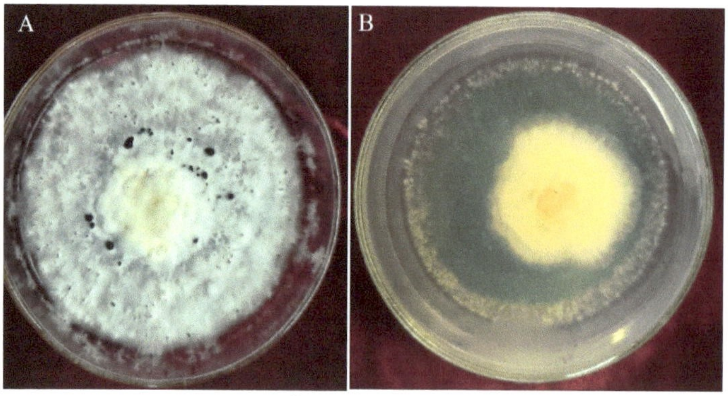

Figure 4. The inhibitory effect of volatile metabolites produced by *T. asperellum* CMT10 against the growth of the fungal pathogen *N. clavispora*. (**A**) PDA plate inoculated with *N. clavispora* and (**B**) the PDA plate inoculated with *N. clavispora* was placed on top of the PDA plate inoculated with *T. asperellum* CMT10 for 7 d at 28 °C, and the colony diameter was measured.

3.3.2. Inhibition Rates of Soluble Metabolites from *Trichoderma* CMT10 on *N. clavispora*

The antifungal activity of soluble metabolites produced by *T. asperellum* CMT10 was assessed against the fungal pathogen CMGF3. The results demonstrated that the soluble metabolites of CMT10 had a strong inhibitory effect against the growth of the fungal pathogen CMGF3 on PDA plates. After 7 d of incubation at 28 °C, the colony diameter of the CMT10-treated fungal growth was reduced significantly, compared to the untreated control (Figure 5). The IR of soluble metabolites produced by *T. asperellum* CMT10 was 79.67% against CMGF3.

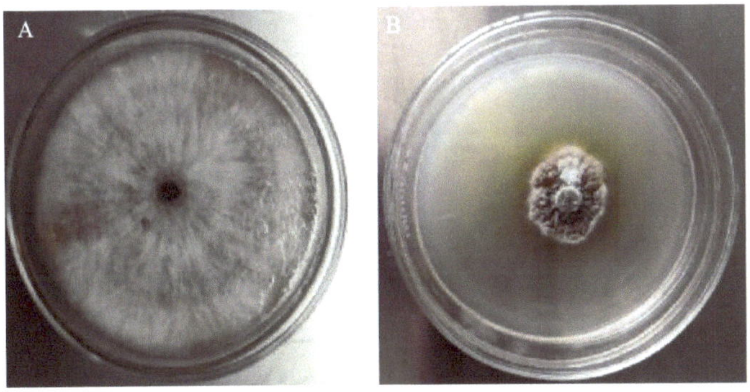

Figure 5. The inhibitory effect of soluble metabolites produced by *T. asperellum* CMT10 against the growth of the fungal pathogen *N. clavispora*. (**A**) inoculated with *N. clavispora* on the PDA plate mixed with sterile water and (**B**) inoculated with *N. clavispora* on the PDA plate mixed with the sterile filtrate of *T. asperellum* CMT10 for 7 d at 28 °C, and the colony diameter was measured.

3.3.3. Hyperparasitism of *Trichoderma* CMT10 on *N. clavispora*

Microscopic observation of the hyphal interaction between *T. asperellum* CMT10 and the pathogen CMGF3 revealed that the mycelia of both strains began to contact each other, but the antagonistic effect between them was not evident after 48 h (Figure 6A). CMT10 mycelia were attached to CMGF3 after 72 h (Figure 6B). After 96 h, CMT10 mycelia grew along and entwined with CMGF3 mycelia, causing the contraction of CMGF3 mycelia (Figure 6C,D). Moreover, CMGF3 mycelia were observed being penetrated and were embedded by CMT10 mycelia (Figure 6E). The results showed that *T. asperellum* CMT10 indicates a strong hyperparasitic effect against the strawberry root rot pathogen *N. clavispora* CMGF3.

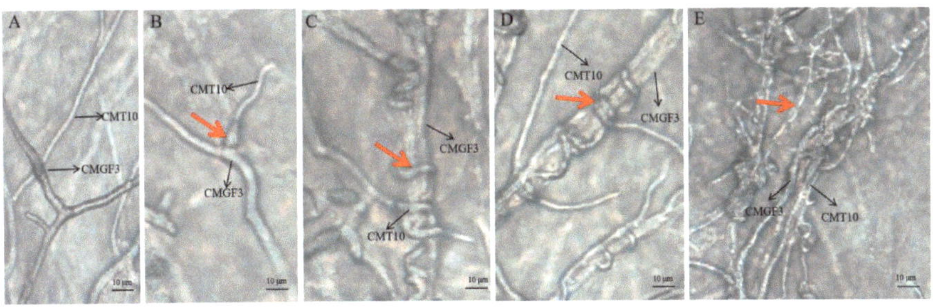

Figure 6. The hyperparasitic effects of *T. asperellum* CMT10 against *N. clavispora*. (**A**) Healthy mycelium morphology of CMT10 and healthy mycelium morphology of CMGF3; (**B**) the hyphae of CMGF3 were attached by CMT10 hyphae (as shown by the red arrow); (**C**,**D**) the hyphae of CMGF3 were entangled by CMT10 hyphae (as shown by the red arrow); and (**E**) the hyphae of CMGF3 were penetrated by CMT10 hyphae (as shown by the red arrow). The hyphae of CMT10 and CMG3 were determined primarily by the difference in mycelia diameter under microscopic conditions.

3.4. Determination of Biochemical Properties of T. asperellum CMT10

The results of the biochemical properties' determination revealed that the mixed solution of *T. asperellum* CMT10 with the Salkowski reagent did not turn pink, indicating that CMT10 could not produce indole-3-acetic acid (IAA), but it could grow on inorganic phosphate medium (Figure 7A). In the nitrogen-fixing medium, the mycelia were sparse, sporulation was limited, and spore distribution exhibited a spotty pattern (Figure 7B). Moreover, CMT10 grew on siderophore medium, demonstrating its ability to produce siderophores (Figure 7C).

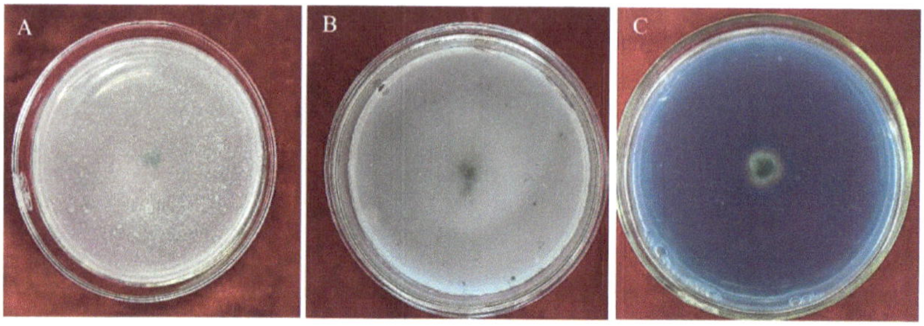

Figure 7. The biochemical properties of *T. asperellum* CMT10. (**A**) The ability of inorganic phosphorus solution; (**B**) the ability of nitrogen fixation solution; and (**C**) the ability of siderophore production.

3.5. Biocontrol Efficiency of T. asperellum CMT10 against Strawberry Root Rot

The incidence of strawberry root rot on each treatment was investigated after inoculation for 60 days (Table 4, Figure 8). The results revealed that the inoculation with *T. asperellum* CMT10 and the water control did not exhibit disease symptoms in strawberries. Treatment with the inoculation of *N. clavispora* CMGF3 showed the most severe disease symptoms, with a disease index of 84.00, which was significantly higher than that of the other treatments ($p \leq 0.05$). The disease index for treatment with *N. clavispora* CMGF3 + *T. asperellum* CMT10 was 31.00 and its biocontrol efficiency against strawberry root rot reached 63.09%, indicating that *T. asperellum* CMT10 effectively controlled the occurrence of potted strawberry root rot.

Figure 8. The control efficiency of *T. asperellum* CMT10 against strawberry root rot.

Table 4. The control efficiency of *T. asperellum* CMT10 against strawberry root rot.

Treatments	Disease Index	Control Efficiency (%)
CMGF3	84.00 ± 0.04 a	-
CMT10	0.00 ± 0.00 c	-
CMGF3 + CMT10	31.00 ± 0.61 b	63.09 ± 0.07 a
Control	0.00 ± 0.00 c	-

Note: water inoculation was used as a control. Data were mean ± SD. Different letters in the same column indicate a significant difference at the 0.05 level using Duncan's new multiple range test.

3.6. Growth-Promoting Efficiency of T. asperellum *CMT10 on Strawberry Seedlings*

The growth-promoting effects of *T. asperellum* CMT10 on strawberry seedlings were investigated after 60 days of inoculation. There was a significant increase in plant height, root length, total fresh weight, root fresh weight, stem fresh weight, and root dry weight compared with the water control. The growth-promoting rates were 20.09%, 22.39%, 87.11%, 101.58%, 79.82%, and 72.33%, respectively (Table 5, Figures 9 and 10).

Table 5. Growth-promoting effects of *T. asperellum* CMT10 on strawberry seedings.

Treatments	Plant Height (cm)	Root Length (cm)	Total Fresh Weight (g)	Root Fresh Weight (g)	Stem Fresh Weight (g)	Root Dry Weight (g)
CMT10	12.57 ± 1.35 a	23.75 ± 2.18 a	13.55 ± 3.53 a	7.18 ± 3.37 a	6.37 ± 2.08 a	2.66 ± 1.00 a
Control	10.53 ± 1.41 b	19.67 ± 2.70 b	7.61 ± 1.66 b	3.87 ± 1.59 b	3.74 ± 0.61 b	1.56 ± 0.50 b

Note: water inoculation was used as a control. Data were mean ± SD. Different letters in the same column indicate a significant difference at the 0.05 level using Duncan's new multiple range test.

Figure 9. Growth-promoting effects of *T. asperellum* CMT10 on strawberry seedlings.

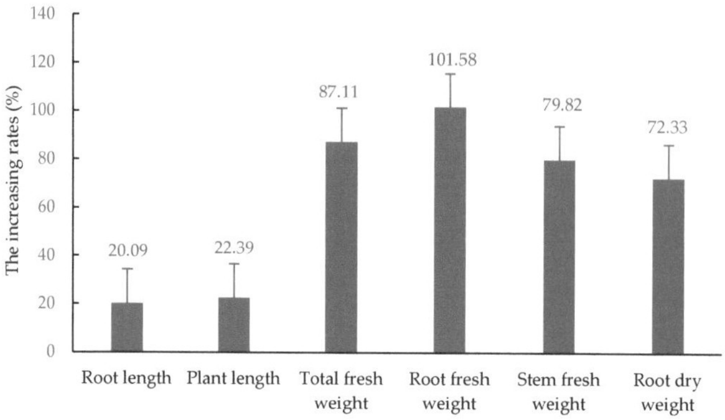

Figure 10. The increasing rates of *T. asperellum* CMT10 on the biomass of strawberry seedlings.

4. Discussion

4.1. Significance of Exploring Biocontrol Resources for Strawberry Root Rot

The prevention of strawberry root rot is complicated due to its diverse composition of pathogens, making it difficult to control. This disease is responsible for significant economic losses to the strawberry industry under greenhouse cultivation [31]. *Trichoderma* spp., recognized as crucial biocontrol resources, have been widely utilized in the disease control of various crops [32]. *Trichoderma* spp. have played a pivotal role in the prevention and control of strawberry root rot. However, due to the diverse composition of root rot pathogens, most studies have focused on *Trichoderma* against *Fusarium* spp. and *Rhizoctonia solani* [33,34]. Studies focusing on *N. clavispora*, a pathogen associated with strawberry root rot, are scarce. In this study, *N. clavispora*, an important pathogen causing strawberry root rot, was specifically selected as a target pathogenic fungus and obtained a strain of *T. asperellum* CMT10 with significant inhibition activity against *N. clavispora*. This study highlighted the remarkable effectiveness of *T. asperellum* against strawberry root rot and its ability to promote growth in strawberry seedlings. These findings contributed to the development of biocontrol resources for managing strawberry root rot and broadened the potential applications of *T. asperellum*.

4.2. Biocontrol Mechanism of T. asperellum CMT10

Most studies had demonstrated that *Trichoderma* strains could inhibit pathogenic fungi through nutrient and spatial competition, hyperparasitism, and the production of antimycotic secondary metabolites, while they could also promote plant growth and enhance plant stress resistance [35]. Ecological niche competition is a crucial mechanism of biocontrol microorganisms for preventing disease in biocontrol. *Trichoderma*, a biocontrol agent, is able to rapidly occupy ecological niches in environments with low concentrations of nutrients, which can cause pathogenic fungi to lose their ability to thrive and survive [36]. Risoli et al. [37] found that the growth rate of *Trichoderma* was 2.0–4.2 times that of *Botrytis cinerea*, indicating a significantly faster growth of *Trichoderma* compared to the pathogen, impeding the growth and reproduction of the pathogen. The results of this study indicated that *T. asperellum* CMT10 could significantly inhibit the growth and reproduction of *N. clavispora*. In the initial phase, CMT10 exhibited rapid growth and a strong competitiveness, and it quickly occupied nutritional and ecological spatial sites and produced an inhibition zone. In the later stages of cultivation, the *N. clavispora* colony completely disappeared and was replaced by dark green conidia of *T. asperellum*.

Trichoderma employes the mechanism of antibiosis in its biological control. Metabolites produced by *Trichoderma*, both volatile and soluble, have been reported to restrict the growth of various pathogenic fungi [38]. The metabolites included triohodexrmin, gliotoxin, viridin, and peptide antimycotics [39]. Naglot et al. [40] found that metabolites of *Trichoderma* significantly inhibited *F. oxysporum* with an inhibition rate of up to 54.81%. Manganiello et al. [41] discovered that volatile secondary metabolites secreted by *T. viride* TG050609 caused irregular growth, fragmentation, and even dissolution of *Phytophthora nicotianae*. By determining the inhibitory effects of the soluble and volatile metabolites of *T. asperellum* CMT10 on the *N. clavispora*, causing strawberry root rot, it was found that after 7 days of cultivation on CMT10-fermented metabolite plates, the inhibition rate reached 79.67% and the inhibition rate of volatile metabolites against *N. clavispora* reached 69.84%. This suggests that CMT10 metabolites have a strong inhibitory effect on the *N. clavispora* that causes strawberry root rot. However, the metabolites responsible for this effect have not been identified in our study and require further investigation.

Hyperparasitism was a vital mechanism employed by *Trichoderma* for its biological control. *Trichoderma* recognized lectins on the mycelia of a pathogenic fungi and engages in processes such as identification, contact, wrapping, penetration, parasitism, and dissolution of the fungi [42]. Hewedy et al. [43] found that *T. harzianum* Th6 could adhere to, invade, and disrupt the mycelia of *F. graminearum*. Larran et al. [44] observed that *T. harzianum* could form adhesive structures on the mycelia of *F. sudanense*, leading to curling, wrinkling, and dissolution of the mycelia of *F. sudanense*. The present study also found that *T. asperellum* CMT10 exhibited hyperparasitism against *N. clavispora*. It could recognize, contact, wrap, and parasitize the mycelia of the pathogen. However, mycelium dissolution, protoplasm leakage, or cell disintegration were not observed, which may be related to the observation time during cultivation. It is believed that the cell wall hydrolytic enzymes secreted by *Trichoderma* played a crucial role in its hyperparasitic activity, such as chitinases, glucanases, and proteases, which can dissolve the cell walls of pathogenic fungi, allowing *Trichoderma* to parasitize, absorb nutrients, and ultimately cause the death of the pathogenic fungi [45]. Whether *T. asperellum* CMT10 can secrete enzymes with lytic effects was not investigated in our study and is a direction for future research.

4.3. Practical Application of T. asperellum CMT10

Currently, the production of *Trichoderma* generally involves the simultaneous or sequential action of several disease prevention mechanisms. *Trichoderma* can utilize different antagonistic mechanisms at different stages to biocontrol effects [46]. This study demonstrated that *T. asperellum* CMT10 exerted competitive, antibiosis, and hyperparasitic effects against the pathogenic fungus causing strawberry root rot. *T. asperellum* CMT10 could effectively control the occurrence of strawberry root rot. However, the biocontrol mechanisms at different stages of interaction between *Trichoderma* and the pathogenic fungi in plants still need further exploration, which may provide a theoretical foundation for the practical application of *T. asperellum* CMT10 in strawberry production. Therefore, future research should focus on the field disease control effect and the interactive relationships among *T. asperellum* CMT10, the pathogenic fungi causing root rot, and the host plant. In addition, this study specifically evaluated the growth-promoting effects of *T. asperellum* CMT10 on strawberry seedlings by measuring parameters such as plant height, root length, total fresh weight, root fresh weight, stem fresh weight, and root dry weight. It is essential to conduct more field experiments to fully understand the growth-promoting effects of CMT10 on strawberry plants, as well as to investigate the impact of *T. asperellum* CMT10 on strawberry fruit size, yield, and quality.

5. Conclusions

In summary, *T. asperellum* CMT10 was obtained among 10 *Trichoderma* strains as a potent biocontrol agent against *N. clavispora*, the pathogenic fungi causing strawberry root rot. The results of the pot experiment demonstrated that *T. asperellum* CMT10 effectively inhibited root rot and significantly enhanced the growth of strawberry seedlings. These findings indicate that *T. asperellum* CMT10 has great potential as a biocontrol resource for preventing and controlling strawberry root rot, making it a promising candidate for future development.

Author Contributions: Conceptualization, R.Y.; methodology, P.L., Z.W., W.R. and R.Y.; software, D.W., Y.M., W.Y. and W.R.; formal analysis, P.L., W.R. and R.Y.; writing—original draft preparation, R.Y.; visualization, D.W., Y.M. and W.R.; project administration, R.Y. and W.Y.; funding acquisition, W.Y. and R.Y. All authors have read and agreed to the published version of the manuscript.

Funding: This research was funded by the Science and Technology Program of Henan province of China (232102320111), the Science and Technology Planning Major Project of Fujian province of China (2022N0010), and the Natural Resources Science and Technology Innovation Project of Fujian province of China (KY-090000-04-2022-016).

Data Availability Statement: The original contributions presented in the study are included in the article, further inquiries can be directed to the corresponding author.

Conflicts of Interest: The authors declare no conflicts of interest.

References

1. Basu, A.; Nguyen, A.; Betts, N.M.; Lyons, T.J. Strawberry as a functional food: An evidence-based review. *Crit. Rev. Food Sci. Nutr.* **2014**, *54*, 790–806. [CrossRef]
2. Ji, Y.; Li, X.; Gao, Q.H.; Geng, C.; Duan, K. *Colletotrichum* species pathogenic to strawberry: Discovery history, global diversity, prevalence in China, and the host range of top two species. *Phytopathol. Res.* **2022**, *4*, 42. [CrossRef]
3. Wang, M.Y.; Du, Y.Q.; Cai, W.W.; Wang, Z.H.; Zhu, J.Q. Effect of complex antagonistic bacteria on controlling strawberry root rot. *J. Agric. Sci. Technol.* **2020**, *22*, 100–110.
4. Lazcano, C.; Boyd, E.; Holmes, G.; Hewavitharana, S.; Pasulka, A.; Ivors, K. The rhizosphere microbiome plays a role in the resistance to soil-borne pathogens and nutrient uptake of strawberry cultivars under field conditions. *Sci. Rep.* **2021**, *11*, 3188. [CrossRef]
5. Iqbal, M.; Jamshaid, M.; Zahid, M.A.; Andreasson, E.; Vetukuri, R.A.; Stenberg, J.A. Biological control of strawberry crown rot, root rot and grey mould by the beneficial fungus *Aureobasidium pullulans*. *Bio-Control* **2021**, *66*, 535–545. [CrossRef]
6. Baggio, J.S.; Cordova, L.G.; Peres, N.A. Sources of inoculum and survival of 22 *Macrophomina phaseolina* in Florida strawberry fields. *Plant Dis.* **2019**, *103*, 2417–2424. [CrossRef]
7. Feliziani, E.; Romanazzi, G. Postharvest decay of strawberry fruit: Etiology, epidemiology, and disease management. *J. Berry Res.* **2016**, *6*, 47–63. [CrossRef]
8. Hong, S.; Kim, T.Y.; Won, S.J.; Moon, J.H.; Ajuna, H.B.; Kim, K.; Ahn, Y.S. Control of fungal diseases and fruit yield improvement of strawberry using *Bacillus velezensis* CE100. *Microorganisms* **2022**, *10*, 365. [CrossRef] [PubMed]
9. Hu, Y.J.; Yang, H.M.; Jin, J.; Yan, H.H.; Wang, J.P.; Zhang, R.Q. Synergistic activity of antagonistic *Trichoderma* spp. and *Rhizoctonia solani* increases disease severity on strawberry petioles. *Eur. J. Plant Pathol.* **2022**, *164*, 375–389. [CrossRef]
10. Abdel-lateif, K.S. *Trichoderma* as biological control weapon against soil borne plant pathogens. *Afr. J. Biotechnol.* **2017**, *16*, 2299–2306.
11. Benítez, T.; Rincón, A.M.; Limón, M.C.; Codón, A.C. Biocontrol mechanisms of *Trichoderma* strains. *Int. Microbiol.* **2004**, *7*, 249–260.
12. Ferreira, F.V.; Musumeci, M.A. Trichoderma as biological control agent: Scope and prospects to improve efficacy. *World J. Microbiol. Biotechnol.* **2021**, *37*, 90. [CrossRef]
13. Zhang, H.; Du, G.D.; Song, Y.N.; Lu, X.F.; Ying, N. Screening, identification and the effect validation of *Trichoderma* against the root rot of strawberry. *J. Shenyang Agric. Univ.* **2015**, *46*, 654–660.
14. Mercado, J.A.; Barceló, M.; Pliego, C.; Rey, M.; Caballero, J.L.; Muñoz-Blanco, J.; Ruano-Rosa, D.; López-Herrera, C.; de Los Santos, B.; Romero-Muñoz, F.; et al. Expression of the β-1,3-glucanase gene bgn13.1 from *Trichoderma harzianum* in strawberry increases tolerance to crown rot diseases but interferes with plant growth. *Transgenic Res.* **2015**, *24*, 979–989. [CrossRef]
15. Rees, H.J.; Drakulic, J.; Cromey, M.G.; Bailey, A.M.; Foster, G.D. Endophytic *Trichoderma* spp. can protect strawberry and privet plants from infection by the fungus *Armillaria mellea*. *PLoS ONE* **2022**, *17*, e0271622. [CrossRef]
16. Mirzaeipour, Z.; Bazgir, E.; Zafari, D.; Darvishnia, M. Selection and biocontrol efficiency of *Trichoderma* isolates against *Rhizoctonia* root rot and their growth promotion effects on strawberry plants. *J. Plant Pathol.* **2023**, *105*, 1563–1579. [CrossRef]
17. Debbi, A.; Boureghda, H.; Monte, E.; Hermosa, R. Distribution and genetic variability of *Fusarium oxysporum* associated with tomato diseases in Algeria and a biocontrol strategy with indigenous *Trichoderma* spp. *Front. Microbiol.* **2018**, *9*, 282. [CrossRef] [PubMed]
18. Pimentel, M.F.; Arnão, E.; Warner, A.J.; Subedi, A.; Rocha, L.F.; Srour, A.; Bond, J.P.; Fakhoury, A.M. *Trichoderma* isolates inhibit *Fusarium virguliforme* growth, reduce root rot, and induce defense-related genes on soybean seedlings. *Plant Dis.* **2020**, *104*, 1949–1959. [CrossRef] [PubMed]
19. Shaigan, S.; Seraji, A.; Moghaddam, S.A. Identification and investigation on antagonistic effect of *Trichoderma* spp. on tea seedlings white foot and root rot (*Sclerotium rolfsii* Sacc.) in vitro condition. *Pak. J. Biol. Sci.* **2008**, *11*, 2346–2350. [CrossRef] [PubMed]
20. Yang, H. *Classification and Identification of* Trichoderma; China Land Press: Beijing, China, 2009; pp. 14–20.

21. White, T.J.; Bruns, T.; Lee, S.; Taylor, J. Amplification and direct sequencing of fungal ribosomal RNA genes for phylogenetics. In *PCR Protocols: A Guide to Methods and Applications*; Academic Press: San Diego, CA, USA, 1990; pp. 315–322.
22. O'donnell, K.; Kistler, H.C.; Cigelnik, E.; Ploetz, R. Multiple evolutionary origins of the fungus causing Panama disease of banana: Concordant evidence from nuclear and mitochondrial gene genealogies. *Proc. Natl. Acad. Sci. USA* **1998**, *95*, 2044–2049. [CrossRef]
23. Damodaran, T.; Rajan, S.; Muthukumar, M.; Ram, G.; Yadav, K.; Kumar, S.; Ahmad, I.; Nidhi, K.; Mishra, V.K.; Vinay, K.; et al. Biological management of banana Fusarium wilt caused by Fusarium oxysporum f. sp. cubense tropical race 4 using antagonistic fungal isolate CSR-T-3 (Trichoderma reesei). *Front. Microbiol.* **2020**, *11*, 595845. [CrossRef] [PubMed]
24. Kexiang, G.; Xiaoguang, L.; Yonghong, L.; Tianbo, Z.; Shuliang, W. Potential of Trichoderma harzianum and T. atroviride to Control Botryosphaeria berengeriana f. sp. piricola, the cause of apple ring rot. *J. Phytopathol.* **2002**, *150*, 271–276. [CrossRef]
25. Nautiyal, C.S. An efficient microbiological growth medium for screening phosphate solubilizing microorganisms. *FEMS Microbiol. Lett.* **1999**, *170*, 265–270. [CrossRef]
26. Shin, S.H.; Lim, Y.; Lee, S.E.; Yang, N.W.; Rhee, J.H. CAS agar diffusion assay for the measurement of siderophores in biological fluids. *J. Microbiol. Methods* **2001**, *44*, 89–95. [CrossRef] [PubMed]
27. Liaqat, F.; Eltem, R. Identification and characterization of endophytic bacteria isolated from in vitro cultures of peach and pear rootstocks. *3 Biotech* **2016**, *6*, 120. [CrossRef]
28. Yao, C.X.; Li, X.J.; Li, Q.; Xing, G.Z.; Fang, W.Y.; Li, C.H.; Zhang, Y.Y.; Yao, C.X.; Xu, M.; Li, F.F.; et al. Screening and identification of antagonistic bacteria against tobacco *Fusarium* root rot and evaluation of their effects on growth promoting and disease control. *Chin. J. Biol. Control* **2021**, *37*, 1066–1072.
29. Brick, J.M.; Bostock, R.M.; Silversone, S.E. Rapid in situ assay for indole acetic acid production by bacteria immobilized on nitrocellulose membrane. *Appl. Environ. Microbiol.* **1991**, *57*, 535–538. [CrossRef]
30. Vestberg, M.; Kukkonen, S.; Saari, K.; Parikka, P.; Huttunen, J.; Tainio, L.; Devos, N.; Weekers, F.; Kevers, C.; Thonart, P.; et al. Microbial inoculation for improving the growth and health of micropropagated strawberry. *Appl. Soil Ecol.* **2004**, *27*, 243–258. [CrossRef]
31. Zhang, M.; Kong, Z.; Fu, H.; Shu, X.; Xue, Q.; Lai, H.; Guo, Q. Rhizosphere microbial ecological characteristics of strawberry root rot. *Front. Microbiol.* **2023**, *14*, 1286740. [CrossRef] [PubMed]
32. Elad, Y.; Chet, I.; Henis, Y. Biological control of *Rhizoctonia solani* in strawberry fields by *Trichoderma harzianum*. *Plant Soil* **1981**, *60*, 245–254. [CrossRef]
33. Hernández-Muñiz, P.; Borrero, C.; Ordóñez-Martín, J.; Pastrana, A.M.; Avilés, M. Optimization of the use of industrial wastes in Anaerobic soil disinfestation for the control of *Fusarium* wilt in strawberry. *Plants* **2023**, *12*, 3185. [CrossRef] [PubMed]
34. Tyśkiewicz, R.; Nowak, A.; Ozimek, E.; Jaroszuk-Ściseł, J. *Trichoderma*: The current status of its application in agriculture for the biocontrol of fungal phytopathogens and stimulation of plant growth. *Int. J. Mol. Sci.* **2022**, *23*, 2329. [CrossRef]
35. Yao, X.; Guo, H.; Zhang, K.; Zhao, M.; Ruan, J.; Chen, J. *Trichoderma* and its role in biological control of plant fungal and nematode disease. *Front. Microbiol.* **2023**, *14*, 1160551. [CrossRef]
36. Mohiddin, F.A.; Padder, S.A.; Bhat, A.H.; Ahanger, M.A.; Shikari, A.B.; Wani, S.H.; Bhat, F.A.; Nabi, S.U.; Hamid, A.; Bhat, N.A.; et al. Phylogeny and optimization of *Trichoderma harzianum* for Chitinase production: Evaluation of their antifungal behaviour against the prominent soil borne phyto-pathogens of temperate India. *Microorganisms* **2021**, *9*, 1962. [CrossRef] [PubMed]
37. Risoli, S.; Cotrozzi, L.; Sarrocco, S.; Nuzzaci, M.; Pellegrini, E.; Vitti, A. *Trichoderma*-induced resistance to *Botrytis cinerea* in Solanum species: A meta-analysis. *Plants* **2022**, *11*, 180. [CrossRef]
38. Kottb, M.; Gigolashvili, T.; Großkinsky, D.K.; Piechulla, B. *Trichoderma* volatiles effecting Arabidopsis: From inhibition to protection against phytopathogenic fungi. *Front. Microbiol.* **2015**, *6*, 995. [CrossRef] [PubMed]
39. Khan, R.A.A.; Najeeb, S.; Hussain, S.; Xie, B.; Li, Y. Bioactive secondary metabolites from *Trichoderma* spp. against phytopathogenic fungi. *Microorganisms* **2020**, *29*, 817. [CrossRef]
40. Naglot, A.; Goswami, S.; Rahman, I.; Shrimali, D.D.; Yadav, K.K.; Gupta, V.K.; Veer, V. Antagonistic potential of native *Trichoderma viride* strain against potent tea fungal pathogens in north east India. *Plant Pathol. J.* **2015**, *31*, 278–289. [CrossRef]
41. Manganiello, G.; Sacco, A.; Ercolano, M.R.; Vinale, F.; Lanzuise, S.; Pascale, A.; Napolitano, M.; Lombardi, N.; Lorito, M.; Woo, S.L. Modulation of tomato response to *Rhizoctonia solani* by *Trichoderma harzianum* and its secondary metabolite harzianic acid. *Front. Microbiol.* **2018**, *9*, 1966. [CrossRef]
42. Shaw, S.; Le Cocq, K.; Paszkiewicz, K.; Moore, K.; Winsbury, R.; de Torres Zabala, M.; Studholme, D.J.; Salmon, D.; Thornton, C.R.; Grant, M.R. Transcriptional reprogramming underpins enhanced plant growth promotion by the biocontrol fungus *Trichoderma hamatum* GD12 during antagonistic interactions with *Sclerotinia sclerotiorum* in soil. *Mol. Plant Pathol.* **2016**, *17*, 1425–1441. [CrossRef]
43. Hewedy, O.A.; Abdel Lateif, K.S.; Seleiman, M.F.; Shami, A.; Albarakaty, F.M.; El-Meihy, R.M. Phylogenetic diversity of *Trichoderma* strains and their antagonistic potential against soil-borne pathogens under stress conditions. *Biology* **2020**, *9*, 189. [CrossRef] [PubMed]
44. Larran, S.; Santamarina Siurana, M.P.; Roselló Caselles, J.; Simón, M.R.; Perelló, A. In vitro antagonistic activity of *Trichoderma harzianum* against *Fusarium sudanense* causing seedling blight and seed rot on wheat. *ACS Omega* **2020**, *5*, 23276–23283. [CrossRef] [PubMed]

45. Sood, M.; Kapoor, D.; Kumar, V.; Sheteiwy, M.S.; Ramakrishnan, M.; Landi, M.; Araniti, F.; Sharma, A. *Trichoderma*: The secrets of a multitalented biocontrol agent. *Plants* **2020**, *9*, 762. [CrossRef] [PubMed]
46. Jogaiah, S.; Abdelrahman, M.; Tran, L.P.; Ito, S.I. Different mechanisms of *Trichoderma* virens-mediated resistance in tomato against *Fusarium* wilt involve the jasmonic and salicylic acid pathways. *Mol. Plant Pathol.* **2018**, *19*, 870–882. [CrossRef]

horticulturae

MDPI

Article

Exploring the Potential Biocontrol Isolates of *Trichoderma asperellum* for Management of Collar Rot Disease in Tomato

C. Shanmugaraj [1], Deeba Kamil [1,*], Aditi Kundu [2], Praveen Kumar Singh [3], Amrita Das [1], Zakir Hussain [4], Robin Gogoi [1], P. R. Shashank [5], R. Gangaraj [1] and M. Chaithra [1]

1 Division of Plant Pathology, ICAR-Indian Agricultural Research Institute, New Delhi 110012, India; spcshanmugaraj@gmail.com (C.S.); amrita.das@icar.org.in (A.D.); rgogoi@iari.res.in (R.G.); gangaeverest.123@gmail.com (R.G.); chaithram06@gmail.com (M.C.)

2 Division of Agricultural Chemicals, ICAR-Indian Agricultural Research Institute, New Delhi 110012, India; aditi@iari.res.in

3 Division of Centre for Protected Cultivation Technology (CPCT), ICAR-Indian Agricultural Research Institute, New Delhi 110012, India; pksingh128@iari.res.in

4 Division of Vegetable Science, ICAR-Indian Agricultural Research Institute, New Delhi 110012, India; zakir.hussain1@icar.gov.in

5 Division of Entomology, ICAR-Indian Agricultural Research Institute, New Delhi 110012, India; shashank@iari.res.in

* Correspondence: deeba.kamil@icar.gov.in

Abstract: Bio-control agents are the best alternative to chemicals for the successful management of plant diseases. Among them, *Trichoderma* is commonly used as a biological control agent in plant disease management due to its ability to suppress soil-borne plant pathogens. In the present study, 20 *Trichoderma asperellum* isolates were collected from different geographical locations and confirmed using morphological characteristics and molecular phylogenetic inferences based on combined ITS and β-tubulin sequences. All twenty isolates were screened for their antagonism against the collar rot pathogen under in vitro and in planta conditions. The isolates were evaluated through dual culture and volatile methods in an in vitro study. Isolate A10 inhibited the test pathogen *Agroathelia rolfsii* at 94.66% in a dual culture assay and 70.95% in a volatile assay, followed by the isolates A11 and A17, which recorded 82.64% and 81.19% in dual culture assay and 63.75% and 68.27% in the volatile assay respectively. An in planta study was conducted under greenhouse conditions in tomato var. pusa ruby by pre- and post-inoculation of *T. asperellum* isolates in the *A. rolfsii* infected soil to evaluate their antagonistic potential against the disease. The A10 isolate was found effective under both pre- and post-inoculation conditions, with a disease inhibition percent of 86.17 and 80.60, respectively, followed by the isolates A11 and A17, which exhibited inhibition of 77.80% and 75.00% in pre-inoculation and 72.22% and 69.44% in post-inoculation, respectively. Further, biochemical analysis was conducted to determine the specific activity of hydrolytic enzymes produced by *T. asperellum* during interaction with *A. rolfsii*. We found that isolate A10 produces more hydrolytic enzymes with the specific activity of 174.68 IU/mg of β-1,3 glucanase, 183.48 IU/mg of β-1, 4 glucanase, 106.06 IU/mg of protease, followed by isolate A17, A11 respectively. In GC-MS analysis, we observed maximum anti-microbial volatile organic compounds from the isolate A10, including 2H-Pyran-2-one (17.39%), which was found to be most abundant, followed by dienolactone (8.43%), α-pyrone (2.19%), and harziandione (0.24%) respective retention time of 33.48, 33.85, 33.39, and 64.23 min, respectively, compared to other isolates. In the TLC assay, we observed that a greater number of bands were produced by the A10 and A17 isolates in the Hexane: Ethyl Acetate (1:1) solvent system than in the 9:1 solvent system, which represents the presence of major metabolites in the ethyl acetate extract.

Keywords: bio-control; bio-efficacy; GC-MS; hydrolytic enzymes; *Agroathelia rolfsii*; secondary metabolites; *Trichoderma asperellum*

1. Introduction

Agroathelia rolfsii (Sacc.) Redhead and Mullineux (Agaricomycetes: Amylocorticiales) is a destructive, necrotrophic, soil-borne plant pathogen that causes collar rot or southern blight disease [1]. It was first reported in Florida in 1892 by Rolfs on tomatoes. They have an extremely wide host range that includes more than 600 plant species across 100 plant families, such as groundnut, green bean, lima bean, onion, garden bean, pepper, potato, sweet potato, tomato, and watermelon [2–4], causing huge losses worldwide. It causes significant losses in various crops in India, including peanuts, soybeans, tomatoes, peppers, and many others. In peanut and tomato crops, the losses due to *A. rolfsii* in India have been estimated to be around 30–40% of the total crop yield [5–7]. Management is not successful in the field because of the soil-borne nature of the pathogen.

Moreover, it produces prolific growth and can produce persistent sclerotia, which plays a key role in the disease cycle. Like many other soil-borne fungal diseases, *A. rolfsii* continues to be a difficult pathogen to control due to its wide host range, abundant persistent sclerotia, and genetic variability among populations [8]. Biological control is an environmentally sound and effective means of managing plant diseases. Recently, the use of potential biocontrol agents has gained importance for the management of pests and diseases because of their cost-effectiveness, sustainability, and eco-friendly nature.

The genus *Trichoderma* is a group of fungi that is well-known for their biocontrol potential against a range of plant pathogens, including *A. rolfsii* [9–12]. *Trichoderma* spp. are natural antagonists that can compete with plant pathogens via various mechanisms, *viz.*, space and nutrients, production of enzymes, and secondary metabolites that can inhibit or kill the pathogens [13,14] and confrontation through mycoparasitism or either by inducing resistance and plant defense reactions [15–17].

However, *Trichoderma asperellum* (Sordariomycetes; Hypocreaceae) possesses a dual activity of both biocontrol and plant growth-promoting properties [18–20]. The colonization of the *Arabidopsis* root by *T. asperellum* SKT-1 and its culture filtrate elicit an ISR (Induced Systemic Resistance) against *Pseudomonas syringae* pv. *tomato* DC3000 through the increased expression of JA (Jasmonic Acid)/ET (Ethylene) and SA (Salicylic Acid) inducing genes [21]. Similar to Plant Growth Promoting Rhizobacteria (PGPR), *T. asperellum*- ISR is activated by the JA/ET signaling pathway [22]. For instance, *T. asperellum* was used by a number of nations to stop crown rot, root rot, and damping off [23]. Swollenin production is induced in plants as a result of *T. asperellum* colonization of the roots to provide a local defense [24]. The ethylene and H_2O_2-mediated plant defense responses against *Rhizoctonia solani* attacks were induced by the synthesis of *T. asperellum* xylanases in plants [25].

There were many secondary metabolites like 6-pentyl-2H-pyran-2-one (6-PP), 2,4-ditert-butyl phenol, propenyl phenyl methyl ester, heptanes, viridin, and harzianolides were reported from *Trichoderma* spp. [26]. More importantly, metabolites, like 6-pentyl-2H-pyran-2-one, have antifungal activity against various plant pathogens [27]. The isolates of *T. asperellum* were known to produce several hydrolytic enzymes like cellulase, β-1,3-glucanases, β-1,4-glucanases, chitinase, and protease with antifungal ability by degrading the cell wall components of the various plant pathogens [28–31].

Therefore, the present study was carried out to evaluate *T. asperellum* isolates for their antagonistic activity against the collar rot pathogen under in vitro and in planta conditions and to characterize the bioactive compounds produced by this bio-control agent. This information will facilitate extensive applications of formulation in the field of bio-control in the future for the successful management of collar rot disease.

2. Materials and Methods

2.1. Collection and Isolation of the Fungus

In 2022, from March to October, 20 soil and rootsamples were collected from the nine states of India (New Delhi, Andhra Pradesh, Rajasthan, West Bengal, Gujarat, Karnataka, Maharashtra, Tamil Nadu, and Uttar Pradesh). Isolation of the potential fungal biocontrol isolates from soil samples was performed on potato dextrose agar (PDA) according to the procedure described by [32]. Plates were incubated at 28 ± 2 °C for 5–7 days. The *Trichoderma* growth was obtained after 5–7 days and further purified by sub-culturing. The purified fungal isolates were maintained on PDA slants at 4°C. All the twenty obtained isolates were deposited in the Indian Type Culture Collection (ITCC) at the Indian Agricultural Research Institute, New Delhi. A highly virulent isolate *A. rolfsii* (Sr38), obtained during our previous work (ITCC Code: 8665, Accession number: OR192927), was used to evaluate the biocontrol potential of the obtained twenty *T. asperellum* isolates.

2.2. Morphological Observations

Morphological identification of the twenty potential biocontrol *T. asperellum* isolates was performed under the compound microscope along with camera attachment (ProgRes capture pro2.7- JENOPTIK). Isolates were identified based on microscopic features, including conidiophores branching, the shape and size of the phialides, and the conidia.

2.3. Molecular Identification and Phylogenetic Analysis

DNA from all the twenty isolates was extracted using a CTAB (Cetyl Trimethyl Ammonium Bromide), procedure given by Cullings [33]. The amplification of the ITS region was carried out using universal primers ITS1 5′-TCCGTAGGTGAACCTGCGG-3′ and ITS4 5′-TCCTCCGCTTATTGATATGC-3′ [34], B-tubf1-F 5′-CAGCTCGAGCGTATGAACGTCT-3′and B-tubr1-R 5′-TGTACCAATGCAAGAAAGCCTT-3′ primers were used for the amplification of β-tubulin region [35]. A total volume of 25 μL PCR mixture consisted of 12.5 μL DreamTaq Green PCR master mix (Thermo Scientific, Pune, India) (including 0.25 mM each of dNTP, 2mM $MgCl_2$, and Taq DNA polymerase), 9.5 μL nuclease-free water, 1 μL (10 Pmol/ μL of each forward and reverse primer), and 1 μL (100 ng/ μL) of fungal DNA. PCR was performed with initial denaturation for 3 min at 94 °C, 30 cycles of denaturation at 94 °C for 30 s, annealing temperature at 57 °C for 1 min, primer extension at 72 °C for 2 min, and final primer extension at 72 °C for 5 min for the ITS region. The PCR for β-tubulin was performed with initial denaturation for 3 min at 94 °C, 30 cycles of denaturation at 94 °C for 30 s, annealing temperature at 61 °C for 40 s, primer extension at 72 °C for 30 s, and final primer extension at 72 °C for 7 min. The electrophoresis was performed using 1.2% agarose gel with 0.5 mg/μL ethidium bromide in 1xTAE buffer (100 V, 400 mA for 30 min) to visualize the PCR products. The 1 kb DNA marker (Thermo Fisher Scientific, Waltham, MA, USA) was used to estimate the size of the PCR products, and further, the samples were sequenced through outsourcing. The NCBI Nucleotide BLAST was done on all the sequences to confirm the species' identity. The phylogenetic analysis and the tree were constructed by combining sequences of ITS (550 bp) and β-tubulin (1025 bp) using MEGA version 11 with the maximum likelihood method. The ex-type strain of *T. asperellum* CBS 433.97 (NR130668, XM024905825) as a reference sequence, *Chaetomium globosum*CBS 160.62 (NR144851, KT214742) as an outgroup, and other closely related species includes *T. pubescens* (KF294849, KF609264), *T. harzianum* (MK886859, MK895942), *T. hamatum* (LT707583, LT707606), *T. longibrachiatum* (JN039058, KF595268), *T. virens* (HM046563, KF595234) were used to construct the tree. To assess the stability of branches, a bootstrap analysis with 1000 replications was performed.

2.4. In Vitro Antagonistic Assay

The twenty *T. asperellum* isolates were tested for their in vitro antagonistic activity against *A. rolfsii* using a dual culture assay. All the *T. asperellum* isolates, and *A. rolfsii* isolate Sr38 were grown on PDA plates for 7 days at 28 ± 2 °C. On the PDA plate, a 5 mm disc of *T. asperellum* and the test pathogen (*A. rolfsii*) were placed immediately opposite one another, 2 cm from the edge, and cultured for 7 days at 28 ± 2 °C. As a control, separate petri plates containing only the pathogen (*A. rolfsii*) were kept. Three replications of each treatment were used for the whole experiment. Following Garcia's formula [36], which measures Inhibition (%) = 100 [(C − T)/C], the radial growth (mm) and percent inhibition were measured and calculated. C represents the growth of the pathogen in the control plate, and T represents the growth of the pathogen in the dual culture plate (treatment) [37]. The impact of antibiotics generated by antagonistic fungi (*Trichoderma*) was investigated using a volatile assay [38,39]. A 5 mm *T. asperellum* mycelial disc was centered on the PDA plate, and another plate had the pathogen *A. rolfsii* inoculated in the same way. Each petri dish pair was sealed together with cellophane adhesive tape with the antagonist plate down and the pathogen plate on the upside and incubated at 28 ± 2 °C for 7 days. As a control, Petri dishes were used with the pathogen on the top plate and without the antagonist on the bottom plate. According to Garcia, the percentage of mycelial growth inhibition was calculated 7 days after inoculation.

2.5. In Planta Bio-Control Assay

An in planta study was conducted on one-month-old tomato plants (var. Pusa Ruby) by pre- and post-inoculation approach of *T. asperellum* isolates in the *A. rolfsii* infected soil to evaluate their antagonistic potential against the disease. The experiment was carried out at the Center for Protected Cultivation Technology (CPCT), IARI, New Delhi. The experiment was conducted by keeping four treatments for all the isolates, viz., control (treated with distilled water), biocontrol treated, pathogen treated, and pathogen + biocontrol-treated plants to compare the effect.

Sorghum grains were used to multiply the test pathogen *A. rolfsii*. The grains were initially soaked in water overnight and then drained off the excess water, filled 100 g sorghum grains in 500 mL Erlenmeyer flasks, plugged with non-absorbent cotton, and autoclaved at 15 psi at 121 °C for 30 min. After cooling at room temperature, the flasks were inoculated with mycelial discs of the 7-day-old culture of *A. rolfsii* under aseptic conditions and incubated at 28 ± 2 °C for 7 days until the sorghum grains were completely covered with *A. rolfsii* growth. The evaluation was conducted by mixing 10 g of *A. rolfsii* inoculums/ kg of soil to make the infected soil.

The seven-day-old pure culture of *T. asperellum* isolates was used after culturing on PDA plates upon incubation at 28 ± 2 °C. To prepare the spore suspension, 5 mL of sterile distilled water was added to each plate, and the spores were then scraped off with a sterile spatula; the spore suspensions were filtered, and the concentrations were adjusted to 10^8 spores/ ml for all the isolates. The evaluation was carried out through pre- and post-inoculation approaches by applying 100 mL of *T. asperellum* spore suspension per pot as a drench and seedling root dip.

In the case of the post-inoculation approach, each *T. asperellum* isolate's spore suspension was then added separately for each isolate evaluation after 5 days of transplanting plants in the infected soil in the pot. Dipping the seedlings in *T. asperellum* spore suspension for 60 min and then transplanting the plants in the soil, which is already drenched with *T. asperellum* spore suspensions in the infected soil in case of the pre-inoculation approach.

After 10 days, the disease severity index was calculated according to the scale, which includes 0 = no disease symptoms; 1 = disease symptoms without visible fungal outgrowth; 2 = disease symptoms with visible fungal growth; 3 = partial wilting of the plant; and 4 = complete wilting and plant death. The experiment was conducted thrice, with three replications of each treatment, using a Completely Randomized Design (CRD). The data collected from the pathogen + biocontrol treatment from all the isolate's evaluation and the disease severity index (DSI) was calculated for both pre- and post-inoculation conditions. The disease severity index was calculated using the following formula,

$$\text{DSI }(\%) = \frac{\Sigma(\text{Severity score} \times \text{Number of infected plants having the sames core})}{\text{Total number of plants observed} \times \text{Maximum rating scale number}} \times 100$$

2.6. Biochemical Analysis

The specific activity of the hydrolytic enzymes (cellulase, β-1,3 glucanase, β-1,4 glucanase, protease, and chitinase) in association with *A. rolfsii* was investigated for all the isolates of *T. asperellum*. The enzyme assay was conducted in a minimal synthetic medium supplemented with colloidal chitin at 0.05% (*w*/*v*) as sole carbon source by inoculation with 1 mL spore suspension of *T. asperellum* isolates (2×10^8 cfu/mL) and test pathogen (*A. rolfsii*) in a 50 mL medium and incubated in a BOD cum rotary shaker at 150 rpm at 28 ± 2 °C for 7 days. The mycelial mat was separated to obtain the culture filtrate by running it through Whatman no. 42 filter paper before centrifuging at 6000 rpm for 10 min at 4 °C. The supernatant was immediately tested for enzyme activity after being collected into a sterilized conical flask. Enzyme activity was expressed in specific activity as IU/mg protein. The approach utilized by Bradford [40] was used to calculate the protein content of each treatment's culture supernatants. Non-enzymatic controls were performed using boiled enzymes and were subtracted from the enzymatic values. The details of substrates and standards used are given in Table 1. The amount of enzyme required to create one micromole of reducing sugar per minute per milliliter of culture supernatants was the definition of unit activity.

Table 1. Details of the selected enzymes used in this study.

Enzymes	Substrates	Supplier and Product Number	Standards (1 mg/mL Stock)	OD (in nm)
Cellulase	Cellulose (0.5%)	Sigma-Aldrich—435236	Glucose	530
β-1,3 glucanase	Laminarin (3.2 mg/mL distilled water)	Sigma-Aldrich—L9634	Glucose	530
β-1,4 glucanase	Carboxy methyl cellulose (1%)	Himedia—GRM329	Glucose	575
Chitinase	Colloidal chitin (0.5%)	Himedia—GRM1356	NAG	420
Protease	Casein (1%)	Sigma-Aldrich—C9801	Tyrosine	280

2.6.1. Cellulase Assay

A total of 1 mL of 0.5% cellulose was suspended in 50 mM (0.05 M) citrate phosphate buffer (pH 4.8), and 1 mL of culture filtrates from various *T. asperellum* isolates were used separately in the assay mixture taken in 15 mL test tubes. At 50 °C, the reaction mixture was incubated for 30 min. A 3 mL solution of 1% DNS (Dinitrosalicylate) reagent was added to stop the reaction. It was then heated for 10 min at 100 °C to produce the reddish-brown coloration. In place of culture filtrate, distilled water was used to create the blanks. The amount of reducing sugar released was calculated using the standard glucose curve, and the absorbance was measured at 530 nm [41].

2.6.2. β-1,3 Glucanase Assay

The assay mixture contains 0.5 mL laminarin (3.2 mg/mL distilled water), 1 mL of 0.05 M citrate buffer, and 0.5 mL of culture filtrates of various *T. asperellum* isolates separately. At 40 °C, the reaction mixture was incubated for 60 min. The reaction was stopped by mixing 2 mL of DNS and incubating in a boiling water bath for 15 min. The absorbance was measured at 530 nm and calculated using the standard glucose curve [42].

2.6.3. β-1,4 Glucanase Assay

A mixture of 1.0 mL of culture filtrate, 2.0 mL of 0.05 M citrate buffer (pH 4.8), and 1 mL of 1.0% carboxy methyl cellulose was combined and incubated at 55 °C for 30 min in a water bath with periodic shaking. By boiling the mixture and adding 4.0 mL of the dinitro salicylic acid reagent to terminate the reaction, the absorbance was measured at 575 nm, and the enzyme activity was calculated using a standard glucose curve [43].

2.6.4. Chitinase Assay

Preparation of Colloidal Chitin

The colloidal chitin was made from crude chitin from shrimps using the process described by Vessey and Pegg [44]:

A total of 30 g of crude chitin (Himedia-GRM1356) were ground and washed in 500 mL of distilled water. After decanting the water, the residue was combined with 505 mL of acidified ethanol- ether combination (ethanol: diethyl ether: Hcl = 250:250:5 mL). The mixture was centrifuged for 15 min at 5 °C at 12,000 rpm, and the residue was recovered. At 0 °C, 10 mL acetone was added, followed by concentrated HCl until the residues were completely dissolved. The mixture was centrifuged for 15 min at 5 °C at 12,000 rpm, and the supernatant was collected. 1.5 L distilled water and ice was added to the supernatant, and chitin was allowed to precipitate for 2 h at 0 °C. Chitin was washed three times with 1 L of distilled water, and each time it was centrifuged. The last residue was collected, and this colloidal form of chitin was stored in an airtight container at 0 °C for subsequent use.

Enzyme Assay

A total of 1.0 mL of Mcllvaine's buffer (pH 4.0), 0.5 mL of culture filtrate (an enzyme source), and 0.5 mL of colloidal chitin suspension were added to the reaction mixture. This mixture was properly mixed before being incubated at 37 °C for 20 min in a water bath with occasional shaking. By boiling the mixture for three minutes in a water bath, the process was stopped. 3.0 mL of potassium ferric cyanide reagent was added and warmed in a boiling water bath for 15 min. The absorbance of the reaction mixture at 420 nm was used to calculate the amount of N-acetyl glucosamine (NAG) released using the standard NAG curve [45].

2.6.5. Protease Assay

The substrate (1% casein in 50 mM phosphate buffer, pH 7.0) was denatured in a water bath at 100 °C for 15 min before cooling to room temperature. Later a reaction mixture consisting of 1 mL of the substrate and 1 mL of enzyme solution was incubated for 20 min at 37 °C. To that reaction mixture, 3 mL of tri-chloro acetic acid (TCA) was added to settle down. The tubes were left at 4 °C for an hour to allow the precipitation of undigested protein. At 280 nm, the absorbance of liberated tyrosine in the filtrate was measured.

2.7. *Secondary Metabolites Profiling*

2.7.1. Extraction and Separation of Antifungal Metabolites

Seven *T. asperellum* isolates (A8, A10, A11, A12, A15, A17, and A20) were chosen according to their efficiency for the secondary metabolites profiling. They were then cultured in 100 mL of potato dextrose broth (PDB). Following inoculation, the medium was incubated for 15 days at 28 ± 2 °C while being continuously shaken at 160 rpm in an incubator shaker. The mycelial biomass was then separated by passing it through a Whatman no. 1 filter, and the filtrates were then successively extracted three times using a separating funnel with an equal volume of ethyl acetate. The extracted fractions were evaporated to produce a concentrate of ethyl acetate using a rotary evaporator (IKA® RV 10; Staufen, Germany). The remaining polar filtrates were passed through anhydrous sodium sulfate (10 g) to remove the excess water and stored for further analysis. The ethyl acetate concentrate was diluted to prepare a 5 µg/mL solution in gas chromatography-mass spectrometry (GCMS) grade ethyl acetate and subjected to analysis for volatile organic compounds.

2.7.2. GC-MS Analysis

Volatile organic compounds produced by each of the seven tested isolates were analyzed on 6850C Agilent GC-MS (Agilent Technologies®, Santa Clara, CA, USA). Following their separation by an Agilent HP-5MS column (30 m, 0.25 mm, film thickness 0.25 µm), the components was identified by the mass spectrometer based on their retention index and molecular weights. Helium gas (>99.99% purity) was utilized as the carrier gas at a flow rate of 1 mL/min and a pressure of 10 psi. Each sample (1 µL) was injected using a built-in auto-injector with a 20:1 split ratio into the gas chromatograph (GC). A GC-MS temperature program was developed that started at 40 °C, increased by 3 °C each minute to 130 °C, and then held that temperature for 2 min. In addition, the temperature increased by 5 °C every minute until it reached 210 °C and stayed there for two minutes. The temperature was then increased by 10 °C/min to 350 °C. The total run time for each sample was 64 min. The following settings were used for the MS acquisition parameters: a solvent delay of 2 min, an E.M. voltage of 1214 V, an ion source temperature of 200 °C, electron ionization energy of 70 eV, a transfer line temperature of 200 °C, and full scan mode (50–550 amu). Identification of the compounds from their respective spectrum was done using the NIST library, matching with their accurate mass and retention index.

2.8. Thin Layer Chromatography (TLC) Assay

The TLC assay was conducted to identify the number of spots present in the ethyl acetate extract, which could give preliminary knowledge about the major components present in the isolates. For primary component partition, thin layer chromatography was done on Merck TLC Silica gel60 F_{254} plates (Product number: 1.05554.0007). Seven metabolite extracts of *T. asperellum* isolates were spotted using capillary tubes at a height of one inch above the plate's bottom. Two different solvent systems, viz., Hexane: Ethyl Acetate (9:1) and Hexane: Ethyl Acetate (1:1) was used for the analysis to determine the best mobile phase to separate the components. Various spots were detected in the TLC plate, and the number of spots was identified using visualizing agents like iodine and UV light at 350 nm [46].

2.9. Data Analysis

The statistical program WASP 1.0 (Web Agri Stat Package) (https://ccari.icar.gov.in/waspnew.html) (accessed on 15 June 2023) was used to analyze the data using ANOVA to determine whether there were any differences in parameter values. Three replications of each treatment were used in the experiments. At a 5% level of significance, changes between treatments were assessed using Duncan's multiple-range test. The SRplot (Science and Research Online Plot) (https://www.bioinformatics.com.cn/en (accessed on 15 June 2023)) was used to create the heat maps by statistically analyzing the GC-MS data. The NTSYSpc-2.20e version was used to create the dendrogram by analyzing the TLC banding patterns from the metabolite profiles of seven *T. asperellum* isolates.

3. Results

3.1. Morphological Identification of *T. asperellum* Isolates

On the PDA medium, by incubating at 28 ± 2 °C, all the isolates grew faster, and the mycelial colonies were initially white three days after incubation (DAI) and turned greenish after sporulation on 7DAI (Table 2, Figure S1). The microscopic observation resulted in branched, mostly paired conidiophores, ampuliform phialides, and slightly ovoidal to globose light green conidia (Figure 1).

Table 2. Collection, identification, and NCBI GenBank accession numbers of the twenty *T. asperellum* isolates.

Isolates Code	Collection Source and Location	GPS Location		ITS [a]	β-Tubulin [a]	Colony Characteristics
		Latitude	Longitude			
A1	Soil, IARI	28°38′ N	77°10′ E	OR133614	OR193716	Abundant mycelium with dark green spores
A2	Soil, IARI	28°38′ N	77°10′ E	OQ892293	OR193717	Abundant mycelium with yellowish-green spores
A3	Rhizosphere, Guntur	16°18′ N	80°27′ E	OR133699	OR193718	Less aerial mycelium with dark green centered spores

Table 2. *Cont.*

Isolates Code	Collection Source and Location	GPS Location		ITS [a]	β-Tubulin [a]	Colony Characteristics
		Latitude	Longitude			
A4	Groundnut field, TN	11°39′ N	78°12′ E	OR133720	OR193707	Abundant mycelium with yellowish-green spores
A5	Vegetable field, TN	11°3′ N	77°17′ E	OR133722	OR193719	Dense dark green spores with less aerial mycelium form rings
A6	Soil, IIHR, KA	13°7′ N	72°29′ E	OR133723	OR193720	Dense dark green spores with cottony mycelium
A7	Soil, Akola	20°42′ N	76°59′ E	OR133724	OR193721	Cottony mycelium with light green spores
A8	Soil, Tirupathi	13°37′ N	79°25′ E	OR133725	OR193708	Dark green spores forming rings with cottony mycelium
A9	Soil, Navsari	20°57′ N	72°55′ E	OR133728	OR193709	Dense light to dark greenish spores with aerial mycelium
A10	Soil, Navsari	20°57′ N	72°55′ E	OR137590	OR193710	Fast-growing, dense dark green spores with aerial mycelium
A11	Soil, Jaipur	26°55′ N	75°49′ E	OR133981	OR193711	Fast-growing, abundant cottony mycelium forms spores at later
A12	Veg. field, IARI	28°38′ N	77°10′ E	OR134019	OR193712	Dark green-centered spores with aerial mycelium
A13	Chilli field, IARI	28°38′ N	77°10′ E	OR134022	OR193722	Aerial mycelium with abundant yellowish to light green spores
A14	Soil, Lucknow	26°50′ N	80°55′ E	OR134094	OR193723	Abundant aerial mycelium with dense dark green spores
A15	Soil, Varanasi	25°19′ N	82°58′ E	OR134235	OR193713	Less mycelium, slow growing with very less spores
A16	Soil, Barracpore	22°45′ N	88°22′ E	OR134238	OR193714	Dense light green spores with aerial mycelium
A17	Soil, IIHR	13°7′ N	72°29′ E	OR134254	OR193724	Fast-growing, dense aerial mycelium forms spores at later
A18	Soil, Navsari	20°57′ N	72°55′ E	OR134258	OR193725	Dense light green spores abundant throughout the plate
A19	Soil, Navsari	20°57′ N	72°55′ E	OR134337	OR193715	Dense yellowish to light green spores with less mycelium
A20	Soil, NBPGR	28°38′ N	77°10′ E	OR134338	OR193726	Dense dark green spores with less aerial mycelium

[a] GenBank accession number.

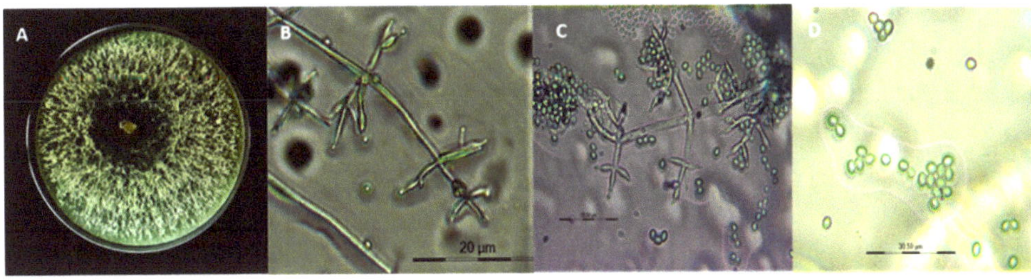

Figure 1. Morphological characteristics of *T. asperellum*. (**A**) Colony growth on potato dextrose agar. (**B**) Microscopic images showing conidiophores structures and branching pattern (magnification at 40×). (**C**) Phialides arrangement with conidia (magnification at 40×). (**D**) Conidia (magnification at 100×).

3.2. Molecular Identification and Phylogenetic Analysis

The amplified products of ITS and β-tubulin gene primers were confirmed using agarose gel electrophoresis (Figures S2 and S3) and then sequenced using the Sanger technique by outsourcing (Anuvanshiki (OPC) Pvt. Ltd.). All sequences were submitted to GenBank, and accession numbers obtained are listed in Table 2. All isolate's taxonomic identities were established by NCBI blast, and phylogenetic analysis of combined partial ITS and β-tubulin sequences was performed using the maximum likelihood method (Figure 2). The phylogenetic tree represents that the isolates A10, A8, A9, A13, A20, A7, A3, A14, A19, A1, A16, and A6 forms a clade. The isolates A5, A11, A15, A18, A4, A17, A2, and A12 form a separate clade, which shows the genetic diversity among the isolates between the above two clades. The outcome revealed that all the 20 isolates share the highest homology with *T. asperellum* Type strain CBS 433.97 followed by *T. pubescens*, *T. harzianum*, *T. hamatum*, *T. virens*, *T. longibrachiatum* with *Chetomium globosum* CBS 160.62 used as an outgroup.

3.3. In Vitro Antagonistic Assay

Under the dual culture assay, the mycelial growth of *A. rolfsii* was significantly inhibited by all the *T. asperellum* isolates, with the percent inhibition ranging from 20.87% to 94.66% ($p < 0.05$). Among them, the A10 isolate showed maximum inhibition of 94.66%, followed by A11 and A17, which exhibited 82.64% and 81.19% inhibition, respectively, 7 days post-inoculation. The volatile compounds produced by the *T. asperellum* isolates showed 11.55% to 70.95% mycelial inhibition. Among them, the A10 isolate showed maximum inhibition of 70.95%, followed by A17 and A11, which showed 68.27% and 63.75% inhibition, respectively (Table 3, Figures 3 and 4). The isolates A12 and A15 showed the least inhibition in both assays, which shows that they have low antagonistic potential, and the remaining isolates showed moderate mycelial growth inhibition. The comparison of the treatment means was evaluated using Duncan's multiple-range test.

Table 3. Effect of *T. asperellum* isolates on radial growth inhibition of *A. rolfsii* on PDA at 28 ± 2 °C at 7 days after inoculation (DAI).

Isolates	Percent Inhibition of *A. rolfsii* Growth									
	Dual Culture Assay					Volatile Assay				
	R1	R2	R3	Average	SD	R1	R2	R3	Average	SD
A1	64.58	64.22	64.15	64.32 ^{f}	0.23	40.61	41.50	41.85	41.32 ^{k}	0.64
A2	64.62	64.50	63.66	64.26 ^{f}	0.52	45.20	44.00	44.50	44.57 ^{j}	0.60
A3	66.69	66.20	65.90	66.26 ^{e}	0.40	48.15	46.20	46.50	46.95 ^{i}	1.05
A4	68.62	67.50	68.22	68.11 ^{d}	0.57	50.16	48.26	48.80	49.07 ^{h}	0.98
A5	68.89	67.82	66.95	67.89 ^{d}	0.97	50.88	49.60	49.22	49.90 ^{h}	0.87

Table 3. *Cont.*

Isolates	Percent Inhibition of *A. rolfsii* Growth									
	Dual Culture Assay					Volatile Assay				
	R1	R2	R3	Average	SD	R1	R2	R3	Average	SD
A6	50.61	51.20	52.33	51.38 ^{h}	0.87	28.06	26.30	26.75	27.04 ^{m}	0.91
A7	67.10	67.85	68.45	67.80 ^{d}	0.68	51.66	50.20	49.24	50.37 ^{gh}	1.22
A8	69.23	68.50	68.00	68.58 ^{d}	0.62	53.24	51.76	51.90	52.30 ^{f}	0.82
A9	62.31	60.50	61.45	61.42 ^{g}	0.91	46.32	46.90	46.15	46.46 ^{i}	0.39
A10	95.23	94.60	94.15	94.66 ^{a}	0.54	72.10	70.15	70.60	70.95 ^{a}	1.02
A11	82.64	80.69	84.59	82.64 ^{b}	1.95	65.72	63.16	62.38	63.75 ^{c}	1.75
A12	30.58	31.22	31.00	30.93 ^{i}	0.33	22.64	22.55	21.80	22.33 ^{n}	0.46
A13	70.54	71.66	71.50	71.23 ^{c}	0.61	60.50	61.25	61.80	61.18 ^{d}	0.65
A14	70.12	71.50	70.88	70.83 ^{c}	0.69	60.12	59.00	59.25	59.46 ^{e}	0.59
A15	22.82	18.92	20.87	20.87 ^{j}	1.95	11.20	12.30	11.15	11.55 ^{op}	0.65
A16	68.23	67.77	68.00	68.00 ^{d}	0.23	59.88	58.20	56.30	58.13 ^{e}	1.79
A17	80.46	82.10	81.00	81.19 ^{b}	0.84	69.21	67.24	68.35	68.27 ^{b}	0.99
A18	65.91	65.25	64.20	65.12 ^{ef}	0.86	52.65	51.60	51.25	51.83 ^{fg}	0.73
A19	68.70	68.45	68.00	68.38 ^{d}	0.35	50.12	49.45	50.00	49.86 ^{h}	0.36
A20	52.35	51.00	53.70	52.35 ^{h}	1.35	31.22	29.45	28.60	29.76 ^{l}	1.34
CD @ 5%				1.497					1.603	
SEm ±				0.823					0.943	
CV (%)				1.411					2.034	

SD—Standard Deviation, CD—Critical Difference, SEm—Standard Error of mean, CV—Coefficient of variation, Different letters after values are significantly different at $p \leq 0.05$.

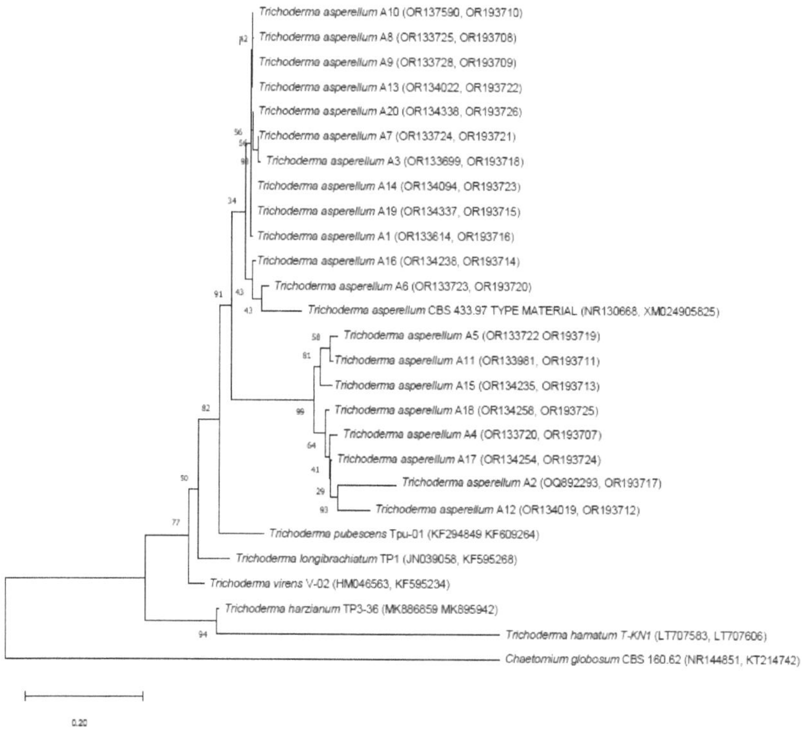

Figure 2. Maximum likelihood phylogenetic tree based on concatenated Internal Transcribed Spacer (ITS) and β-tubulin sequences of the 20 *T. asperellum* test isolates and its related species and reference

T. asperellum Type strain CBS 433.97 retrieved from the NCBI database. The tree was rooted using *Chaetomium globosum* CBS 160.62 as an outgroup.

Figure 3. Effect of *T. asperellum* isolates on *A. rolfsii* growth through the bio efficacy tests. (**A**) Dual culture assay. (**B**) Volatile assay.

3.4. In Planta Bio-Control Assay

In the greenhouse pot assay, twenty *T. asperellum* isolates were evaluated against collar rot in tomatoes using a completely randomized design, and the disease severity index was recorded. The disease was significantly inhibited by all the *T. asperellum* isolates ranging from a maximum of 86.17% and 80.60 % disease inhibition from the A10 isolate to the minimum disease inhibition of 11.11% and 8.33% from the isolate A15 during pre- and post-inoculation treatment. The isolates A10 (13.83%), A11 (22.20%), and A14 (27.67%) treated plants on the *A. rolfsii* infected soil showed lower disease incidence as compared to pathogen alone treated when applied as a pre-inoculation approach against *A. rolfsii*. Pathogen treatment was maintained by inoculating only with *A. rolfsii*, which exhibited complete wilting and plant death, to compare the effect. The efficiency of *T. asperellum* isolates was also assessed by the post-inoculation approach of *T. asperellum* against *A. rolfsii*. The lowest percent of disease incidence was observed from the evaluation of A10 (19.40%), followed by A11 (27.78%) and A14 (33.33%), respectively, in post-inoculation conditions. Notably, the A15 isolate-treated plants on the infected soil recorded a maximum disease incidence of 88.89% in pre-inoculation and 91.67% in post-inoculation. The standard deviation was recorded from each measurement to analyze the variance (Table 4, Figures 5 and 6).

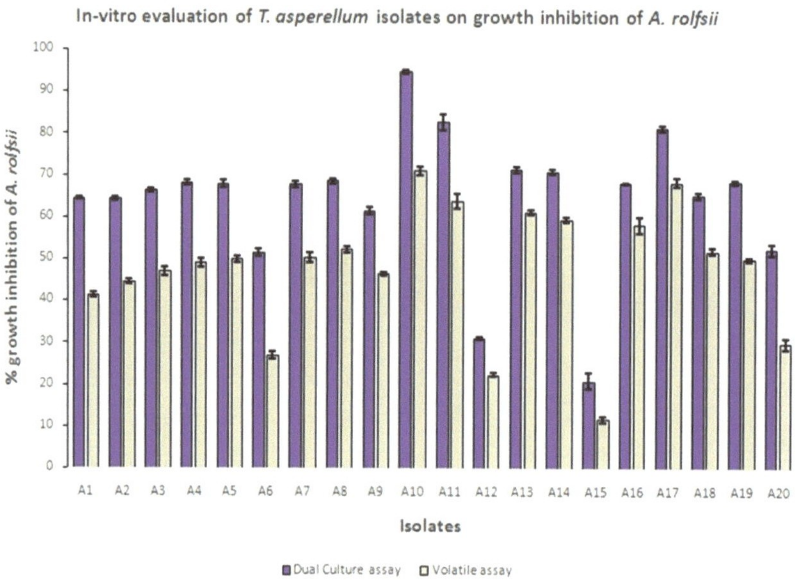

Figure 4. Graphical representation of the effect of *T. asperellum* isolates on the growth inhibition of *A. rolfsii*.

Table 4. Effect of *T. asperellum* isolates against collar rot disease in tomato plants var. pusa ruby in pot condition.

Isolates	% Disease Severity Index (DSI)									
	Pre-Inoculation Treatment					Post-Inoculation Treatment				
	R1	R2	R3	Average	SD	R1	R2	R3	Average	SD
A1	50	41.67	41.67	44.45	4.81	50	50	58.33	52.78	4.81
A2	50	50	41.67	47.22	4.81	50	58.33	58.33	55.55	4.81
A3	41.67	41.67	33.33	38.89	4.82	58.33	41.67	41.67	47.22	9.62
A4	50	50	50	50.00	0.00	58.33	58.33	41.67	52.78	9.62
A5	50	25	33.3	36.11	12.73	58.33	50	33.3	47.22	12.75
A6	66.67	66.67	58.33	63.89	4.82	66.67	66.67	75	69.45	4.81
A7	41.67	41.67	41.67	41.67	0.00	50	50	50	50.00	0.00
A8	50	25	33.3	36.11	12.73	58.33	50	33.37	47.22	12.71
A9	50	41.67	41.67	44.48	4.81	50	58.33	50	52.78	4.81
A10	16.6	8.3	16.6	13.83	4.79	16.6	16.6	25	19.40	4.85
A11	25	25	16.6	22.20	4.85	25	25	33.33	27.78	4.81
A12	66.67	83.33	83.33	77.77	9.62	66.67	91.67	91.67	83.33	14.43
A13	50	41.67	25	38.89	12.73	33.33	50	50	44.44	9.62
A14	25	25	33	27.67	4.62	33.33	33.33	33.33	33.33	0.00
A15	91.67	91.67	83.33	88.89	4.82	91.67	91.67	91.67	91.67	0.00
A16	41.67	41.67	33	38.78	5.01	41.67	41.67	50	44.45	4.81
A17	25	25	25	25.00	0.00	25	25	41.67	30.56	9.62
A18	41.67	41.67	33.33	38.89	4.82	41.67	41.67	50	44.45	4.81
A19	41.67	41.67	41.67	41.67	0.00	41.67	41.67	58.33	47.22	9.62
A20	66.67	66.67	58.33	63.89	4.82	66.67	75	58.33	66.67	8.34

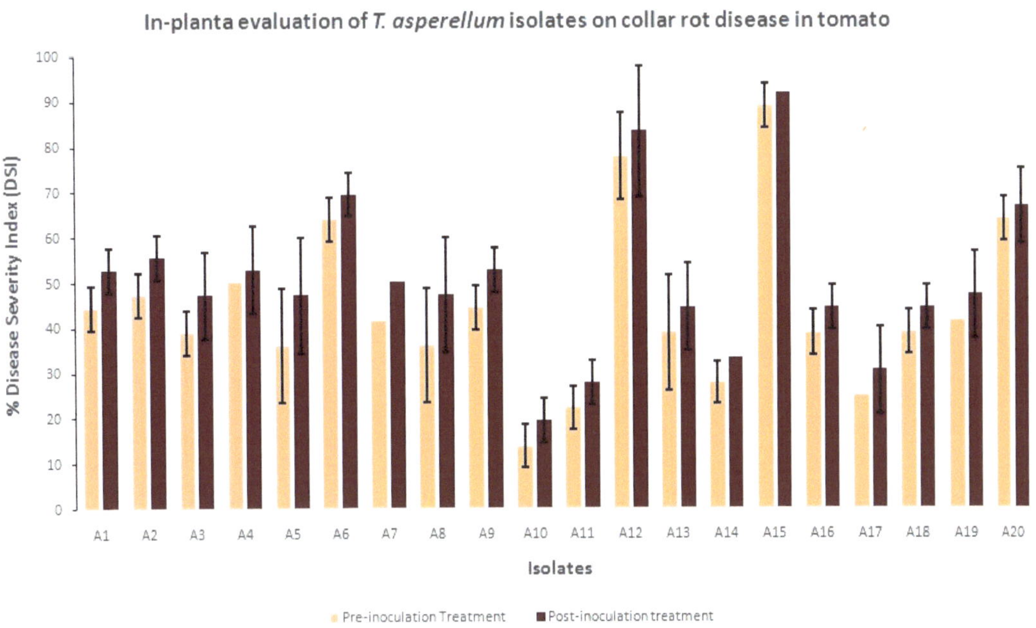

Figure 5. Graphical representation of the effect of *T. asperellum* isolates against collar rot disease in tomato.

Figure 6. Effect of the *T. asperellum* isolate A10 on percent disease inhibition of *A. rolfsii* through the in planta evaluation.

3.5. Enzyme Assay

The results revealed that all the tested isolates of *T. asperellum* significantly produced different hydrolytic enzymes during their interaction with the test pathogen. The isolate A10 (174.68 IU/mg) produced the highest amount of β-1,3 glucanase, followed by A11 (149.08 IU/mg), A14 (134.68 IU/mg), and the least in A15 (24.79 IU/mg). The isolate A10 (183.48 IU/mg) showed the highest production of β-1, 4 glucanase followed by A11 (159.60 IU/mg) and A14 (147.25 IU/mg), and the least production was observed in A15

(51.28 IU/mg). The isolate A17 (80.92 IU/mg) produced the highest production of cellulase, followed by A14 (77.94 IU/mg), A16 (77.82 IU/mg), and the least production in A15 (20.84). In protease production, isolate A10 (106.06 IU/mg) showed the highest production, followed by A17 (103.08 IU/mg), and the least production was recorded in isolate A15 (15.60 IU/mg). In chitinase production, A17 (53.09 IU/mg) recorded the maximum yield, followed by A16 (52.17), and the least was observed in A15 (13.07 IU/mg) (Table 5, Figure 7).

Table 5. Specific activity of hydrolytic enzymes produced by *T. asperellum* isolates during the interaction with *A. rolfsii*.

Isolates	Specific Activity in IU/mg									
	β-1,3 Glucanase		β-1,4 Glucanase		Cellulase		Protease		Chitinase	
	Avg	SD	Avg	SD	Avg	SD	Avg	SD	Avg	SD
A1	78.25 i	2.00	79.60 k	2.30	73.00 cde	2.30	71.53 f	2.80	45.07 bcd	1.28
A2	90.25 h	2.00	79.93 k	2.40	70.94 ef	2.70	74.53 ef	1.60	37.33 fg	2.32
A3	111.00 ef	0.25	123.13 g	1.80	75.44 bcd	2.50	55.81 h	0.80	7.02 j	0.77
A4	112.75 e	0.25	109.56 h	2.90	68.09 fg	1.80	77.83 d	2.40	45.85 bc	2.96
A5	125.50 d	2.00	124.18 fg	1.70	62.47 h	3.00	65.33 g	2.40	45.17 bcd	2.60
A6	59.50 j	2.00	52.10 m	2.20	52.04 i	1.90	44.85 i	0.80	39.88 ef	2.13
A7	111.58 ef	1.50	124.28 fg	2.70	61.92 h	2.30	63.38 g	1.70	36.07 g	1.62
A8	122.92 d	1.80	127.68 ef	2.50	76.04 bc	2.10	66.08 g	0.90	34.85 gh	1.24
A9	62.40 j	1.70	88.25 j	2.10	70.87 efg	2.40	74.53 ef	2.50	44.37 bcd	2.52
A10	174.68 a	2.30	183.48 a	2.40	77.09 b	1.80	106.06 a	1.90	41.87 de	2.54
A11	149.08 b	2.40	159.60 b	1.30	67.26 g	0.70	82.33 bc	2.40	46.82 b	2.39
A12	37.93 l	2.80	30.73 n	0.90	34.99 j	1.60	53.53 h	0.90	37.65 fg	1.76
A13	108.96 fg	1.80	129.58 e	1.40	76.75 b	2.90	85.13 b	2.40	42.40 de	1.77
A14	134.68 c	2.10	147.25 c	2.60	77.94 ab	1.50	42.03 i	1.20	50.81 a	2.32
A15	24.79 m	2.90	51.28 m	1.90	20.84 k	1.90	15.53 j	1.30	13.07 i	1.52
A16	107.64 g	2.50	112.95 h	1.80	77.82 ab	2.60	81.35 c	2.20	52.17 a	1.95
A17	133.18 c	2.40	136.85 d	1.60	80.92 a	2.90	103.08 a	1.57	53.09 a	1.85
A18	60.91 j	1.60	133.43 d	2.40	72.04 de	2.50	75.43 de	2.41	42.12 de	1.88
A19	112.36 e	1.90	102.83 i	1.80	61.75 h	1.80	82.05 c	1.23	31.77 h	1.98
A20	43.10 k	1.80	58.38 l	2.90	72.02 de	1.60	77.15 de	1.26	43.22 cd	1.33
CD @ 5%	3.318		3.543		3.651		3.052		3.315	
SEm ±	4.044		4.609		4.896		3.421		4.035	
CV (%)	2.050		1.992		3.327		2.647		5.082	

Avg—Average value of three replications, SD—Standard Deviation, CD—Critical difference, SEm—Standard error of the mean, CV—Coefficient of variation, Different letters after values are significantly different at $p \leq 0.05$.

3.6. Comparative Analysis of Volatile Organic Compounds of T. asperellum Isolates

In the present investigation, it was revealed that the culture filtrates of the seven isolates of *T. asperellum* (A8, A10, A11, A12, A15, A17 and A20) showed the presence of many secondary metabolites at different retention times (Rt). The anti-microbial compounds were found highest from the isolate A10, 2H-pyran-2-one (17.39%) was found to be most abundant in it, followed by dienolactone (8.43%), α-pyrone (2.19%), and harziandione (0.24%) with the respective Rt of 33.48, 33.85, 33.39, and 64.23 min. 2H-Pyran-2-one (9.87%) was observed to be the most abundant in isolate A11, followed by 2,3-butanediol (7.24%) and harziandione (6.70%) with Rt of 18.65, 42.67 and 63.77 min, respectively. The isolate A17 showed many metabolites, like hexadecanoic acid (9.42%), octadecanoic acid (3.87%), hexadecane (3.83%), docosene (4.07%), and (E)-6-pen-1-enylpyran-2-one (2.18%) with the Rt of 13.67, 23.90, 19.55, 32.55 and 15.35 min respectively. 2,3-butanediol (23.41%), (E)-6-Pent-1-enyl pyran-2-one (5.88%), and 2H-pyran-2-one (2.49%) with the corresponding Rt of 64.5, 13.8 and 15.7 min respectively, were recorded to be the major compounds, followed by long-chain hydrocarbons from the isolate A8. Few anti-microbial compounds, like dimethyl disulfoxide were also observed from the isolate A12 (0.79%) and A20 (0.30%)

with the respective Rt of 33.70 and 24.45 min. No major anti-microbial compounds were detected from isolate A15, suggesting that this isolate does not produce many bioactive volatile metabolites.

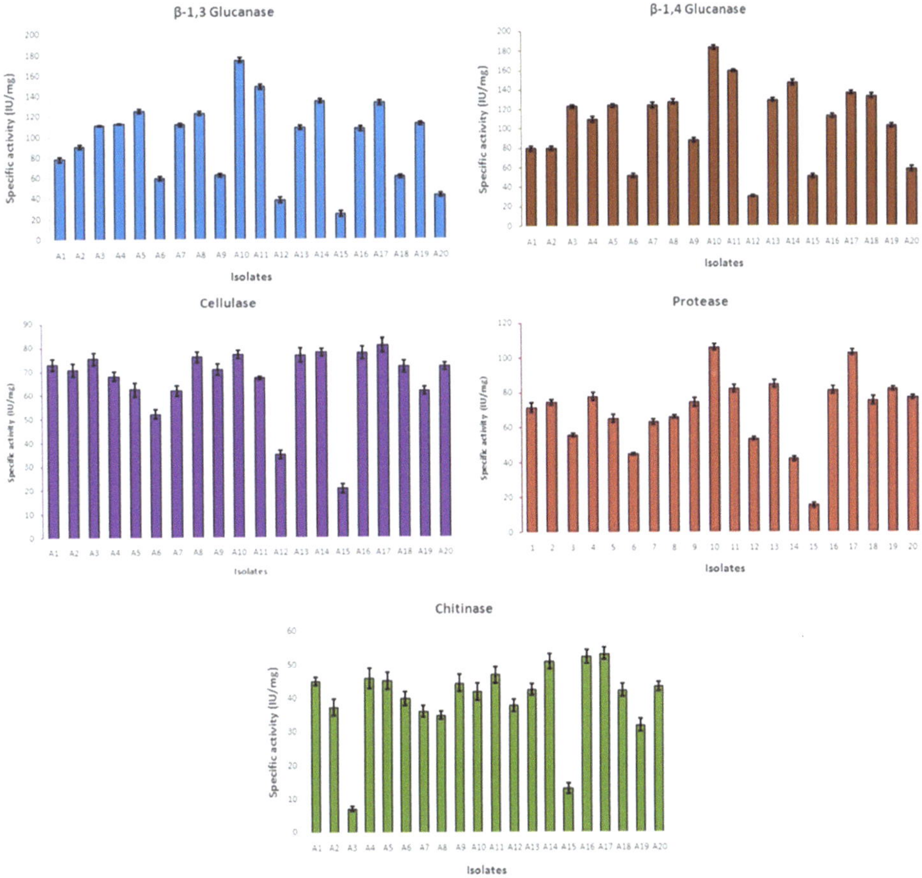

Figure 7. Graphical representation of the specific activity of hydrolytic enzymes produced by *T. asperellum* isolates during the interaction with *A. rolfsii*.

From the ethyl acetate fractions of *T. asperellum* isolates, the predominant compounds were identified and grouped into various functional groups with varying content ranges among isolates, including alcohols (23.68%) from the isolate A8, ketones (20.78%) from isolate A20, and acids (18.23%) from the isolate A17. Among the alcohols, 2, 3-butanediol (23.41%) was observed to be most abundant from the isolate A8. Among the Ketones having anti-microbial compounds, 2H-pyran-2-one (17.39%) was observed as the most abundant from the isolate A10, (E)-6-pent-1-enylpyran-2-one (5.88%) from the isolate A8, and α-pyrone (5.14) from the isolate A20. Among the acids, the highest content of hexadecanoic acid (9.42%) was found in the isolate A17, followed by acetic acid (3.99%) in A10, octadecanoic acid (3.87%) in A17, and 9, 12-octadecadienoic acid (2.83%) in A10. Among the lactones, δ-2,4-dienolactone (8.43%) was recorded at the maximum from isolate A10.

We observed more hydrocarbons viz., dodecane, 1-tetradecene, tetradecane, 1-hexadecene, hexadecane, cyclohexadecane, tetracosane, hexacosane, heptadecane, 1-octadecene, octadecane, docosene, tetracosahexaene from various isolates. Among them, hexadecane (3.83%) and docosene (4.07%) were observed at the highest content in the isolate A17, whereas diterpenes, namely, harziandione (6.70%), were abundant in the isolate A11. The list of identified compounds in the ethyl acetate fraction from various *T. asperellum* isolates is listed in Table 6.

The total ion chromatograms for all the isolates are shown in Figures S4–S10. Furthermore, a heat map was generated to compare the visualization and interpretation of changes in the volatile organic compounds (VOCs) profile among the isolates (Figure 8). The area showing the red color in the heat map indicates the presence of that particular volatile organic compound having the maximum area-wise among the isolates. The most important anti-microbial compounds like 2H-pyran-2-one are abundant among the isolates A11, A12, A10, and A20. The 2,3-butanediol presence is more in the isolates A11, A12, and A8. The isolate A17 has high hexadecanoic acid. The presence of harziandione is more in the isolates A11 and A20. The isolates A20 and A10 contain high α-pyrone and 2,4 dienolactone, respectively.

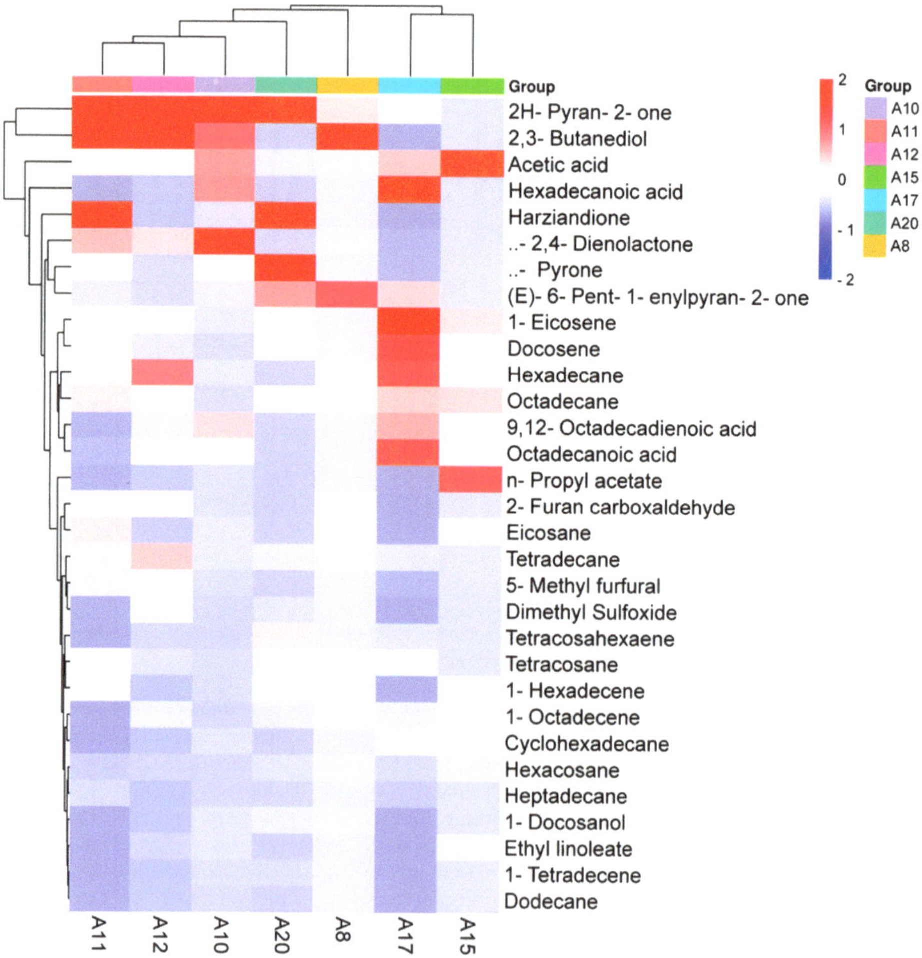

Figure 8. Heat map clustering of Volatile Organic Compounds (VOCs) profiles from *T. asperellum* isolates.

Table 6. Characteristic features of secondary metabolites isolated from *T. asperellum* isolates by GC-MS.

Compounds	Functional Group	Molecular Formula	Isolates													
			A10		A17		A12		A15		A8		A11		A20	
			* RA %	RT **	* RA %	RT **	* RA %	RT **	*RA %	RT **	* RA %	RT **	* RA %	RT **	* RA %	RT **
Acetic acid	Acid	CH_3COOH	3.99 ± 1.12	5.68	2.35 ± 0.42	4.83	0.53 ± 0.03	6.88	1.47 ± 0.39	12.53	0.15 ± 0.02	9.43	1.28 ± 0.62	16.52	0.40 ± 0.02	8.24
n-Propyl acetate	Ester	$C_5H_{10}O_2$	0.21 ± 0.06	5.82	-	-	-	-	0.40 ± 0.02	13.05	-	-	-	-	-	-
2,3-Butanediol	Alcohol	$C_4H_{10}O_2$	4.99 ± 1.07	8.27	-	-	6.57 ± 1.32		-	-	23.41 ± 4.56	64.5	7.24 ± 1.64	42.67	-	-
2-Furan carboxaldehyde	Aldehyde	$C_5H_4O_2$	0.12 ± 0.03	10.55	-	-	0.97 ± 0.05		-	-	0.90 ± 0.02	14.2	1.88 ± 0.51	24.52	-	-
5-Methyl furfural	Aldehyde	$C_6H_6O_2$	0.12 ± 0.01	10.55	0.07 ± 0.02	12.65	0.97 ± 0.06		-	-	-	-	0.94 ± 0.11	18.97	-	-
Dodecane	Hydrocarbon	$C_{12}H_{26}$	0.04 ± 0.01	18.92	-	-	-	-	-	-	-	-	-	-	-	-
1-Tetradecene	Hydrocarbon	$C_{14}H_{28}$	0.12 ± 0.04	27.33	-	-	-	-	-	-	-	-	-	-	0.18 ± 0.01	10.32
Tetradecane	Hydrocarbon	$C_{14}H_{30}$	0.27 ± 0.13	27.65	0.68 ± 0.14	26.45	1.41 ± 0.42	12.35	-	-	0.31 ± 0.01	18.9	0.89 ± 0.07	10.58	0.44 ± 0.02	11.47
α- Pyrone	Ketone	$C_5H_4O_2$	2.19 ± 0.82	33.39	0.11 ± 0.03	12.88	0.12 ± 0.01	18.43	-	-	0.04 ± 0.01	12.7	1.43 ± 0.22	20.34	5.14 ± 1.94	22.25
δ-2,4-Dienolactone	Lactone	C_7H_{12}	8.43 ± 1.53	33.85	0.06 ± 0.01	8.69	1.23 ± 0.47	15.30	-	-	0.21 ± 0.05	8.4	2.76 ± 0.13	12.76	0.13 ± 0.04	7.86
2H-Pyran-2-one	Ketone	$C_5H_4O_2$	17.39 ± 3.45	33.48	1.24 ± 0.32	13.80	5.23 ± 1.94	19.21	-	-	2.49 ± 0.86	15.7	9.87 ± 2.41	18.65	8.89 ± 2.32	20.34
(E)-6-Pent-1-enylpyran-2-one	Ketone	$C_{21}H_{26}O_2$	0.43 ± 0.05	34.55	2.18 ± 0.48	15.35	0.1 ± 0.02	17.33	-	-	5.88 ± 1.23	13.8	0.59 ± 0.04	18.43	2.26 ± 0.91	13.66
1-Hexadecene	Hydrocarbon	$C_{16}H_{32}$	0.12 ± 0.02	35.49	-	-	-	-	0.09 ± 0.01	15.92	0.49 ± 0.03	27.3	1.09 ± 0.32	18.60	1.11 ± 0.07	15.97
Hexadecane	Hydrocarbon	$C_{16}H_{32}$	0.33 ± 0.08	35.74	3.83 ± 0.71	19.55	2.07 ± 0.72	14.73	0.08 ± 0.01	15.95	0.99 ± 0.12	27.6	1.91 ± 0.62	19.71	-	-
Cyclohexadecane	Hydrocarbon	$C_{16}H_{32}$	0.25 ± 0.02	49.10	1.09 ± 0.32	12.75	-	-	0.11 ± 0.03	15.91	-	-	-	-	-	-
1-Eicosene	Ester	$C_{20}H_{40}$	0.25 ± 0.04	49.10	5.63 ± 1.79	5.68	0.86 ± 0.06	28.45	0.19 ± 0.04	5.43	-	-	1.03 ± 0.04	29.56	1.18 ± 0.11	24.70
1-Docosanol	Alcohol	$C_{22}H_{46}O$	0.25 ± 0.01	49.10	-	-	-	-	-	-	0.27 ± 0.07	11.3	-	-	0.46 ± 0.06	26.85
Eicosane	Hydrocarbon	$C_{20}H_{42}$	0.25 ± 0.02	49.28	-	-	-	-	0.14 ± 0.04	8.35	1.40 ± 0.09	8.25	2.10 ± 0.90	28.98	-	-
Hexadecanoic acid	Acid	$C_{16}H_{32}O_2$	4.12 ± 1.23	50.29	9.42 ± 2.96	13.67	-	-	-	-	-	-	-	-	-	-
9,12-Octa decadienoic acid	Acid	$C_{18}H_{32}O_2$	2.83 ± 0.90	55.19	2.59 ± 0.41	24.56	0.37 ± 0.08	8.43	0.04 ± 0.01	18.90	-	-	-	-	-	-
Ethyl linoleate	Ester	$C_{20}H_{36}O_2$	0.35 ± 0.12	54.41	-	-	0.14 ± 0.04	12.56	0.04 ± 0.01	13.50	-	-	-	-	-	-
Tetracosane	Hydrocarbon	$C_{24}H_{50}$	0.11 ± 0.02	60.65	1.06 ± 0.02	13.80	0.26 ± 0.03	13.45	-	-	0.33 ± 0.04	6.76	1.37 ± 0.46	13.08	0.58 ± 0.04	27.59

Table 6. *Cont.*

Compounds	Functional Group	Molecular Formula	Isolates													
			A10		A17		A12		A15		A8		A11		A20	
			* RA %	RT **	* RA %	RT **	* RA %	RT **	*RA %	RT **	* RA %	RT **	* RA %	RT **	* RA %	RT **
Harziandione	Ketone	$C_{20}H_{30}O_2$	0.24 ± 0.01	64.23	-	-	-	-	-	-	-	-	6.70 ± 1.94	63.77	4.49 ± 2.07	58.88
Hexacosane	Hydrocarbon	$C_{26}H_{54}$	0.05 ± 0.01	65.71	0.57 ± 0.12	11.65	0.14 ± 0.01	19.58	0.03 ± 0.01	2.56	0.15 ± 0.03	12.34	0.33 ± 0.04	19.42	0.25 ± 0.03	15.20
Heptadecane	Hydrocarbon	$C_{17}H_{36}$	-	-	0.32 ± 0.08	13.09	-	-	-	-	-	-	0.56 ± 0.05	18.91	0.12 ± 0.02	10.24
1-Octadecene	Hydrocarbon	$C_{18}H_{36}$	-	-	0.83 ± 0.27	12.60	0.41 ± 0.02	13.08	0.09 ± 0.02	14.29	0.44 ± 0.06	14.0	-	-	0.46 ± 0.07	8.65
Octadecane	Hydrocarbon	$C_{18}H_{36}$	-	-	2.23 ± 1.01	12.60	0.62 ± 0.03	15.48	0.20 ± 0.03	9.20	1.91 ± 0.23	13.2	2.16 ± 0.91	16.77	1.07 ± 0.38	18.55
Docosene	Hydrocarbon	$C_{22}H_{44}$	-	-	4.07 ± 1.53	32.55	0.24 ± 0.03	11.35	0.06 ± 0.01	12.11	-	-	1.86 ± 0.11	15.69	0.98 ± 0.23	13.34
Octadecanoic acid	Acid	$C_{18}H_{36}O_2$	0.95 ± 0.23	55.70	3.87 ± 1.25	23.90	0.75 ± 0.11	11.87	0.13 ± 0.02	7.88	-	-	0.07 ± 0.01	7.24	-	-
Tetracosahexaene	Hydrocarbon	$C_{24}H_{38}$	-	-	0.72 ± 0.09	24.76	0.15 ± 0.01	18.68	-	-	-	-	-	-	1.44 ± 0.71	14.21
Dimethyl Sulfoxide	Organosulfur	C_2H_6OS	-	-	-	-	0.79 ± 0.06	33.70	-	-	-	-	-	-	0.30 ± 0.06	24.45
Chemical groups			Content (%)													
Acids			11.89		18.23		1.65		1.64		0.15		1.35		0.40	
Esters			0.81		5.63		1.0		0.63		-		1.03		1.18	
Alcohols			5.24		-		6.57		-		23.68		7.24		0.46	
Aldehydes			0.24		0.07		1.94		-		0.90		2.82		-	
Hydrocarbons			1.54		15.4		5.3		0.80		6.02		12.27		6.63	
Ketones			20.25		3.53		5.45		-		8.41		18.59		20.78	
Lactones			8.43		0.06		1.23		-		0.21		2.76		0.13	
Organosulfur			-		-		0.79		-		-		-		0.30	
Total			48.40		42.92		23.93		3.07		39.37		46.06		29.88	

* RA %—Relative Area %; ** RT—Retention time (min.) of each volatile compound eluted through HP 5MS column in GC-MS.

3.7. TLC Assay

We observed that there were more bands in Hexane: Ethyl Acetate (1:1) solvent system than in the 9:1 solvent system. The A10 isolate produced nine bands in a 1:1 system, followed by the isolates A17 and A11, which produced seven bands that represent the presence of major metabolites in the ethyl acetate extract. The least number of bands observed from isolate A15 was 4 (Figure S11). Using the UPGMA (Unweighted pair group method with arithmetic mean) method in NTSYSpc-2.02e, a dendrogram was performed based on the unique banding patterns from each isolate of the *T. asperellum* metabolome. Two clades were obtained viz., A17, A10, and A11 (more promising isolates) forming one, and A8, A20, A12, and A15 (moderate to less promising isolates) forming the other, as seen in the dendrogram (Figure 9).

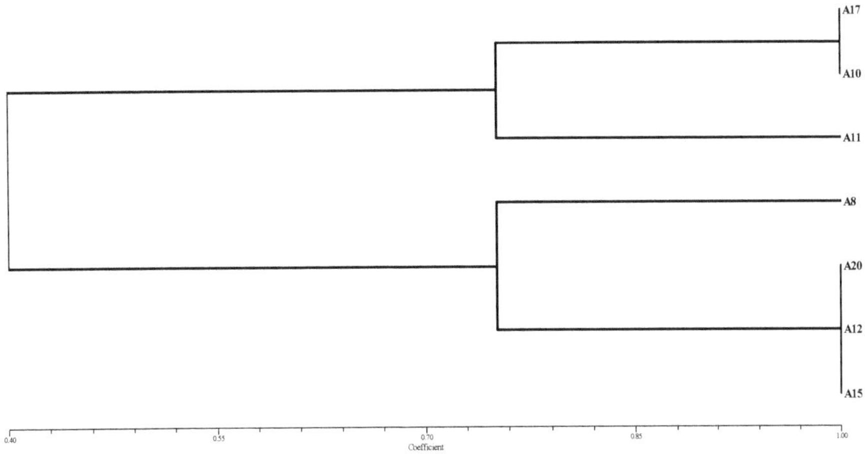

Figure 9. Dendrogram of seven *T. asperellum* isolates based on bioactive compounds using thin layer chromatography (Hexane: Ethyl Acetate (1:1) solvent system and long UV light range).

4. Discussion

In our study, 20 isolates were used as antagonists and were confirmed as *T. asperellum* based on morphological and molecular characteristics. The studied morphological characteristics, such as colony characteristics like a faster growth rate, globose conidia, mostly paired branches, and an ampulliform phialides nature, were also confirmed with the previous reports [47,48]. Conventional morphological approaches may not perform well for accurate identification. So, combining the morphological and molecular strategies is an ideal tool [49]. Using the ITS and β-tubulin markers, the precise species identification of *T. asperellum* was achieved, demonstrating the efficacy of the molecular markers in ensuring reliable species determination.

Further, these isolates were tested against the collar rot pathogen, *A. rolfsii*, indicating that *T. asperellum* is an effective biocontrol agent. In the In vitro antagonistic assay, isolate A10 recorded a maximum inhibition of *A. rolfsii* growth (94.66%) in the dual plate assay and 70.95% in the volatile assay. These results are in parallel with the previous reports, showing that most temperature-tolerant *T. asperellum* isolates showed a 50–68% reduction of hyphal growth of *A. rolfsii* under in vitro [50], and *T. asperellum* caused significant reduction in mycelial growth of 43.57%, 38.16%, and 54.87% obtained for *Pythium aphanidermatum*, *Pythium debaryanum*, and *A. rolfsii* Sr3 respectively [51]. The Tricho05 and Tricho06 isolates also recorded the maximum inhibition of *A. rolfsii* mycelial growth of 70.37% and 63.33%, respectively [52]. The In planta study carried out using tomato plant var. pusa ruby suggests that the application of *T. asperellum* to the soil prior to the development of the disease will reduce the incidence of collar rot significantly, whereas the pathogen alone treated plants showed 100% disease incidence with prominent symptoms. The isolate

A10 application on the infected soil observed lower disease incidence (13.83%) in pre-inoculation and 19.40% in post-inoculation as compared to pathogen-alone treatment. These are correlated with previous studies reports that the presence of *T. asperellum* reduced the severity of disease symptom caused by *A. rolfsii* in onion [53], and *T. asperellum* strains mixture treated cucumber plants showed the highest reduction of Fusarium root and stem rot disease incidence and severity [54].

This indicates that the biocontrol agent, *T. asperellum* inhibits the pathogen by various mechanisms, including the competing nature of space and nutrients and the production of volatile metabolites, thereby restricting the growth of the pathogen [55–57] and providing resistance against pathogens by inducing defense responses in the plants [58–61].

The production of hydrolytic enzymes to dissolve the fungal mycelial cell wall has been attributed to biocontrol activity. According to Inbar and Chet [62], the antagonistic action of *Trichoderma* species against soil-borne diseases may be related to their synthesis of these enzymes. In the present investigation, the *T. asperellum* isolates were evaluated for their potential to produce various extracellular enzymes against *A. rolfsii*. Extracellular enzyme activity was observed in all the isolates. The isolate A10 observed the maximum production of hydrolytic enzymes by the production of β-1,3 glucanase (174.68 IU/mg), β-1,4 glucanase (183.48 IU/mg), and protease (106.06 IU/mg), which correlates with the previous studies showed that *T. asperellum* isolates Ts39, Ts12, Ts42, Ts9, Ts32, Ts36 possessed the wide range of hydrolytic enzyme activities exhibiting chitinase 8.7–10.3 Pmol/s/mL and β-1,3 glucanase (1.4–1.98 nmol/s/mL) [63] and *T. asperellum* CCTCC-RW0014 reported that the production of various hydrolytic enzymes with higher enzyme activity viz., chitinase (87.5 ± 3.21%), protease (52.9 ± 1.23%), cellulase (84.8 ± 2.56%) and β-1.3 glucanase (60.5 ± 2.32%) [64] and *Trichoderma* isolates T2, T8, T9, T10, T11, T16, T17, and T20 recorded the strong production of chitinase, β-1,3 glucanase, protease, and cellulase that exhibited high inhibitory activity against *A. rolfsii* in dual culture [65]. The *T. virens* isolates V-19, and V-21 exhibited the highest enzyme activities like cellulase (30.31 IU/mg, 16.75 IU/mg), β-1,3 glucanase (19.01 IU/mg, 16.46 IU/mg), chitinase (24.21 IU/mg, 22.71 IU/mg), protease (17.13 IU/mg, 18.20 IU/mg) respectively, and *T. harzianum* isolate H10 also possessed the highest enzyme activity viz., cellulase (18.09 IU/mg), β-1,3 glucanase (16.44 IU/mg), β-1,4 glucanase (17.81 IU/mg), Chitinase (88.80 IU/mg), Protease (23.88 IU/mg) recorded during interation with *A. rolfsii* [66].

It is known that *Trichoderma* spp. produces a variety of bioactive secondary metabolites with anti-microbial properties against a variety of phytopathogens, primarily soil-borne pathogens [67]. In the present study, secondary metabolites were profiled using GC-MS analysis from seven *T. asperellum* isolates selected based on the bio-efficacy tested. 2H-Pyran-2-one, dienolactone, α-pyrone, and harziandione are the volatile secondary metabolites that have antifungal activity and are produced by *Trichoderma* spp. From our GC-MS study, we found that the maximum production of antifungal VOCs from isolate A10 is 2H-pyran-2-one, dienolactone, α-pyrone, and harziandione. Among them, 2H-Pyran-2-one is the most important compound, having antifungal activity against various plant pathogens as per the previous studies. The compound 6-pentyl-2H-pyran-2-one (6-PP), an important volatile compound recorded from the *T. asperellum* isolates Ta2, Ta12, Ta17, Ta20, and Ta45 through GC-MS analysis having the peculiar character of coconut odor with anti-microbial activity [68]. The 6-PP recorded from the *T. asperellum* possessing a powerful antifungal compound against *Magnaporthiopsis maydis* and the purified 6-PP compound (30μg/seed) used in seed coating, and the treatments reduced the symptoms (up to 20%), pathogen infection (94–98%) and improved plant biomass by 90–120% and cob weight by 60% [69]. The 6-PP has a significant inhibitory effect on turf dollar spot and recorded good control efficacy recorded both in in-vitro and in-vivo studies [70]. The 6-PP reported from *T. asperellum* PT-15 showed the antifungal effects by applying a nutrient solution containing 25 mg/L 6-PP significantly suppressed the fusarium wilt with 70.71% efficacy and a 27.23% disease index [71].

A TLC assay was conducted to identify the number of bands present in the ethyl acetate extract and to analyze the metabolites produced by the biocontrol agents under two solvent systems [72,73]. TLC results exhibited many bands from tested isolates representing major metabolites present in the ethyl acetate extract. Among the two solvent systems used for the purpose, hexane: ethyl acetate (1:1) was found to be suitable for the separation of the metabolites, which correlates with the previous report confirmed the presence of metabolites using hexane: ethyl acetate (1:1) solvent system and verified the detection of carboxylic acid as it formed as yellow spot when sprayed using bromocresol green in laccase mediated Reactive Black 5 biodegradation by the *T. atroviride* F03 [74].

This solvent system is then recommended for the isolation of the metabolites for further structural elucidation using various spectrometric techniques. However, from the TLC analysis, we observed the formation of two clusters, viz., high promising isolates as one cluster and moderate to low promising isolates as another cluster, in the dendrogram generated based on banding patterns among the isolates. We may conclude from these findings that there is a high correlation between metabolite profiles, hydrolytic enzyme production, space, nutrient competitiveness, priming of defenses in plants, and their potential as biological agents against *A. rolfsii*.

5. Conclusions

In recent years, biological control has received increasing attention as a promising alternative to chemical control of plant pathogens. In the present study, *T. asperellum* isolate A10 was found to be a promising antagonist for the management of *A. rolfsii* when tested under lab and greenhouse conditions. Our study also suggested that the application of *T. asperellum* before the establishment of the disease reduces the incidence of collar rot significantly. GC-MS analysis revealed the production of various antifungal secondary metabolites such as 2H-Pyran-2-one, dienolactone, α-pyrone, harziandione, butanediol, and dimethyl disulfoxide involved in the antagonistic mechanism. We found that isolate A10 produced the 2H-Pyran-2-one with the maximum relative area and retention time, playing a pivotal role in the antagonistic activity. We also observed that isolate A10 produced the maximum production of various hydrolytic enzymes like β-1,3 glucanase, β-1, 4 glucanase, and protease, which are involved in antagonistic effects on *A. rolfsii*. Our findings collectively imply that *T. asperellum* and its anti-microbial substances have great potential to successfully manage collar rot diseases. Since BCAs perform differently under varied climatic conditions, there is a need to evaluate their efficacy at multiple locations and across multiple crops to confirm their robustness and potential.

Supplementary Materials: The following supporting information can be downloaded at: https://www.mdpi.com/article/10.3390/horticulturae9101116/s1, Figure S1: Growth characteristics of different isolates of Trichoderma asperellum on potato dextrose agar (PDA) after 7 days of incubation at 28 ± 2 °C; Figure S2: PCR amplification patterns in ITS1-5.8S-ITS2 with ITS1/ITS4 primers from 20 Trichoderma asperellum isolates; L is 1000-bp DNA ladder; Figure S3: PCR amplification patterns in β-tubulin with B-tubf1-F / B-tubr1-R primers from 20 Trichoderma asperellum isolates; L is 1000-bp DNA ladder; Figure S4–S10: GC-MS chromatogram of secondary metabolites from T. asperellum isolates; Figure S11: TLC plates showing many spots of metabolites (A) Long UV light range (B) Short UV light range in the solvent system: Hexane: Ethyl acetate (9:1) (C) Long UV light range (D) Short UV light range in the solvent system: Hexane: Ethyl acetate (1:1).

Author Contributions: C.S., D.K. and R.G. (Robin Gogoi) were involved in the conceptualization of the project, study design, and critical inputs. C.S. contributed to the lab work and statistical analysis and wrote the first draft. C.S., D.K., A.K., A.D. and R.G. (Robin Gogoi) finalized the outline and prepared schematics. P.K.S., Z.H. and A.D. helped in carry out an in-vivo experiment. C.S. and A.K. carried out GC-MS work and analysis. C.S., P.R.S., M.C. and R.G. (R. Gangaraj) were helped with statistical analysis and editing of the manuscript. All authors contributed to the article and approved the submitted version. All authors have read and agreed to the published version of the manuscript.

Funding: This work was partially funded by the NAHEP-CAAST project on "Genomics assisted crop improvement and management". C. Shanmugaraj received Junior and Senior Research Fellowships from the Indian Council of Agricultural Research (ICAR) for pursuing the Ph.D. program at ICAR-IARI, New Delhi.

Data Availability Statement: The original contributions presented in the study are included in the article/Supplementary material; further inquiries can be directed to the corresponding author.

Acknowledgments: C. Shanmugaraj offers sincere thanks to the ICAR for financial support in the form of Junior and Senior Research Fellowships for the Ph.D. program. C. Shanmugaraj and Deeba Kamil are grateful to NAHEP-CAAST for financial assistance.

Conflicts of Interest: The authors declare that the research was conducted in the absence of any commercial or financial relationships that could be construed as potential conflict of interest.

References

1. Nalim, F.A.; Starr, J.L.; Woodard, K.E.; Segner, S.; Keller, N.P. Mycelial compatibility groups in Texas peanut field populations of *Sclerotium rolfsii*. *Phytopathology* **1995**, *85*, 1507–1512. [CrossRef]
2. Aycock, R. *Stem Rot and Other Diseases Caused by Sclerotium Rolfsii or the Status of Rolf's Fungus after 70 Years*; North Carolina Agricultural Experiment Station Technical Bulletin No.175; North Carolina Agricultural Experiment Station: Salisbury, NC, USA, 1966; pp. 1–202.
3. Cilliers, A.; Herselman, L.; Pretorius, Z. Genetic variability within and among mycelial compatibility groups of *Sclerotium rolfsii* in South Africa. *Phytopathology* **2000**, *90*, 1026–1031. [CrossRef]
4. Harlton, C.E.; Uvesque, C.A.; Punja, Z.K. Genetic diversity in *Sclerotium* (*Athelia*) *rolfsii* and related species. *Phytopathology.* **1995**, *85*, 1269–1281. [CrossRef]
5. Bera, S.K.; Kasundra, S.V.; Kamdar, J.H.; Ajay, B.C.; Lal, C.; Thirumalasmy, P.P. Variable response of interspecific breeding lines of groundnut to *Sclerotium rolfsii* infection under field and laboratory conditions. *Electron. J. Plant Breed.* **2014**, *5*, 22–29.
6. Bosamia, T.C.; Dodia, S.M.; Mishra, G.P.; Ahmad, S.; Joshi, B.; Thirumalaisamy, P.P.; Kumar, N.; Rathnakumar, A.L.; Sangh, C.; Kumar, A.; et al. Unraveling the mechanisms of resistance to *Sclerotium rolfsii* in peanut (*Arachis hypogaea* L.) using comparative RNA-Seq analysis of resistant and susceptible genotypes. *PLoS ONE* **2020**, *15*, e0236823. [CrossRef] [PubMed]
7. Mahato, A.; Biswas, M.K.; Patra, S. Prevalence of collar rot of tomato caused by *Sclerotium rolfsii* (Sacc.) under the red and lateritic zone of West Bengal, India. *Int. J. Curr. Microbiol. Appl. Sci.* **2017**, *6*, 3231–3236. [CrossRef]
8. Punja, Z.K. The biology, ecology and control of *Sclerotium rolfsii*. *Annu. Rev. Phytopathol.* **1985**, *23*, 97–127. [CrossRef]
9. Guzmán-Guzmán, P.; Kumar, A.; de Los Santos-Villalobos, S.; Parra-Cota, F.I.; Orozco-Mosqueda, M.D.C.; Fadiji, A.E.; Hyder, S.; Babalola, O.O.; Santoyo, G. *Trichoderma* species: Our best fungal allies in the biocontrol of plant diseases—A review. *Plants* **2023**, *12*, 432. [CrossRef] [PubMed]
10. Sood, M.; Kapoor, D.; Kumar, V.; Sheteiwy, M.S.; Ramakrishnan, M.; Landi, M.; Araniti, F.; Sharma, A. *Trichoderma*: The secrets of a Multitalented Biocontrol Agent. *Plants* **2020**, *7*, 962. [CrossRef]
11. Verma, M.; Brar, S.K.; Tyagi, R.D.; Surampalli, R.Y.; Valéro, J.R. Antagonistic Fungi, *Trichoderma* Spp.: Panoply of Biological Control. *Biochem. Eng. J.* **2007**, *37*, 1–20. [CrossRef]
12. Woo, S.L.; Hermosa, R.; Lorito, M.; Monte, E. *Trichoderma*: A multipurpose, plant-beneficial microorganism for eco-sustainable agriculture. *Nat. Rev. Microbiol.* **2023**, *21*, 312–326. [CrossRef] [PubMed]
13. Hjeljord, L.; Tronsmo, A.; Harman, G.E.; Kubicek, C.P. Trichoderma and *Gliocladium* in biological control: An overview. In *Trichoderma and Gliocladium, Enzymes, Biological Control and Commercial Applications*; Kubicek, C.P., Harman, G.E., Eds.; Taylor and Francis: London, UK, 1998; pp. 131–151. [CrossRef]
14. Kubicek, C.P.; Pentilla, M.E. Regulation of production of plant polysaccharide degrading enzymes by *Trichoderma*. In *Trichoderma and Gliocladium. Vol.2. Enzymes, Biological Control and Commercial Applications*; Harman, G.E., Kubicek, C.P., Eds.; Taylor and Francis Ltd.: London, UK, 1998; pp. 49–71.
15. Howell, C.R. The role of antibiosis in bio-control. In *Trichoderma and Gliocladium*; Harman, G.E., Kubicek, C.P., Eds.; Taylor and Francis: London, UK, 1998; Volume 2, pp. 173–184.
16. Howell, C.R. Mechanisms employed by *Trichoderma* species in the biological control of plant diseases: The history and evolution of current concepts. *Plant Dis.* **2003**, *87*, 4–10. [CrossRef]
17. Paparizas, G.C. *Trichoderma* and *Gliocladium*: Biology, ecology and potential for bio-control. *Annu. Rev. Phytopathol.* **1985**, *23*, 23–54. [CrossRef]
18. Narasimha Murthy, K.; Nirmala Devi, D.; Srinivas, C. Efficacy of *Trichoderma asperellum* against *Ralstonia solanacearum* under greenhouse conditions. *Ann. Plant Sci.* **2013**, *2*, 342–350.
19. Wu, Q.; Sun, R.; Ni, M.; Yu, J.; Li, Y.; Yu, C.; Dou, K.; Ren, J.; Chen, J. Identification of a novel fungus, *Trichoderma asperellum* GDFS1009, and comprehensive evaluation of its bio-control efficacy. *PLoS ONE* **2017**, *12*, e0179957. [CrossRef]

20. Moussa, Z.; Alanazi, Y.F.; Khateb, A.M.; Eldadamony, N.M.; Ismail, M.M.; Saber, W.I.; Darwish, D.B.E. Domiciliation of *Trichoderma asperellum* Suppresses *Globiosporangium ultimum* and Promotes Pea Growth, Ultrastructure, and Metabolic Features. *Microorganisms* **2023**, *11*, 198. [CrossRef]
21. Yoshioka, Y.; Ichikawa, H.; Naznin, H.A.; Kogure, A.; Hyakumachi, M. Systemic resistance induced in *Arabidopsis thaliana* by *Trichoderma asperellum* SKT-1, a microbial pesticide of seed borne diseases of rice. *Pest Manag. Sci.* **2011**, *68*, 60–66. [CrossRef]
22. Shoresh, M.; Yedidia, I.; Chet, I. Involvement of jasmonic acid/ethylene signaling pathway in the systemic resistance induced in cucumber by *Trichoderma asperellum* T203. *Phytopathology* **2005**, *95*, 76–84. [CrossRef]
23. Herrera, W.; Valbuena, O.; Pavone-Maniscalco, D. Formulation of *Trichoderma asperellum* TV190 for biological control of *Rhizoctonia solani* on corn seedlings. *Egypt. J. Biol. Pest Control.* **2020**, *30*, 44. [CrossRef]
24. Brotman, Y.; Briff, E.; Viterbo, A.; Chet, I. Role of swollenin, an expansin-like protein from *Trichoderma*, in plant root colonization. *Plant Physiol.* **2008**, *147*, 779–789. [CrossRef]
25. Guo, R.; Ji, S.; Wang, Z.; Zhang, H.; Wang, Y.; Liu, Z. *Trichoderma asperellum* xylanases promote growth and induce resistance in poplar. *Microbiol. Res.* **2021**, *248*, 126767. [CrossRef]
26. Srinivasa, N.; Sriram, S.; Singh, C.; Shivashankar, K.S. Secondary metabolites approach to study the bio-efficacy of *Trichoderma asperellum* isolates in India. *Int. J. Curr. Microbiol. Appl. Sci.* **2017**, *6*, 1105–1123. [CrossRef]
27. Degani, O.; Khatib, S.; Becher, P.; Gordani, A.; Harris, R. *Trichoderma asperellum* secreted 6-Pentyl-α-Pyrone to control *Magnaporthiopsis maydis*, the maize late wilt disease agent. *Biology* **2021**, *10*, 897. [CrossRef] [PubMed]
28. Geraldine, A.M.; Lopes, F.A.C.; Carvalho, D.D.C.; Barbosa, E.T.; Rodrigues, A.R.; Brandão, R.S.; Ulhoa, C.J.; Junior, M.L. Cell wall-degrading enzymes and parasitism of sclerotia are key factors on field bio-control of white mold by *Trichoderma* spp. *Biol. Control.* **2013**, *67*, 308–316. [CrossRef]
29. Qualhato, T.F.; Lopes, F.A.C.; Steindorff, A.S.; Brandao, R.S.; Jesuino, R.S.A.; Ulhoa, C.J. Mycoparasitism studies of *Trichoderma* species against three phytopathogenic fungi: Evaluation of antagonism and hydrolytic enzyme production. *Biotechnol. Lett.* **2013**, *35*, 1461–1468. [CrossRef]
30. Saravanakumar, K.; Yu, C.; Dou, K.; Wang, M.; Li, Y.; Chen, J. Synergistic effect of *Trichoderma*-derived antifungal metabolites and cell wall degrading enzymes on enhanced bio-control of *Fusarium oxysporum* f. sp. *cucumerinum*. *Biol. Control.* **2016**, *94*, 37–46. [CrossRef]
31. Silva, B.D.S.; Ulhoa, C.J.; Batista, K.A.; Yamashita, F.; Fernandes, K.F. Potential fungal inhibition by immobilized hydrolytic enzymes from *Trichoderma asperellum*. *J. Agric. Food Chem.* **2011**, *59*, 8148–8154. [CrossRef]
32. Snehal, D.F. Isolation and identification of fungi from contaminated soil to build biological resource as biocontrol activity. *Macromol. Ind. J.* **2017**, *12*, 105.
33. Cullings, K.W. Design and testing of a plant-specific PCR primer for ecological and evolutionary studies. *Mol. Ecol.* **1992**, *1*, 233–240. [CrossRef]
34. White, T.J.; Bruns, T.D.; Lee, S.B.; Taylor, J.W. Amplification and Direct Sequencing of Fungal Ribosomal RNA Genes for Phylogenetics. In *PCR Protocols: A Guide to Methods and Applications*; Innis, M.A., Gelfand, D.H., Sninsky, J.J., White, T.J., Eds.; Academic Press: New York, NY, USA, 1990; pp. 315–322.
35. Thokala, P.; Narayanasamy, P.; Kamil, D.; Choudhary, S.P. Polyphasic taxonomy of Indian *Trichoderma* species. *Phytotaxa* **2021**, *502*, 1–27. [CrossRef]
36. Garcia, E.F. Screening of fungal antagonist to control *Sclerotium cepivorum*. In *New Approaches in Biological Control of Soil Borne Diseases*; Jensen, D.F., Ed.; IOBCM/PRS Bull.: Copenhagen, Denmark, 1991; pp. 79–81.
37. Morton, D.T.; Stroube, N.H. Antagonistic and stimulatory effect of microorganism upon *Sclerotium rolfsii*. *Phytopathology* **1955**, *45*, 419–420.
38. Dennis, C.; Webster, J. Antagonistic properties of species groups of *Trichoderma*. II. Production of volatile antibiotics. *Trans. Brit. Mycol. Soc.* **1971**, *57*, 363–369. [CrossRef]
39. Quimio, A.J.; Cumagun, C.J. *Workbook on Tropical fungi: Collection, Isolation and Identification*; The Mycological Society of the Philippines, Inc.: Laguna, Philippines, 2001.
40. Bradford, M.M. A rapid and sensitive method for the quantitation of microgram quantities of protein utilizing the principle of protein-dye binding. *Anal. Biochem.* **1976**, *72*, 248–254. [CrossRef]
41. Miller, G.L. Use of dinitrosalicylic acid reagent for determination of reducing sugar. *Anal. Chem.* **1959**, *31*, 426–428. [CrossRef]
42. Nelson, N.A. A photometric adaptation of the Somogyi method for the determination of glucose. *J. Biol. Chem.* **1944**, *153*, 375. [CrossRef]
43. Thrane, C.; Jensen, D.F.; Tronsmo, A. Substrate colonization, strain competition, enzyme production in vitro, and bio-control of *Pythium ultimum* by *Trichoderma* spp. isolates P1 and T3. *Eur. J. Plant Pathol.* **2000**, *106*, 215–225. [CrossRef]
44. Vessey, J.C.; Pegg, G.F. Autolysis and chitinase production in cultures of *Verticillium albo-atrum*. *Trans. Br. Mycol. Soc.* **1973**, *60*, 133–143. [CrossRef]
45. Sahai, A.S.; Manocha, M.S. Chitinases of fungi and plants: Their involvement in morphogenesis and host-parasite interaction. *FEMS Microbiol. Rev.* **1993**, *11*, 317–338. [CrossRef]
46. Chaithra, M.; Prameeladevi, T.; Prasad, L.; Kundu, A.; Bhagyasree, S.N.; Subramanian, S.; Kamil, D. Metabolomic diversity of local strains of *Beauveria bassiana* (Balsamo) Vuillemin and their efficacy against the cassava mite, *Tetranychustruncatus* Ehara (Acari: Tetranychidae). *PLoS ONE* **2022**, *17*, e0277124. [CrossRef]

47. Pandian, R.T.P.; Raja, M.; Kumar, A.; Sharma, P. Morphological and molecular characterization of *Trichoderma asperellum* strain Ta13. *Indian Phytopathol.* **2016**, *69*, 297–303.
48. Sriram, S.; Savitha, M.J.; Rohini, H.S.; Jalali, S.K. The most widely used fungal antagonist for plant disease management in India, *Trichoderma viride* is *Trichoderma asperellum* as confirmed by oligonucleotide barcode and morphological characters. *Curr. Sci.* **2013**, *2023*, 1332–1340.
49. Chaithra, M.; Prameeladevi, T.; Bhagyasree, S.N.; Prasad, L.; Subramanian, S.; Kamil, D. Multilocus sequence analysis for population diversity of indigenous entomopathogenic fungus *Beauveria bassiana* and its bio-efficacy against the cassava mite, *Tetranychus truncatus* Ehara (Acari: Tetranychidae). *Front. Microbiol.* **2022**, *13*, 1007017. [CrossRef]
50. Poosapati, S.; Ravulapalli, P.D.; Tippirishetty, N.; Vishwanathaswamy, D.K.; Chunduri, S. Selection of high temperature and salinity tolerant *Trichoderma* isolates with antagonistic activity against *Sclerotium rolfsii*. *SpringerPlus* **2014**, *3*, 641. [CrossRef]
51. Manjunath, M.; Singh, A.; Tripathi, A.N.; Prasanna, R.; Rai, A.B.; Singh, B. Bioprospecting the fungicides compatible *Trichoderma asperellum* isolate effective against multiple plant pathogens in vitro. *J. Environ. Biol.* **2017**, *38*, 553. [CrossRef]
52. Akash, A.U.; Ramya, V.; Uma Devi, G.; Pushpavalli, S.N.C.V.L.; Triveni, S. Antagonist activities of native rhizosphere micro-flora against groundnut stem rot pathogen, *Sclerotium rolfsii* Sacc. *Egypt. J. Biol. Pest Control.* **2022**, *32*, 133. [CrossRef]
53. Guzman-Valle, P.; Bravo-Luna, L.; Montes-Belmont, R.; Guigon-Lopez, C.; Sepulveda-Jimenez, G. Induction of resistance to *Sclerotium rolfsii* in different varieties of onion by inoculation with *Trichoderma asperellum*. *Eur. J. Plant Pathol.* **2014**, *138*, 223–229. [CrossRef]
54. El-Komy, M.H.; Al-Qahtani, R.M.; Ibrahim, Y.E.; Almasrahi, A.A.; Al-Saleh, M.A. Soil application of *Trichoderma asperellum* strains significantly improves Fusarium root and stem rot disease management and promotes growth in cucumbers in semi-arid regions. *Eur. J. Plant Pathol.* **2022**, *162*, 637–653. [CrossRef]
55. Ayyandurai, M.; Akila, R.; Manonmani, K.; Harish, S.; Mini, M.L.; Vellaikumar, S. Deciphering the mechanism of *Trichoderma* spp. consortia possessing volatile organic compounds and antifungal metabolites in the suppression of *Sclerotium rolfsii* in groundnut. *Physiol. Mol. Plant Pathol.* **2023**, *125*, 102005. [CrossRef]
56. Harman, G.E. Overview of mechanisms and uses of *Trichoderma* spp. *Phytopathology* **2006**, *96*, 190–194. [CrossRef]
57. Zhang, Y.; Xiao, J.; Yang, K.; Wang, Y.; Tian, Y.; Liang, Z. Transcriptomic and metabonomic insights into the bio-control mechanism of *Trichoderma asperellum* M45a against watermelon *Fusarium* wilt. *PLoS ONE* **2022**, *17*, e0272702. [CrossRef]
58. Degani, O.; Rabinovitz, O.; Becher, P.; Gordani, A.; Chen, A. *Trichoderma longibrachiatum* and *Trichoderma asperellum* confer growth promotion and protection against late wilt disease in the field. *J. Fungi.* **2021**, *7*, 444. [CrossRef]
59. El-Komy, M.H.; Saleh, A.A.; Ibrahim, Y.E.; Hamad, Y.K.; Molan, Y.Y. *Trichoderma asperellum* strains confer tomato protection and induce its defense-related genes against the *Fusarium* wilt pathogen. *Tropical Plant Pathol.* **2016**, *41*, 277–287. [CrossRef]
60. Karuppiah, V.; He, A.; Lu, Z.; Wang, X.; Li, Y.; Chen, J. *Trichoderma asperellum* GDFS1009-mediated maize resistance against *Fusarium graminearum* stalk rot and mycotoxin degradation. *Biol. Control.* **2022**, *174*, 105026. [CrossRef]
61. Rivera-Mendez, W.; Obregon, M.; Moran-Diez, M.E.; Hermosa, R.; Monte, E. *Trichoderma asperellum* bio-control activity and induction of systemic defenses against *Sclerotium cepivorum* in onion plants under tropical climate conditions. *Biol. Control.* **2020**, *141*, 104145. [CrossRef]
62. Inbar, J.; Chet, I. The role of recognition in the induction of specific chitinases during mycoparasitism by *Trichoderma harzianum*. *Microbiology* **1995**, *141*, 2823–2829. [CrossRef]
63. El-Komy, M.H.; Saleh, A.A.; Eranthodi, A.; Molan, Y.Y. Characterization of novel *Trichoderma asperellum* isolates to select effective bio-control agents against tomato *Fusarium* wilt. *Plant Pathol. J.* **2015**, *31*, 50. [CrossRef] [PubMed]
64. Samuels, G.J.; Lieckfeldt, E.L.K.E.; Nirenberg, H.I. *Trichoderma asperellum*, a new species with warted conidia, and redescription of *T. viride*. *Sydowia* **1999**, *51*, 71–88.
65. Kangjam, V.; Pongener, N.; Singh, R.; Banik, S.; Daiho, L.; Ao, N.T. In vitro Screening for Potential Antagonistic Activity of *Trichoderma* Isolates against *Sclerotium rolfsii* Causing Collar Rot of French Bean. *Environ. Ecol.* **2023**, *41*, 623–632.
66. Srinivasa, N.; Kamil, D.; Singh, C.; Singode, A.; Gupta, D. Molecular and Biochemical Characterization of Potential Isolates of *Trichoderma* Species Effective against Soil-Borne Pathogens. *Int. J. Curr. Microbiol. Appl. Sci.* **2017**, *6*, 3132–3149. [CrossRef]
67. Bhardwaj, N.R.; Kumar, J. Characterization of volatile secondary metabolites from *Trichoderma asperellum*. *J. Appl. Nat. Sci.* **2017**, *9*, 954–959. [CrossRef]
68. Scudeletti, D.; Crusciol, C.A.C.; Bossolani, J.W.; Moretti, L.G.; Momesso, L.; Servaz Tubana, B.; De Castro, S.G.Q.; De Oliveira, E.F.; Hungria, M. *Trichoderma asperellum* inoculation as a tool for attenuating drought stress in sugarcane. *Front. Plant Sci.* **2021**, *12*, 645542. [CrossRef] [PubMed]
69. Degani, O.; Gordani, A. New Antifungal Compound, 6-Pentyl-α-Pyrone, against the Maize Late Wilt Pathogen, *Magnaporthiopsis maydis*. *Agronomy* **2022**, *12*, 2339. [CrossRef]
70. Man, L.I.U.; Qichen, N.I.U.; Shuxia, Y.I.N.; Ziyue, W.A.N.G. Inhibitory activity of 6-pentyl-2H-pyran-2-one against the pathogenesis fungi of dollar spot and its control efficacy. *J. Pestic. Sci.* **2023**, *25*, 104–116. [CrossRef]
71. Hao, J.; Wuyun, D.; Xi, X.; Dong, B.; Wang, D.; Quan, W.; Zhang, Z.; Zhou, H. Application of 6-Pentyl-α-Pyrone in the Nutrient Solution Used in Tomato Soilless Cultivation to Inhibit *Fusarium oxysporum* HF-26 Growth and Development. *Agronomy* **2023**, *13*, 1210. [CrossRef]
72. Aggarwal, R.; Tewari, A.K.; Srivastava, K.D.; Singh, D.V. Role of antibiosis in the biological control of spot blotch (*Cochliobolus sativus*) of wheat by *Chaetomium globosum*. *Mycopathologia* **2004**, *157*, 369–377. [CrossRef] [PubMed]

73. Saputra, H.; Puspota, F.; Tjandrawati, T. Production of an antibacterial compound against the plant pathogen *Erwinia carotovora* subsp. *carotovora* by the bio-control strain *Gliocladium* sp. TN C73. *J. Agric. Tech.* **2013**, *9*, 1157–1165.
74. Adnan, L.A.; Sathishkumar, P.; Yusoff, A.R.M.; Hadibarata, T. Metabolites characterisation of laccase mediated Reactive Black 5 biodegradation by fast growing ascomycete fungus *Trichoderma atroviride* F03. *Int. Biodeterior. Biodegrad.* **2015**, *104*, 274–282. [CrossRef]

horticulturae

Article

The Isolation, Identification, and Insecticidal Activities of Indigenous Entomopathogenic Nematodes (*Steinernema carpocapsae*) and Their Symbiotic Bacteria (*Xenorhabdus nematophila*) against the Larvae of *Pieris brassicae*

Preety Tomar [1], Neelam Thakur [1,*], Avtar Kaur Sidhu [2], Boni Amin Laskar [2], Abeer Hashem [3], Graciela Dolores Avila-Quezada [4] and Elsayed Fathi Abd_Allah [5]

1 Department of Zoology, Akal College of Basic Sciences, Eternal University, Baru Sahib, Sirmour, Himachal Pradesh 173101, India; preetytomar9@gmail.com
2 Zoological Survey of India, High Altitude Regional Center (HARC)-Solan, Himachal Pradesh 173212, India; avtarkaur2000@gmail.com (A.K.S.); boniamin.laskar@gmail.com (B.A.L.)
3 Botany and Microbiology Department, College of Science, King Saud University, P.O. Box 2460, Riyadh 11451, Saudi Arabia; habeer@ksu.edu.sa
4 Facultad de Ciencias Agrotecnológicas, Universidad Autónoma de Chihuahua, Chihuahua 31350, Chihuahua, Mexico; gdavila@uach.mx
5 Plant Production Department, College of Food and Agricultural Sciences, King Saud University, P.O. Box 2460, Riyadh 11451, Saudi Arabia; eabdallah@ksu.edu.sa
* Correspondence: neelamthakur@eternaluniversity.edu.in; Tel.: +91-8351871241

Abstract: The cabbage butterfly, *Pieris brassicae* Linnaeus (Lepidoptera: Pieridae), is an oligophagous and invasive insect pest of various economically important cole crops. Recently, there have been reports about an increase in the incidence and damaging activities of cabbage butterflies, signifying that the existing control methods fail to meet the grower's expectations. Entomopathogenic nematodes (EPNs) and their endosymbiotic bacteria have immense potential for the control of a wide range of insect pests. In this investigation, the EPN species *Steinernema carpocapsae* and its associated bacterial species, *Xenorhabdus nematophila*, were isolated and identified through morphological and molecular techniques. The laboratory bioassay experiment was performed using *S. carpocapsae* and *X. nematophila* against the 3rd instar larvae of *P. brassicae* (25 ± 1 °C; RH = 60%). The efficacy of EPN suspension (30, 60, 90, 120, 150 IJs/mL) and bacterial suspension (1×10^4, 2×10^4, 3×10^4, 4×10^4, and 5×10^4 CFU/mL) via contact and oral routes showed significant mortality among the larvae. Surprisingly, 100% insect mortality within 48 h was recorded in the bacterial inoculum 5×10^4 CFU/mL. However, in the case of EPNs (*S. carpocapsae*), 150 IJs/mL caused the highest, 92%, larval mortality rate after 96 h. The results signify that both indigenous EPNs and their associated bacteria can provide efficient control against *P. brassicae* larvae and could effectively contribute to IPM programs. However, further analyses are required to authenticate their effectiveness in field conditions.

Keywords: diversity; entomopathogenic nematodes; infective juveniles; *Steinernema*

1. Introduction

Insects are phytophagous organisms, with >85 percent of species containing more than 9 lakh identified insects [1]. Numerous agricultural insect pests, along with forest encroaching species, contribute to losses of approximately 76.9 billion USD per year, and it is a need of the hour to control the effects of insect bio-invasion worldwide [2]. Currently, phytophagous insect invasion has intensified due to global warming projects [3] that might contribute to the expansion of pest populations, empirical exceedance in outbreaks, and geographical spreading of several species, resulting in huge economic impairments and a decline in food safety [4]. Among insect arthropods, lepidopteran insect pests are causing substantial damage to agricultural produce. *Pieris brassicae* (the cabbage butterfly) is a destructive polyphagous pest that is responsible for huge economic losses to the crucifers [5] and is one of the limiting factors in crop production worldwide. Along with the crucifers, they also feed upon plants of the Tropaeolaceae, Resedaceae, and Capparaceae families [6], causing over 50% losses in production every year [7].

The management of cabbage butterflies is challenging due to their wide host range, feeding habits, and high reproduction potential. A wide spectrum of chemical-based insecticides and pesticides have been employed by farmers worldwide to protect their crops from invasive damage. The pesticides used are quite expensive and unsafe. The consumption of pesticides not only decays soil prolificacy but also affects the intrinsic habitats of advantageous organisms and causes resistance among insect pests [8–10]. Additionally, education and consciousness concerning the risks associated with pesticide consumption, including pollution, contamination, resistance, pest revival, and effect on other non-target creatures has risen. The consequences raised by these chemical-based formulations have drawn attention to a better and safer alternative. Continuous research in this field is being carried out, and an alternative means of insect pest control without affecting natural entities has emerged as "Biological control". The utilization of bio-agents, such as indigenous EPNs, is the best way to tackle insect pest problems.

The word "Entomopathogenic" is integrated from two different words: first *entomon*, related to insects, and *pathogen*, related to causing disease. Thus, EPNs are responsible for causing diseases in various insects [11]. Phoretic, communalistic, symbiotic, obligatory, and facultative interactions have been recorded among nematodes and insects [12]. The EPN order Rhabditida belongs to two families, i.e., Heterorhabditidae [13] and Steinernematidae [14]. Genera *Steinernema* (Steinernematidae) and *Heterorhabditis* (Heterorhabditidae) exhibit a total of 117 species of EPNs, with 100 species described from *Steinernema* and 17 species from the genus *Heterorhabditis* [15]. The nematodes showed mutualistic alliances with the bacterial species that inhabit their alimentary canals and are responsible for forming nemato-bacterial complexes [16]. The EPNs of genus *Steinernema* showed mutualistic association with the bacterial genus *Xenorhabdus* whereas *Heterorhabditis* showed mutual interrelation with the bacterial genus *Photorhabdus*, respectively [17–19]. They are well-established biocontrol agents (BCAs) that play an imperative role in virulence and have been successfully employed against agricultural insect pests in different management strategies [20–22]. Third-stage juveniles, or infective juveniles (IJs), or dauer juveniles, are actually responsible for invading and parasitizing the host insect [23]. The symbiotic bacteria of EPNs releasing toxic substances upon invasion into the hemolymph and multiply there [24].

The correct and precise recognition of EPNs is the initial and prime exigency to accomplish any biological control attribute. Additionally, they are eco-friendly, economic, and conservative in nature [25–27]. Keeping this in mind, a survey was conducted to explore EPN incidence in Himachal Pradesh and to isolate the bacteria associated with these EPNs in order to evaluate their virulence capacity towards cabbage butterfly larvae under laboratory conditions.

2. Materials and Methods

2.1. Survey and Sample Collection

Soil-inhabiting entomopathogenic nematodes are usually found naturally in all types of soils, but their incidence is high in undistributed soil. A survey study was conducted during 2018–2021, and soil samples were collected from the undisturbed land of forests and fruit orchards of district Shimla (altitude range between 1021 and 2276 m), Himachal Pradesh, as this region is known as the "fruit bowl" of India and is chiefly a horticulture state that produces and exports apples, peaches, plums, apricots, cherries, and pears to other states. The soil was sampled from the vicinity of plant roots at a depth of 15–30 cm with the help of a spade and hand shovel. The samples were brought into the experimental laboratory after proper labelling of details such as host plant name, locality, date, district, village, type of soil, and altitude [28]. The isolation of EPNs was done via the soil baiting technique [29]. For this, stones, detritus, and leaf litter were removed, and soil was filled into the plastic containers after proper labelling [30]. The soil baiting technique requires insect baits, so *Galleria mellonella* and *Corcyra cephalonica* were used as bait. In this technique, 5–10 last instar larvae of bait insects were kept in plastic containers with soil. The containers were searched for larval mortality after every 48 h up to 120 h. The dead insect cadavers were separated and removed regularly from the containers. These cadavers were investigated, and the existence of nematodes was explored using the white trap method [31]. The emerging nematodes were collected from the white trap and stored in the flask before being transferred to the roux bottle. The incidence of EPNs was assessed using the formula given by Norton [32].

2.2. In-Vivo Mass Multiplication of EPNs

The isolated nematodes were further mass multiplied under laboratory conditions using *G. mellonella* larvae, followed by the white trap methodology for nematode extraction. The isolated nematodes were stored in two forms: (i) directly as suspension in the roux bottle and (ii) in a sterilized synthetic sponge. They were stored at 12–15 °C.

2.3. Morphological Observations of Entomopathogenic Nematode

The isolated nematodes were examined live as well as after being sacrificed, fixed, and mounted. Nematodes were sacrificed and fixed with formalin (4%) fixative, and mounting was carried out using anhydrous glycerine [33]. A total of 10 nematode specimens from each life stage were collected randomly and observed for morphological analysis. The measurements were taken using a compound microscope (Leica DM750) equipped with Leica Application Suite (LAS) version 4.12. The morphological studies were carried out in accordance with the taxonomic keys [34], and the body ratio depiction was made on the basis of the formula given by de Man [35].

2.4. Isolation of Bacteria Associated with Entomopathogenic Nematode

The bacterium associated with the EPNs was also isolated following the methodology given by Thakur et al. [36]. To isolate bacterial endosymbionts from EPNs, 50–100 IJs were taken into an Eppendorf tube (1.5 mL). In this tube, 1 mL of 1% sodium hypochlorite solution was added for surface sterilization (4–6 min). The suspension was spun in the microcentrifuge for 5 min at 5000 rpm for the collection of sterilized IJs. The IJs pellet was resuspended in double distilled water and washed 3–5 times. After washing, the IJs were placed into a sterilized mortar and pestle and crushed properly. After crushing, a few drops of crushed nematodes were placed onto the Petri plate containing Nutrient Bromothymol Blue Tetrazolium Chloride Agar (NBTA) medium. The drops were spread over the plate with the help of a sterile spreader. The plates were then placed into the bacteriological incubator at 28 ± 1 °C for 24–48 h. Lawns of bacterial colonies appeared on the Petri plate after 48 h. A single colony of bacteria was chosen and re-streaking was carried out using a sterile loop until a pure uniform colony was achieved. The culture was stored under the nutrient broth (NB) media and used for DNA isolation as well as for the bioassay study.

2.5. Molecular Characterization of Entomopathogenic Nematode and Bacteria

Molecular characterization was carried out in the molecular laboratory at ZSI, HARC-Solan. The genomic DNA from the nematodes and bacteria was procured using the Qiazen's Dneasy® blood and tissue kit-based method. Before proceeding to DNA isolation, 500 nematodes juveniles were stored in ethanol inside the Eppendorf tube (2 mL) and centrifuged at 12,000 rpm for 3 min. The supernatant was discarded, and the pellet was kept for drying, which was then transferred into the deep freezer (−20 °C). The nematode pellet was crushed using a micropestle inside the tube. The DNA was isolated, and the quantification of the extracted DNA was carried out using 0.8% agarose gel electrophoresis. The DNA quality was confirmed by visualizing the gel under the gel documentation unit (Alpha-Imager). The amplification of the isolated nematodes DNA was carried out, and the ITS region of the rDNA was amplified using the forward primer (18S: 5′-TTGATTACGTCCCTGCCCTTT-3′) and the reverse primer (26S: 5′-TTTCACTCGCCGTTACTAAGG-3′).

After that, the isolation of DNA from bacteria was carried out, and bacterial colonies were inoculated into 2 mL microcentrifuge tubes containing nutrient broth media and incubated for 24 h prior to extraction. The tubes were centrifuged, and the bacterial pallet was used for the isolation of DNA. The amplification of 16S rRNA was carried using pA/pH primers: forward primer pA (5′-AGAGTTTGATCCTGGCTCAG-3′) and reverse primer pH (5′-AAGGAGGTGATCCAGCCGCA-3′).

For the PCR amplification of nematode DNA, 25 μL reaction tube was prepared that contained genomic DNA 2.0 μL, Taq buffer (10X) 4.0 μL, $MgCl_2$ (25 mM) 1.0 μL, dNTPs (1 mM) 1.0 μL, primer forward 1.0 μL, primer reverse 1.0 μL, Taq polymerase 1.0 μL and Mili Q water 14 μL. The amplifications were accomplished in a gradient thermal cycler machine (Applied Biosys VeritiPro Thermal Cycle). The first step consisted of pre-heating the thermal cycler machine to 95 °C, followed by denaturation for 3 min at 94 °C. The third step involved annealing for 1 min at 55 °C and then for an extension of 1 min 30 s at 72 °C. The process was repeated for up to five cycles and was again followed by 30 s of denaturation at 94 °C; double annealing at 55 °C for 3 min; extension at 72 °C for 1 min; and, finally, extension of 5 min at 72 °C for up to 35 cycles to ensure the full length of amplified fragments. The amplified product was stored at 4 °C, which was later transferred to −20 °C.

Similarly, in bacterial DNA amplification, a 25 μL reaction tube was prepared and contained genomic DNA 2.0 μL, Taq buffer (10X) 4.0 μL, $MgCl_2$ (25 mM) 1.0 μL, dNTPs (1 mM) 1.0 μL, primer forward 1.0 μL, primer reverse 1.0 μL, Taq polymerase 1.0 μL, and Mili Q water 14 μL. The optimized PCR conditions include initial denaturation at 95° for 2 min, followed by denaturation at 94 °C for 30 s, followed by annealing at 54° for 30 s, then extension at 72 °C for 2 min, and final extension at 72 °C for 10 min for up to 35 cycles, and then storage at 4 °C.

The products after the PCR reactions, using primers, were observed on 1.2% agarose gels and their sizes were determined by comparison with the DNA ladder (100 bp). The amplified PCR products of the nematode genome were sent to the Eurofins Analytical Services India Pvt. Ltd. (Bengaluru, India) for sequencing. After sequencing, a phylogenetic tree of the obtained sequences was constructed, and the reverse sequence was antisense reversed. Using these sequences, 10 corresponding or similar sequences were downloaded from GenBank, NCBI. Multiple sequence alignment (MSA) of all the sequences was carried out using ClustalW (with a gap opening penalty for multiple alignments of 12 and an extension of 10) using MEGA v 11.0. A phylogenetic tree was constructed using maximum likelihood in bootstrapping 100, using the Tamura 3-parameter model. Gaps and missing data were treated as complete deletions [37]. The partial sequences were submitted to the National Centre for Biotechnology Information (NCBI) and accession numbers were assigned.

2.6. Evaluation of EPNs Infectivity

Rearing of *Pieris brassicae* was carried out in the laboratory following the methodology given by Tomar et al. [38]. A petri plate bioassay experiment was performed to evaluate the infectivity of *S. carpocapsae* and *X. nematophila* against the 3rd instar larvae of *P. brassicae* under laboratory conditions. Two bioassay experiments were performed separately, in which one experiment contained EPNs suspension and another contained bacterial suspension. The nematodes were used at concentrations of 30, 60, 90, 120, 150 IJs/mL as well as a control. The bacterial suspension was applied at a rate of 1×10^4, 2×10^4, 3×10^4, 4×10^4, and 5×10^4 CFU/mL with absolute control. Healthy, laboratory-reared 3rd instar larvae of *P. brassicae* were kept in the petri plates at a rate of 10 larvae per petri plate along with food (cabbage leaves). Each treatment was replicated five times, and in the control, only 1–2 mL of distilled water was applied. The Petri dish was incubated at 26 ± 1 °C. Insect mortality was checked after 24, 48, and 96 h of bacterial inoculation, and the data were recorded daily. The experiment was conducted in a completely randomized design. The data recorded on the insect mortality under laboratory conditions were subjected to statistical analysis, and the significance of the results was determined. The corrected percent mortality was analyzed using Abbott's formula [39]. The analysis of variance (ANOVA) was evaluated (three factorial analysis) using arcsine transformation on software (Op Stat) developed by HAU, Haryana. A Tukey's post hoc test was executed for pairwise comparisons between the EPNs and bacterial treatments.

3. Results

3.1. Entomopathogenic Nematode Occurrence in (Surveyed) Soils

For the soil sampling survey, a total of 31 soil samples were gathered from Chhupari, Rampur, Badiyara, Shimla, Fagu, and Jabbal and processed using the soil baiting technique with *G. mellonella* and *C. cephalonica* larvae (Table 1). Out of these 31 samples, only 5 were observed positive for EPNs, which included 4 *Heterorhabditis* and 1 *Steinernema* species based upon the infected cadaver appearance. The overall frequency of *Steinernema* sp. occurrence was 25% from an apple orchard in the Fagu region. Furthermore, the isolated *Steinernema* sp. strain EUPT-R2 was mass multiplied under laboratory conditions using *G. mellonella* and *C. cephalonica* larvae. The average infective juveniles (IJs) produced by *G. mellonella* were 66,036 (55,120–67,154) and *C. cephalonica* 48,320 (42,532–50,795) IJs/larva.

Table 1. Distribution and frequency of occurrence of entomopathogenic nematodes.

Location/ Villages	Vegetations	Total No. of Samples	Samples Having EPNs	Samples without EPNs	Frequency of Occurrence (%)
Chhupari	Peach, Apple, Pear, Cucumber and Plum	05	01	04	25
Rampur	Apple and Persimmon	04	-	04	-
Badiyara	Peach and Apple	06	01	05	16.67
Shimla	Apricot, Plum, Cherry, Apple, Pear, Peach	05	-	05	-
Fagu	Apple	05	01	04	25
Jabbal	Pear, Apple and Apricot	06	02	04	33.33

The isolated nematodes were identified on the basis of their morphological observations based on the species-specific diagnostic keys.

3.2. Morphological and Molecular Characterization of EPNs

The morphometrics of infective juveniles, adult males, and adult females corresponded with *Steinernema carpocapsae*, Weiser. The 10 individual infective juveniles, female and male, were observed morphologically. The IJs were much narrower, 508.00–598.00 µm in length, 23.64–26.58 µm in width, and the esophagus and intestinal region were collapsed. The nerve ring distance from the anterior end was 72.00–86.00 µm, the mouth was closed, the tail was pointed, and the lateral line fields were clearly visible. The excretory pore distance from the anterior end was 30.00–35.00 µm, the pharynx distance from the anterior end was 111.50–123.00 µm, and the tail length was 48.00–54.00 µm.

Females (n = 10) were somewhat larger in size, 3210.00–4060.00 µm in length and 183.00–211.00 µm, with a smooth cuticle and slightly rounded head, united lips, and a collapsed stoma. Pharynx distance from the anterior end was 180.00–196.00 µm, with a Rhabditoid esophagus. Anterior-to-basal bulb nerve ring was present surrounding the isthmus, and reflexed amphidelphic gonads and the vulval region showed protuberance. The tail was conical with a spiny tip and was 42.00–49.00 µm in length. Males (n = 10) are anteriorly similar to females, with a truncated, slightly rounded head, smooth cuticle, 1080.00–1658.00 µm in length, and 78.00–113.00 µm in width. The nerve ring distance from the anterior end was 93.00–124.00 µm, with a partially collapsed stoma, an absent collar, and the presence of a small amphid. The excretory pore distance from the anterior end was 48.00–72.00 µm, and the pharynx distance from the anterior end was 138.00–164.00 µm. The testis was reflexed, and the paired spicule was slightly curved and symmetrical. Spicule has a wider head, tapered gubernaculums, and a tail bearing genital papillae. The tail was 28.00–34.00 µmwithmucron and no bursa (Figure 1).

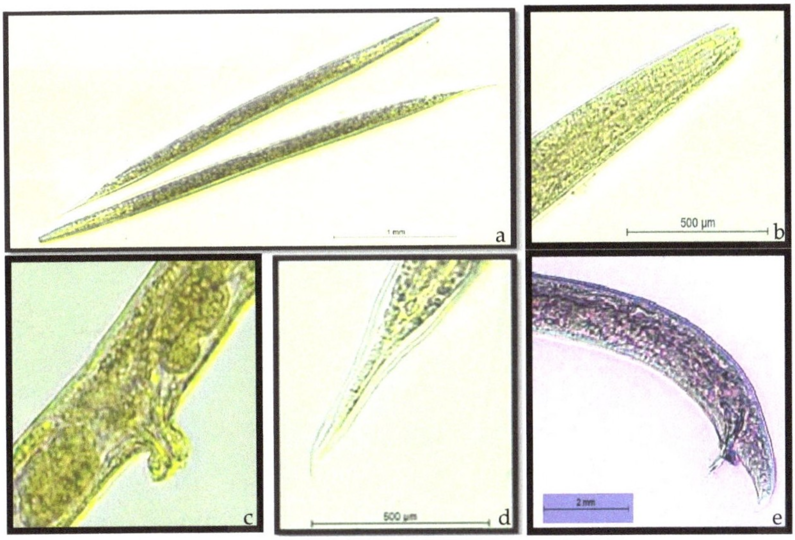

Figure 1. Morphological observations of *S. carpocapsae* Weiser, 1955 (**a**) Entire infective stage juvenile; (**b**) Anterior end of female; (**c**) Vulval region with epiptigma; (**d**) Female tail region; (**e**) Male tail region with mucron.

The results obtained through amplified PCR products showed a band size of 850 bp in the agarose gel electrophoresis run with a 100 bp ladder. The blast comparison based upon the ITS region amplification showed that the *Steinernema* sp. strain EUPT-R2 exhibited 99.38% sequence similarity and 99% query cover with *S. carpocapsae* (MK977607). The sequence was submitted to the gene bank with accession number OP295355. The calculated patterns of nucleotide substitution in the isolated taxon were T = 37.8; C = 16.3; A = 23.6; and G = 22.3 (Figure 2).

3.3. Identification and Molecular Characterization of the Endosymbiotic Bacteria

Initially, the identification of bacteria was carried out based on morphological characteristics. The bacterial colony appeared maroon on the blue agar, and, after Gram staining, it was found to be Gram-negative, rod-shaped, a facultative anaerobe, and motile. The genomic DNA of a bacterium was isolated and amplified using 16S and 18S gene sequencing. The blast comparison of 16S gene amplification in EUDPTB-R2 attributes 99.35% sequence similarity and 100% query cover with *X. nematophila* (MW619913). The sequence was submitted to the gene bank with accession number OQ842895. The calculated patterns of nucleotide substitution in the isolated taxon were T = 18.6; C = 22.1; A = 24.7; and G = 34.6. Based on gene amplification, the phylogenetic tree was constructed using the neighbor-joining method (Figure 3).

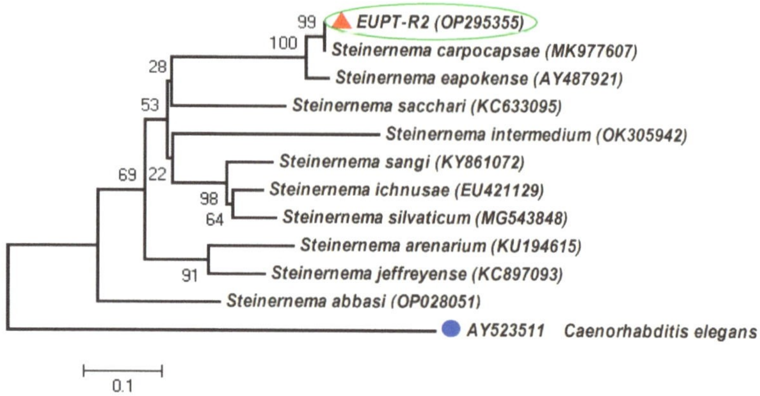

Figure 2. Phylogenetic relationships of nematode isolates with 10 isolated of *Steinernema* species based on ITS-rDNA sequences using neighbor joining (NJ) method. The *Caenorhabditis elegans* (AY523511) marked with blue circle was used as out group. The presented values at the nodes are in the form of bootstrap in ML. The green frame with red triangle denotes the nematode isolate identified during this study.

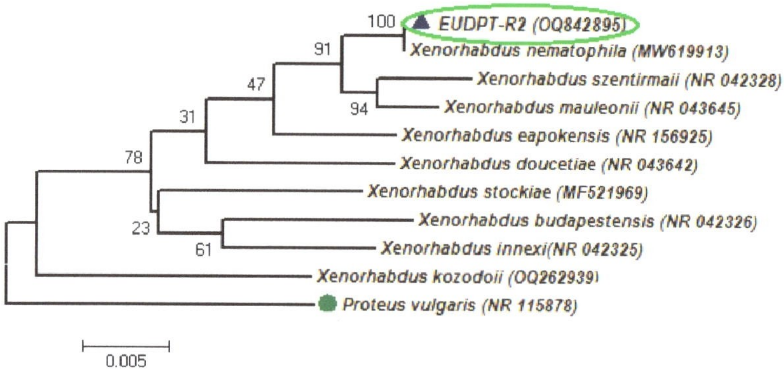

Figure 3. Phylogenetic relationships of bacterial isolates with 10 isolates of *Xenorhabdus* species based on 16S-rRNA gene sequences using neighbor joining (NJ) method. The *Proteus vulgaris* (NR 115878) marked with green circle was used as out group. The presented values at the nodes are in the form of bootstrap in ML. The green frame with blue triangle denotes the bacterial isolate identified during this study.

3.4. Entomopathogenic Capacity of Isolated EPN Species and Associated Bacterium

During the present investigation, the entomopathogenicity of EPNs (30, 60, 90, 120, and 150 IJs/mL) and their associated bacteria (1×10^4, 2×10^4, 3×10^4, 4×10^4, and 5×10^4 CFU/mL) was evaluated against the 3rd instar larvae of *P. brassicae*. A bioassay experiment showed the highest mortality (92%) was recorded at the highest nematode inoculum concentration (150 IJs/mL) after 96 h of infection (Figure 4). Considerable variations have been observed in insect mortality at various inoculums that vary from 45, 58, 64, 82 to 92% after 96 h of nematode inoculation (F =13.647; DF = 5; $p < 0.05$). A significant increase in larval mortality rate was recorded as exposure time increased.

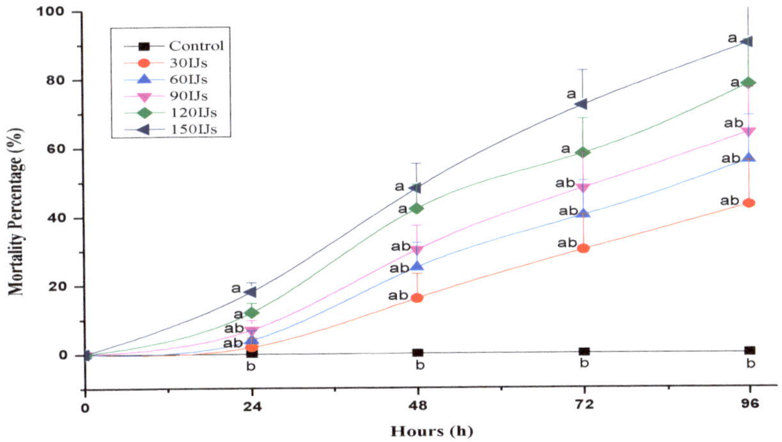

Figure 4. Pathogenicity caused by different concentrations of *Steinernema carpocapsae* in 3rd instar larvae of *Pieris brassicae;* Letters in front of bars showed all pairwise comparisons with statistical ranking from Tukey's post hoc test n = 5 for each nematode treatment. Error bars corresponds to the standard error.

The EPN-associated bacterium *X. nematophila* also caused significant mortality among insect larvae. The highest concentration of bacterial suspension (5×10^4 CFU/mL) resulted in 100% larval mortality within 48 h. A noticeable difference in the mortality rate was observed after the application of different concentrations of bacterial suspension. The mortality ranges vary from 94, 96, 100 to 100% after 96 h of bacterial inoculation ($F = 258.01$; $DF = 5$; $p < 0.05$). Amongst both of these treatments, applications of different concentrations of *X. nematophila* caused rapid infectivity, followed by insect mortality (Figure 5).

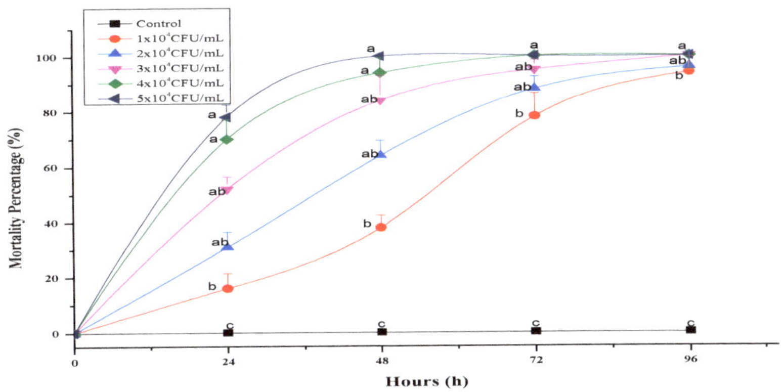

Figure 5. Pathogenicity caused by different concentrations of *Xenorhabdus nematophila* in 3rd instar larvae of *Pieris brassicae;* Letters in front of bars showed all pairwise comparisons with statistical ranking from Tukey's post hoc test n = 5 for each bacterial treatment. Error bars corresponds to the standard error.

4. Discussion

The isolation and identification of infinitesimal EPNs is quite difficult and thus necessitates greater attention during these processes. The isolation of EPNs and mass multiplication using the insect baiting technique has been mentioned by many researchers. During the present investigation, EPN *S. carpocapsae* was isolated and mass multiplied using *C.*

cephalonica and *G. mellonella* larvae, showing that the average infective juveniles (IJs) produced by *G. mellonella* were 66,036 (55,120–67,154) and *C. cephalonica* 48,320 (42,532–50,795) IJs/larva. The results were also supported by an earlier study, where 54,347, 44,697, and 36,860 IJs of *S. carpocapsae* were recorded on large, medium, and small-sized larvae of *C. cephalonica* [40]. The highest 71,532.0 IJs/larva were recovered during the in vivo mass multiplication of *S. carpocapsae* on *G. mellonella* larvae [41].

The taxonomic identification of genus level has been considered unstable due to overlapped morphological characteristics. With the advancement of molecular techniques, morphological observations can be accompanied with molecular observations that have been considered stable. The analysis of nucleotide sequences has also been effective for providing the phylogenetic relationship amongst the nematodes. On the basis of morphological and morphometrical analysis, the juveniles and adults of isolated EPNs resembled *S. carpocapsae* [42,43]. Earlier, *H. bacteriophora, Steinernema* sp. and *S. feltiae* were recovered from the apple orchards of Himachal Pradesh, India [44]. *Steinernema glaseri, Steinernema thermophilum,* and *H. indica* recovered from Meghalaya, India [45]. Based upon the amplification of the ITS region, the EPNs for *Steinernema* sp. strain EUPT-R2 attribute 99.38% sequence similarity with *S. carpocapsae.* Earlier, *H. bacteriophora* and *S. carpocapsae* were reported from Uttar Pradesh (India) using morphological and molecular analyses [46,47]. *S. carpocapsae* was identified on the basis of morphological and molecular characterization in Italy (Veneto) [48].

The nematodes genus *Steinernema* showed mutualistic alliance with the bacterial genus *Xenorhabdus.* The mutualism between both partners is highly specialized and both of them are mutually benefitted. Here, in this study, *S. carpocapsae*-associated bacteria, *X. nematophila,* was isolated and identified morphologically as well as molecularly. Similar observations were recorded by researchers in Cauca-Colombia [49]. *S. carpocapsae* (MK350941), *S. monticolum* (MK503674), and *Rhabditis blumi* (MN453373) were isolated and identified on the basis of 18S rDNA sequencing from the tropical and subtropical agro-ecosystems of Tamil Nadu, India [50]. The nematode-associated bacterial isolates *X. nematophila* were also identified morphologically, biochemically, and molecularly. Earlier, *X. nematophilus* from *S. carpocapsae* was isolated in Korea [51].

EPNs and their symbionts have been known for their virulence activities against insect pests since ancient times. During this investigation, a laboratory bioassay experiment was conducted for the estimation of the insecticidal potential of *S. carpocapsae* and *X. nematophila* against the 3rd instar larvae of *P. brassicae.* Amongst both of these treatments, applications of different concentrations of *X. nematophila* caused rapid infectivity followed by insect mortality. The observations of the current findings are supported by previous observations that showed 100% larval mortality in the early instars after 96 h [45,52–54]. It was also suggested that increased exposure time significantly increased larval mortality and vice versa [55]. In total, 72.08% and 85.38% larval mortality was recorded among 2nd and 4th instar larvae upon treatment with *H. bacteriophora* [56]. The mortality rate of 67.42 and 69.50% was recorded among the 2nd and 4th instars upon treatment with *S. feltiae.* High insect mortality was recorded among the 3rd and 4th instar larvae with the inoculum of 200 IJs, which is similar to the observations of this investigation [57]. In total, 100 percent mortality was recorded among all larval stages of the cabbage butterfly after exposure to *S. glaseri* and *H. bacteriophora* [58].

High mortality in *P. brassicae* larvae was recorded by several researchers upon treatment with native EPN isolates from Himachal Pradesh [56,59]. The virulence of *Xenorhabdus* sp. was also reported against mushroom mites [60]. EPNs and their associated bacteria were found to be highly effective in managing the 3rd and 4th larval instars of *Pieris rapae* and *Pentodonal gerinus* [61]. The *Xenorhabdus* and *Photorhabdus* bacteria (cell suspensions and cell-free supernatant) applied for the management of *Agrotis ipsilon* and good management potential was recorded [62]. The biocontrol potential of *Steinernema* sp. and their symbiontic bacteria, *Xenorhabdus* sp., was also found to be effective for the management of *Ephestia cautella* larvae [63].

5. Conclusions

Entomopathogens have been known to manage insect populations for years. *Steinernema carpocapsae* (OP295355) has been isolated from the apple (*Malus domestica*) orchards of the district of Shimla, Himachal Pradesh. The bacteria associated with this EPN species were recognized as *X. nematophila* (OQ842895). The laboratory bioassay experiments were performed using different concentrations of *S. carpocapsae* and *X. nematophila* against the 3rd instar larvae of *P. brassicae*. Among both of these treatments, applications of different concentrations of *X. nematophila* resulted in rapid infection and faster mortality when compared with EPNs. It can be concluded from the present investigation that the soil environment of the fruit orchards of Himachal Pradesh provides an excellent habitat to proliferate the entomopathogenic nematodes naturally in the environment. Further survey studies are required to explore nematode diversity and evaluate their biocontrol attributes in this region.

Author Contributions: N.T. gave the concept. P.T. performed the experiment, wrote the manuscript and did the statistical analysis. A.K.S. and B.A.L. helped in the molecular part (DNA isolation, PCR amplification). A.H., G.D.A.-Q. and E.F.A. also helped in the manuscript writing (review and editing). All authors have read and agreed to the published version of the manuscript.

Funding: The authors would like to extend their sincere appreciation to the Researchers Supporting Project Number (RSP2023R356), King Saud University, Riyadh, Saudi Arabia.

Data Availability Statement: Not applicable.

Acknowledgments: The study was carried out in the Zoology laboratory at Eternal University Baru Sahib, Himachal Pradesh. The authors are also thankful to Vice Chancellor, Eternal University, Baru Sahib and Zoological survey of India (HARC-Solan) for providing necessary laboratory facilities required to carry this investigation. The authors would like to extend their sincere appreciation to the Researchers Supporting Project Number (RSP2023R356), King Saud University, Riyadh, Saudi Arabia.

Conflicts of Interest: The authors declare no conflict of interest.

References

1. Dhaliwal, G.; Jindal, V.; Mohindru, B. Crop losses due to insect pests: Global and Indian scenario. *Indian J. Entomol.* **2015**, *77*, 165–168. [CrossRef]
2. Bradshaw, C.J.; Leroy, B.; Bellard, C.; Roiz, D.; Albert, C.; Fournier, A.; Barbet-Massin, M.; Salles, J.-M.; Simard, F.; Courchamp, F. Massive yet grossly underestimated global costs of invasive insects. *Nat. Commun.* **2016**, *7*, 12986. [CrossRef] [PubMed]
3. *IPCC (Intergovernmental Panel on Climate Change) Working Group I Contribution to the Fifth Assessment Report of the Intergovernmental Panel on Climate Change*; U.C.U.P.: Cambridge, UK, 2013.
4. Thackeray, S.J.; Henrys, P.A.; Hemming, D.; Bell, J.R.; Botham, M.S.; Burthe, S.; Helaouet, P.; Johns, D.G.; Jones, I.D.; Leech, D.I. Phenological sensitivity to climate across taxa and trophic levels. *Nature* **2016**, *535*, 241–245. [CrossRef] [PubMed]
5. Brown, K.; Phillips, C.; Broome, K.; Green, C.; Toft, R.; Walker, G. Feasibility of eradicating the large white butterfly (*Pieris brassicae*) from New Zealand: Data gathering to inform decisions about the feasibility of eradication. In *Island Invasives: Scaling Up to Meet the Challenge*; Veitch, C.R., Clout, M.N., Martin, A.R., Russell, J.C., West, C.J., Eds.; Occasional Paper SSC; IUCN: Gland, Switzerland, 2019; Volume 62, pp. 364–369.
6. Feltwell, J. *Large White Butterfly: The Biology, Biochemistry and Physiology of Pieris brassicae (Linnaeus)*; Springer Science & Business Media: Dordrecht, The Netherlands, 2012; Volume 18.
7. Abbas, W.; Javed, N.; Haq, I.-U.; Ahmed, S. Virulence potential of two entomopathogenic nematodes, their associated bacteria, and its metabolites to larvae of *Pieris brassicae* L.(Lepidoptera, Pieridae) in cabbage under greenhouse and field bioassays. *Int. J. Trop. Insect Sci.* **2022**, *42*, 557–563. [CrossRef]
8. Meszka, B.; Broniarek-Niemiec, A.; Bielenin, A. The status of dodine resistance of *Venturia inaequalis* populations in Poland. *Phytopathol. Pol.* **2008**, *47*, 57–61.
9. Matson, M.E.; Small, I.M.; Fry, W.E.; Judelson, H.S. Metalaxyl resistance in *Phytophthora* infestans: Assessing role of RPA190 gene and diversity within clonal lineages. *Phytopathology* **2015**, *105*, 1594–1600. [CrossRef]
10. Singh, A.; Shukla, N.; Kabadwal, B.; Tewari, A.; Kumar, J. Review on plant-Trichoderma-pathogen interaction. *Int. J. Curr. Microbiol. Appl. Sci.* **2018**, *7*, 2382–2397. [CrossRef]
11. Gozel, U.; Gozel, C. Entomopathogenic nematodes in pest management. In *Integrated Pest Management (IPM): Environmentally Sound Pest Management*; Gill, H.K., Goyal, G., Eds.; IntechOpen: London, UK, 2016; Volume 55.

12. Mracek, Z. Use of entomoparasitic nematodes (EPANs) in biological control. In *Advances in Microbial Control of Insect Pests*; Upadhyay, R.K., Ed.; Springer: Boston, MA, USA, 2003; pp. 235–261.
13. Poinar, G.O., Jr. Description and biology of a new insect parasitic Rhabditoid, *Heterorhabditis bacteriophora* N. Gen., N. Sp.(Rhabditida; Heterorhabditidae N. Fam.). *Nematologica* **1975**, *21*, 463–470. [CrossRef]
14. Chitwood, B.G.; Chitwood, M.B. An introduction to nematology. *Nature* **1937**, *139*, 654.
15. Bhat, A.H.; Chaubey, A.K.; Askary, T.H. Global distribution of entomopathogenic nematodes, *Steinernema* and *Heterorhabditis*. *Egypt. J. Biol. Pest Control* **2020**, *30*, 31. [CrossRef]
16. Tomar, P.; Thakur, N.; Yadav, A.N. Endosymbiotic microbes from entomopathogenic nematode (EPNs) and their applications as biocontrol agents for agro-environmental sustainability. *Egypt. J. Biol. Pest Control* **2022**, *32*, 80. [CrossRef]
17. Machado, R.A.; Wüthrich, D.; Kuhnert, P.; Arce, C.C.; Thönen, L.; Ruiz, C.; Zhang, X.; Robert, C.A.; Karimi, J.; Kamali, S. Whole-genome-based revisit of *Photorhabdus* phylogeny: Proposal for the elevation of most *Photorhabdus* subspecies to the species level and description of one novel species *Photorhabdus bodei* sp. nov., and one novel subspecies *Photorhabdus laumondii* subsp. clarkei subsp. nov. *Int. J. Syst. Evol. Microbiol.* **2018**, *68*, 2664–2681.
18. Sajnaga, E.; Kazimierczak, W. Evolution and taxonomy of nematode-associated entomopathogenic bacteria of the genera *Xenorhabdus* and *Photorhabdus*: An overview. *Symbiosis* **2020**, *80*, 1–13. [CrossRef]
19. Singh, A.K.; Kumar, M.; Ahuja, A.; Vinay, B.; Kommu, K.K.; Thakur, S.; Paschapur, A.U.; Jeevan, B.; Mishra, K.; Meena, R.P. Entomopathogenic nematodes: A sustainable option for insect pest management. In *Biopesticides*; Rakshit, A., Meena, V.S., Abhilash, P.C., Sarma, B.K., Singh, H.B., Fraceto, L., Parihar, M., Singh, A.K., Eds.; Elsevier: Amsterdam, The Netherlands, 2022; pp. 73–92.
20. Georgis, R.; Koppenhöfer, A.; Lacey, L.; Bélair, G.; Duncan, L.; Grewal, P.; Samish, M.; Tan, L.; Torr, P.; Van Tol, R. Successes and failures in the use of parasitic nematodes for pest control. *Biol. Control.* **2006**, *38*, 103–123. [CrossRef]
21. Thakur, N.; Tomar, P.; Kaur, S.; Kumari, P. Virulence of native entomopathogenic nematodes against major lepidopteran insect species of tomato (*Solanum lycopersicum* L.). *J. Appl. Biol. Biotechnol.* **2022**, *10*, 6–14. [CrossRef]
22. Tomar, P.; Thakur, N.; Yadav, A.N. Indigenous entomopathogenic nematode as biocontrol agents for insect pest management in hilly regions. *Plant Sci. Today* **2021**, *8*, 51–59. [CrossRef]
23. Kepenekci, I.; Hazir, S.; Lewis, E.E. Evaluation of entomopathogenic nematodes and the supernatants of the in vitro culture medium of their mutualistic bacteria for the control of the root-knot nematodes *Meloidogyne incognita* and *M. arenaria*. *Pest Manag. Sci.* **2016**, *72*, 327–334. [CrossRef] [PubMed]
24. Snyder, H.; Stock, S.P.; Kim, S.-K.; Flores-Lara, Y.; Forst, S. New insights into the colonization and release processes of *Xenorhabdus nematophila* and the morphology and ultrastructure of the bacterial receptacle of its nematode host, *Steinernema carpocapsae*. *Appl. Environ. Microbiol.* **2007**, *73*, 5338–5346. [CrossRef]
25. Thakur, N.; Kaur, S.; Tomar, P.; Thakur, S.; Yadav, A.N. Microbial biopesticides: Current status and advancement for sustainable agriculture and environment. In *New and Future Developments in Microbial Biotechnology and Bioengineering*; Rastegari, A.A., Yadav, A.N., Yadav, N., Eds.; Elsevier: Amsterdam, The Netherlands, 2020; pp. 243–282.
26. Thakur, N.; Tomar, P.; Kaur, S.; Jhamta, S.; Thakur, R.; Yadav, A.N. Entomopathogenic soil microbes for sustainable crop protection. In *Soil Microbiomes for Sustainable Agriculture: Functional Annotation*; Yadav, A.N., Ed.; Springer: Cham, Switzerland, 2021; pp. 529–571.
27. Thakur, N.; Tomar, P.; Sharma, S.; Kaur, S.; Sharma, S.; Yadav, A.N.; Hesham, A.E.-L. Synergistic effect of entomopathogens against *Spodoptera litura* (Fabricius) under laboratory and greenhouse conditions. *Egypt. J. Biol. Pest Control* **2022**, *32*, 39. [CrossRef]
28. Orozco, R.A.; Lee, M.-M.; Stock, S.P. Soil sampling and isolation of entomopathogenic nematodes (Steinernematidae, Heterorhabditidae). *J. Vis. Exp.* **2014**, *89*, e52083.
29. Bedding, R.; Akhurst, R. A simple technique for the detection of insect paristic rhabditid nematodes in soil. *Nematologica* **1975**, *21*, 109–110. [CrossRef]
30. Nickle, W.R. *Manual of Agricultural Nematology*; CRC Press: Boca Raton, FL, USA, 2020.
31. White, G. A method for obtaining infective nematode larvae from cultures. *Science* **1927**, *66*, 302–303. [CrossRef]
32. Norton, D.C. *Ecology of Plant Parasitic Nematode*; John Willey and Sons: New York, NY, USA, 1978.
33. Seinhorst, J. Killing nematodes for taxonomic study with hot fa 4: 1. *Nematologica* **1966**, *12*, 178. [CrossRef]
34. Hominick, W.; Briscoe, B.; del Pino, F.G.; Heng, J.; Hunt, D.; Kozodoy, E.; Mracek, Z.; Nguyen, K.; Reid, A.; Spiridonov, S. Biosystematics of entomopathogenic nematodes: Current status, protocols and definitions. *J. Helminthol.* **1997**, *71*, 271–298. [CrossRef] [PubMed]
35. De Man, J.G. *Die, frei in der reinen Erde und im Süssen Wasser lebenden Nematoden der Niederländischen Fauna: Eine Systematisch-faunistische Monographie*; EJ Brill: Leiden, The Netherlands, 1884; Volume 1.
36. Thakur, N.; Tomar, P.; Kaur, J.; Kaur, S.; Sharma, A.; Jhamta, S.; Yadav, A.N.; Dhaliwal, H.S.; Thakur, R.; Thakur, S. Eco-friendly management of *Spodoptera litura* (Lepidoptera: Noctuidae) in tomato under polyhouse and field conditions using *Heterorhabditis bacteriophora* Poinar, their associated bacteria (*Photorhabdus luminescens*), and *Bacillus thuringiensis* var. kurstaki. *Egypt. J. Biol. Pest Control* **2023**, *33*, 7.
37. Tamura, K.; Dudley, J.; Nei, M.; Kumar, S. MEGA4: Molecular evolutionary genetics analysis (MEGA) software version 4.0. *Mol. Biol. Evol.* **2007**, *24*, 1596–1599. [CrossRef]

38. Tomar, P.; Thakur, N.; Sharma, A. Infectivity of entomopathogenic nematode against the cabbage butterfly (*Pieris brassicae* L.) in polyhouse and in field condition. *Egypt. J. Biol. Pest Control* **2022**, *32*, 38. [CrossRef]
39. Abbott, W.S. A method of computing the effectiveness of an insecticide. *J. Econ. Entomol.* **1925**, *18*, 265–267. [CrossRef]
40. Dhaliwal, A. Biocontrol of Maize Stem Borer (*Chilo partellus*) Using Entomopathogenic Nematodes. Master Thesis, Submitted to Maharana Pratap University of Agriculture & Technology (MPUAT), Udaipur, India, 2006.
41. Chand, P.; Parihar, A.; Maru, A.K.; Sharma, S. Mass production (in vivo) of the entomopathogenic nematode, *Steinernema carpocapsae* on greater wax moth, *Galleria mellonella* and rice moth, *Corcyra cephalonica*. *Biodiversitas J. Biol. Divers.* **2019**, *20*, 1344–1349.
42. Weiser, J. *Neoaplectana carpocapsae* n. sp. (Anguillulata, Steinernematinae), novy cizopasník housenek obalece jablecného, *Carpocapsa pomonella* L. *Vestn. Ceskoslovenske Spol. Zool.* **1955**, *19*, 44–52.
43. Poinar, G. Description and taxonomic position of DD-136 nematode (Steinernematidae Rhabditoidea) and its relationship to *Neoaplectana carpocapsae* Weiser. *Proc. Helminthol. Soc. Wash.* **1967**, *34*, 199.
44. Singh, M.; Gupta, P. Occurrence of entomopathogenic nematodes in Himachal Pradesh, India and their pathogenicity against various insect species. *Pest Manag. Econ. Zool.* **2006**, *14*, 179–189.
45. Lalramliana, Y.A.; Kumar, A. Occurrence of entomopathogenic nematodes (Rhabditida: Steinernematidae and Heterorhabditidae) in Meghalaya, NE India. *Sci. Vis.* **2010**, *10*, 89–100.
46. Rana, A.; Bhat, A.H.; Chaubey, A.K.; Shokoohi, E.; Machado, R.A. Morphological and molecular characterization of *Heterorhabditis bacteriophora* isolated from Indian soils and their biocontrol potential. *Zootaxa* **2020**, *4878*, 77–102. [CrossRef] [PubMed]
47. Pervez, R.; Eapen, S.J.; Devasahayam, S.; Jacob, T. Natural occurrence of entomopathogenic nematodes associated with ginger (*Zingiber officinale* Rosc.) ecosystem in India. *Indian J. Nematol.* **2014**, *44*, 238–246.
48. Torrini, G.; Landi, S.; Benvenuti, C.; De Luca, F.; Fanelli, E.; Troccoli, A.; Tarasco, E.; Bazzoffi, P. Morphological and molecular characterization of a *Steinernema carpocapsae* (Nematoda Steinernematidae) strain isolated in Veneto region (Italy). *Redia* **2014**, *97*, 89–94.
49. Neira-Monsalve, E.; Wilches-Ramírez, N.C.; Terán, W.; del Pilar Márquez, M.; Mosquera-Espinosa, A.T.; Sáenz-Aponte, A. Isolation, identification, and pathogenicity of and its bacterial symbiont in Cauca-Colombia. *J. Nematol.* **2020**, *52*, e2020-89. [CrossRef] [PubMed]
50. Lalitha, K.; Venkatesan, S.; Balamuralikrishnan, B.; Shivakumar, M.S. Isolation and biocontrol efficacy of entomopathogenic nematodes *Steinernema carpocapsae*, *Steinernema monticolum* and *Rhabditis blumi* on lepidopteran pest *Spodoptera litura*. *Biocatal. Agric. Biotechnol.* **2022**, *39*, 102291. [CrossRef]
51. Park, Y.; Kim, Y.; Yi, Y. Identification and characterization of a symbiotic bacterium associated with *Steinernema carpocapsae* in Korea. *J. Asia-Pac. Entomol.* **1999**, *2*, 105–111. [CrossRef]
52. Mantoo, M.A.; Zaki, F. Biological control of cabbage butterfly, *Pieris brassicae*, by a locally isolated entomopathogenic nematode, *Heterorhabditis bacteriophora* SKUASTK-EPN-Hr-1 in Kashmir. *SKUAST J. Res.* **2014**, *16*, 66–70.
53. Gorgadze, O.; Fanelli, E.; Lortkhipanidze, M.; Troccoli, A.; Burjanadze, M.; Tarasco, E.; De Luca, F. *Steinernema borjomiense* n. sp.(Rhabditida: Steinernematidae), a new entomopathogenic nematode from Georgia. *Nematology* **2018**, *20*, 653–669. [CrossRef]
54. Tomar, P.; Thakur, N. Biocidal potential of indigenous isolates of Entomopathogenic Nematodes (EPNs) against tobacco cutworm, *Spodoptera litura* Fabricius (Lepidoptera: Noctuidae). *Egypt. J. Biol. Pest Control* **2022**, *32*, 107. [CrossRef]
55. Sabry, A.; Metwallya, H.; Abolmaatyb, S. Compatibility and efficacy of entomopathogenic nematode, *Steinernema carpocapsae* all alone and in combination with some insecticides against *Tuta absoluta*. *Der. Pharm. Let.* **2016**, *8*, 311–315.
56. Kasi, I.K.; Singh, M.; Waiba, K.M.; Monika, S.; Waseem, M.; Archie, D.; Gilhotra, H. Bio-efficacy of entomopathogenic nematodes, *Steinernema feltiae* and *Heterorhabditis bacteriophora* against the Cabbage butterfly (*Pieris brassicae* [L.]) under laboratory conditions. *Egypt. J. Biol. Pest Control* **2021**, *31*, 125. [CrossRef]
57. Askary, T.H.; Ahmad, M.J. Efficacy of entomopathogenic nematodes against the cabbage butterfly (*Pieris brassicae* (L.)(Lepidoptera: Pieridae) infesting cabbage under field conditions. *Egypt. J. Biol. Pest Control* **2020**, *30*, 39. [CrossRef]
58. Abbas, W.; Javed, N.; Haq, I.U.; Ahmed, S. Pathogenicity of Entomopathogenic nematodes against cabbage butterfly (*Pieris brassicae*) Linnaeus (Lepidoptera: Pieridae) in laboratory conditions. *Int. J. Trop. Insect Sci.* **2021**, *41*, 525–531. [CrossRef]
59. Tomar, P.; Thakur, N. Isolation and evaluation of *Heterorhabditis bacteriophora* strain-S26 as biocontrol agents against *Pieris brassicae* L. under laboratory conditions. *Indian J. Nematol.* **2022**, *52*, 49–58. [CrossRef]
60. Sobanboa, S.; Bussaman, P.; Chandrapatya, A. Efficacy of *Xenorhabdus* sp.(X1) as biocontrol against for controlling mushroom mites (*Luciaphorus* sp.). *Asian J. Food Agro-Ind.* **2009**, *2*, S145–S154.
61. Elbrense, H.; Elmasry, A.M.; Seleiman, M.F.; Al-Harbi, M.S.; Abd El-Raheem, A.M. Can symbiotic bacteria (*Xenorhabdus* and *Photorhabdus*) be more efficient than their entomopathogenic nematodes against *Pieris rapae* and *Pentodon algerinus* larvae? *Biology* **2021**, *10*, 999. [CrossRef]

62. Ünal, M.; Yüksel, E.; Canhilal, R. Biocontrol potential of cell suspensions and cell-free superntants of different *Xenorhabdus* and *Photorhabdus* bacteria against the different larval instars of *Agrotis ipsilon* (Hufnagel) (Lepidoptera: Noctuidae). *Exp. Parasitol.* **2022**, *242*, 108394. [CrossRef]
63. Yüksel, E.; Ormanoğlu, N.; İmren, M.; Canhilal, R. Assessment of biocontrol potential of different *Steinernema* species and their bacterial symbionts, *Xenorhabdus* species against larvae of almond moth, *Ephestia cautella* (Walker). *J. Stored Prod. Res.* **2023**, *101*, 102082. [CrossRef]

horticulturae

MDPI

Article

Two Bacterial Bioagents Boost Onion Response to *Stromatinia cepivora* and Promote Growth and Yield via Enhancing the Antioxidant Defense System and Auxin Production

Hanan E. M. Osman [1,†], Yasser Nehela [2,*,†], Abdelnaser A. Elzaawely [2], Mohamed H. El-Morsy [3,4] and Asmaa El-Nagar [2]

1 Biology Department, Faculty of Applied Science, Umm-Al-Qura University, Makkah 21955, Saudi Arabia
2 Department of Agricultural Botany, Faculty of Agriculture, Tanta University, Tanta 31527, Egypt
3 Deanship of Scientific Research, Umm Al-Qura University, Makkah 24243, Saudi Arabia
4 Plant Ecology and Range Management Department, Desert Research Center, Cairo 11753, Egypt
* Correspondence: yasser.nehela@agr.tanta.edu.eg
† These authors contributed equally to this work.

Abstract: White rot, caused by *Stromatinia cepivora* (Anamorph: *Sclerotium cepivorum* Berk), is a serious soil-borne disease of the onion that restricts its cultivation and production worldwide. Herein, we isolated and characterized a plant growth-promoting rhizobacterium *Stenotrophomonas maltophilia* from healthy onion roots and an endophytic bacterium *Serratia liquefaciens* from healthy bean leaves. Both isolates showed strong fungistatic activity against *S. cepivora* using the dual culture and culture filtrate methods. This effect might be due to the presence of several volatile compounds, especially menthol in both culture filtrates as shown with a GC-MS analysis. Additionally, the root drench application of cell-free culture filtrates of *S. maltophilia* and *S. liquefaciens* significantly reduced the incidence and severity of white rot disease on treated onion plants, which was associated with the activation of both enzymatic (POX and PPO) and non-enzymatic (phenolics and flavonoids) antioxidant defense machineries of *S. cepivora*-infected onion plants. Moreover, the culture filtrates of both bacterial bioagents remarkably enhanced the growth (as expressed by root length, plant height, and number of leaves) and yield parameters (as indicated by bulb circumference, fresh weight of the bulb, and bulb yield per plot) of treated onion plants under field conditions during two successive seasons (2020/2021 and 2021/2022). This might be because of a reduced disease severity and/or the accumulation of the main auxin, indole-3-acetic acid (IAA), and its precursor, the amino acid tryptophan. Our findings suggest that both bioagents might be utilized as eco-friendly alternative control measures to reduce the utilization of chemical fungicides entirely or partially for the safer production of onion in *S. cepivora*-infested soils.

Keywords: onion; biocontrol; white rot; *Sclerotium*; *Stromatinia*; *Stenotrophomonas*; *Serratia*; dual culture; cell-free culture filtrate; menthol; IAA

1. Introduction

An onion (*Allium cepa* L.) is one of the oldest vegetable crops known to humans, as its cultivation dates back more than 5000 years [1], and, nowadays, it is one of the most cultivated vegetables all over the world. It is an important crop grown in Egypt and worldwide, whether for local consumption or exportation. Onions are rich in chemical compounds that have medical importance, such as flavonoids [2]. The global production of onions reached 106.59 million tons and the harvested area was 5.78 million hectares in 2021 [3]. Additionally, Egypt cultivated 94,457 hectares of onions, yielding 35.0684 tons per hectare, with a total production of 3.31 million tons [3]. An onion is attacked by several phytopathogens such as viruses, bacteria, fungi, and nematodes, which cause serious diseases and economic losses [4]. However, white rot disease caused by *Sclerotium cepivorum* Berk (teleomorph: *Stromatinia cepivora*) is the most serious disease and a limiting factor for onion production worldwide [5]. *S. cepivora* is a soil-borne fungus that can attack onion roots from seedlings to the harvest stage, resulting in plant death during a season [6].

The ascomycetous *S. cepivora* is a necrotrophic fungus that can infect numerous susceptible *Allium* species, particularly onions, leeks, and garlic, causing white rot disease. Although *S. cepivora* does not produce any known asexual spores, it survives and overwinters as sclerotia (the survival stage) [7]. Sclerotia are small, black, globular, solid structures that are produced by the pathogen at the end of its life cycle and can resist unfavorable environmental conditions and remain dormant in the soil for years even without a host [7]. It was reported previously that an unpredictable quantity of sclerotia produced naturally on infected hosts might decay shortly after the formation for unknown reasons; however, sclerotia that survive beyond the decay period are probable to persist viably for 20 years in the soil under field conditions in the absence of host plants [8]. Although sclerotia can spread from field to field through unsuccessful sanitation practices such as soil movement and/or contaminated water, only *Allium* root exudates stimulate sclerotia germination, which remains as the host range limited to *Allium* species [9].

Management of onion white rot is incredibly challenging and requires multi-pronged strategies due to the above-listed characteristics of sclerotia. These strategies include but are not limited to cultural controls such as crop rotation with nonhost plants, sanitation such as soil solarization [10], biological control using sclerotia germination stimulants, composted onion waste [11], or fungal and bacterial bioagents [12,13]. Although an individual control method does not give the desired level of disease management, chemical fungicides are considered the most effective means used to control this disease [14]. Despite the effectiveness of synthetic fungicides, the extensive and repeated use of these fungicides results in several environmental problems and has undesirable effects on humans, animals, and non-targeted microorganisms. Moreover, fungicides might disrupt natural biological systems via the development of fungicide-resistant fungal strains [15,16]. Consequently, it is necessary to prompt an intensive search for cheap eco-friendly alternatives that are safer for humans, animals, and the environment.

Biological control of soil-borne phytopathogens is a potential alternative to reduce the use of hazardous chemical fungicides entirely or partially. Biological control is one of the most sustainable alternative methods in controlling plant diseases due to its low cost and environmental friendliness [17,18]. Direct mechanisms/mode-of-actions of biological control against soil-borne phytopathogenic fungi involve, but are not limited to, antibiosis, cross-protection, hyperparasitism, predation, soil amendments, induced systemic resistance (ISR), as well as competition for a site and nutrient [17,18]. Moreover, plant growth-promoting rhizobacteria (PGPR) are considered one of the most important biological control agents for plant diseases that colonize the roots of many plants [19]. *Stenotrophomonas maltophilia* is a widespread PGPR that is isolated from the rhizosphere of many plants [20] such as cruciferous, corn, and beets [21]. Additionally, Garbeva et al. found that *S. maltophilia* colonizes and persists inside the tissues of potato plants [22]. *S. maltophilia* was effective in inhibiting the mycelial growth of *Rhizoctonia solani*, *Verticillium dahliae* [23], *Sclerotium rolfsii* [24], and *Pyricularia oryzae* (sexual morph *Magnaporthe oryzae*) [25]. Moreover, it had an antagonistic effect against *Ralstonia solanacearum* in vitro and on potato plants under greenhouse conditions [20].

On the other hand, endophytic bacteria could be isolated from different parts of a plant including roots, stems, leaves, flowers, fruits, and seeds, and be used in biological control for plant diseases [26]. Endophytic bacteria colonize the internal tissues of plants without causing any negative effects or disease symptoms [27]. In addition, they were identified as effective biological control agents for a variety of plant diseases [26]. *Serratia liquefaciens* is an endophytic bacterium isolated from *Pinellia ternata* for the first time by Liu et al. [28]. Additionally, *S. liquefaciens* and *S. proteamaculans* were investigated for their antifungal activity against several phytopathogenic fungi including *Sclerotinia sclerotiorum Fusarium oxysporum*, *Rhizoctonia solani*, *Botrytis cinerea*, *and Alternaria alternata* by Michail et al. [29].

In the current study, we isolated and characterized *S. maltophilia* and *S. liquefaciens*, and investigated their potential antifungal activity against *S. cepivora*, the causal agent of onion white rot disease, in vitro and in vivo under field conditions. We hypothesized that both bioagents may affect the growth and yield components of *S. cepivora*-infected onion plants. Moreover, we suggest that the plant growth-promoting properties of both bioagents are correlated with the activation of enzymatic and non-enzymatic antioxidant defense machinery and the accumulation of phytohormones such as auxins. Gas chromatography–mass spectrometry (GC-MS) was used to identify the chemical composition of culture filtrates of *S. maltophilia* and *S. liquefaciens* to better understand the function of these bacterial secretions, as well as their involvement in suppressing the growth of *S. cepivora* and reducing the disease severity of onion white rot.

2. Materials and Methods

2.1. Isolation of the Causal Agent of Onion White Rot Disease

The causal agent *S. cepivora* was isolated from onion bulbs and roots that exhibited typical symptoms of white rot disease. Samples were collected from different sites at EL-Gharbia governorate Egypt. Isolates of *S. cepivora* were obtained by scraping the mycelium or sclerotia from onion bulbs and roots then placed on a Potato Dextrose Agar (PDA) medium and incubated at 20 ± 2 °C for 7 days. Isolates were initially identified based on their cultural, morphological, and macroscopic characteristics and then tested for their pathogenicity.

2.1.1. Pathogenicity Test

The pathogenicity of seven isolates of *S. cepivora* was tested on 50-day-old healthy onion seedlings of the susceptible cultivar Giza 20. To prepare the fungal inoculum, 500 mL glass bottles containing 100 g of barley grains and 50 mL of water were autoclaved; five discs of each isolate were inoculated into the barley medium and incubated at 20 ± 2 °C for 30 days. Plastic pots (30 cm) were filled with sterilized sand–clay soil 1:1 (v/v) and infected with fungal inoculum 14 days before transplanting at a rate of 15 $g{\cdot}kg^{-1}$. Plants were watered as needed and other agricultural practices were conducted as recommended. Onions were uprooted at 150 days post transplanting (dpt) to evaluate the disease incidence and severity. The disease incidence was assessed by counting onion plants that had a visible white rot mycelium and/or sclerotia on the roots and scored as white rot-infected. The disease incidence (%) was calculated using Equation (1):

$$\text{Disease incidence } (\%) = \frac{\text{Number of infected plants}}{\text{Number of total plants}} \times 100 \quad (1)$$

Whereas the severity of onion white rot disease was evaluated according to a 5-degree symptom scale according to Tian and Bertolini [30] with slight modifications as follows: 0: Healthy; 1: Bulb covered with mycelium but not rotted; 2: Bulb covered with mycelium, and 1–25% of the bulb rotted; 3: Bulb covered with mycelium, 25–50% of the bulb rotted; 4: Bulb covered with mycelium, 50–75% of the bulb rotted; and 5: Bulb covered with mycelium, 75–100% of the bulb rotted.

Disease severity scores were converted into percentages according to Equation (2) as described by Zewide et al. [31] as follows:

$$\text{Disease severity } (\%) = \frac{\text{Total of all ratings}}{\text{Total number of plants} \times \text{maximum score}} \times 100 \quad (2)$$

Based on the pathogenicity test, isolate #4 was the most aggressive isolate and showed the highest pathogenicity among the tested isolates, so it was selected for molecular identification based on the sequence of its internal transcribed spacer (ITS) region. Moreover, it was selected for all subsequent experiments.

2.1.2. Molecular Identification of *S. cepivora*

The most aggressive isolate of *S. cepivora* (isolate #4) was subjected to molecular identification [32,33]. Briefly, this isolate was grown on a sterilized potato dextrose broth (PDB) and incubated at 20 ± 2 °C for 10 days. Subsequently, the mycelium and sclerotia were collected and filtered using cheesecloth, washed twice with sterilized deionized water, and dried using filter paper. Approximately 0.1 g of the mycelium and sclerotia were ground to a fine powder using liquid nitrogen. The total DNA of the pathogenic fungus was extracted using a Quick-DNA™ Fungal/Bacterial Miniprep Kit according to the manufacturer's instructions and then purified, and the targeted sequences of the ITS region (ITS-5.8S rDNA) were amplified using PCR. The purified PCR products were sent for sequencing (Aoke Dingsheng Biotechnology Co., Beijing, China). Sanger sequencing was used to perform the two-directional sequencing of the ITS-5.8S rDNA sequences. DNABASER software (Heracle BioSoft S.R.L., Arges, Romania) was used to process and assemble consensus sequences. Subsequently, a Nucleotide-Nucleotide Basic Local Alignment Search Tool (BLASTn) was used to compare the assembled sequence with the most recent available data in GenBank and the national center for biotechnology information website (NCBI, http://www.ncbi.nlm.nih.gov/gene/; accessed on 7 February 2023).

2.1.3. Phylogenetic Analysis

An evolutionary analysis and phylogenetic trees were constructed using the assembled sequence of the ITS-5.8S rDNA based on the Maximum Likelihood method and Tamura-Nei model [34] using Molecular Evolutionary Genetics Analysis–Version 11 (MEGA 11) software using 500 bootstrap replications [35]. In addition to the query sequence, about 20 reference strains/isolates (Tables S1–S3 in Supplementary Material) were selected and used for multiple sequence alignment using ClustalW multiple sequence alignment algorithms.

2.2. Isolation of Plant Growth-Promoting Rhizobacterium and Endophytic Bacterium

A plant growth-promoting rhizobacterium (PGPR) was isolated from the rhizosphere soil of healthy onion plants growing next to diseased plants using the plate of soil dilution method as described by [36]. The soil attached to healthy plants' roots was removed, collected, air dried, and mixed well. In total, 10 g of homogenized soil was placed in conical flasks, 100 mL of sterilized distilled water (SDW) was added—well shaken for 10 min (150 rpm at 30 °C)—and serial dilutions were prepared. Subsequently, 100 µL of the last dilution was spread on surfaces of Petri dishes containing a nutrient agar (NA) medium with a sterilized Drigalski glass triangle; the dishes were incubated at 28 ± 2 °C for 48 h. Growth in incubated dishes was examined and purified using a single colony.

Furthermore, an endophytic bacterium was isolated from fresh healthy leaves of common bean plants. Firstly, leaves were washed with tap water, air dried, and surface sterilized in 70% ethanol for 3 min, then in 2% sodium hypochlorite for 2 min, and finally rinsed with SDW 3 times. The samples were crushed in a sterilized mortar with 5 mL of an aqueous saline solution of 0.9% NaCl. In total, 100 µL of a suspension was spread on the surface of the Petri dishes containing the nutrient agar (NA) medium according to [37]; the dishes were incubated at 30 °C for 48 h.

Molecular Identification of *S. maltophilia* and *S. liqufaciens*

The tested isolates of bacteria were subjected to molecular identification through 16S rRNA sequencing. Total genomic DNA was extracted from a 2-day-old culture, using a Quick-DNA™ Fungal/Bacterial Miniprep Kit according to the manufacturer's instructions. The PCR amplification and sequencing of the DNA extracts were performed. The PCR products were examined using agarose gel electrophoresis, and products that had shape bands were sent to a sequencing company (Aoke Dingsheng Biotechnology Co., Beijing,

China) for sequencing. BLASTn was used to compare the raw fast sequence data with the NCBI nucleotide sequence database after being processed as described above.

2.3. In Vitro Antifungal Activity of S. maltophilia and S. liqufaciens

2.3.1. Dual Culture Assay

The dual culture assay was used to investigate the in vitro antifungal activity of *S. maltophilia* and *S. liqufaciens* isolates against *S. cepivora*. Briefly, bacterial isolates were separately streaked at 1 cm on one side of the outer edge of the Petri dishes containing a PDA medium and incubated at 28 ± 2 °C for 48 h. Then, a 6 mm mycelial disc of *S. cepivora* obtained from a freshly growing culture was placed on the other side of the plates. Plates inoculated with *S. cepivora* discs from one side and streaked with a line of sterile deionized water on the other side were used as a negative control. Likewise, Folicure 25% EC (Tebuconazole 25%), a triazole fungicide, was used as a positive control by dipping a 0.4 × 3.5 cm bar of filter papers in a solution of the recommended dose (25 mL per liter) and placed on the surface of the PDA on one side and a 6 mm mycelial disc of *S. cepivora* on the other side. All inoculated plates were incubated at 20 ± 2 °C for 10 days until the mycelial growth of the pathogenic fungus covered the entire control plates. The inhibition zone was calculated according to [38] as the distance between the bioagent bacteria and the edge of the pathogenic fungal mycelium.

2.3.2. Culture Filtrate Assay

A cell-free cultural filtrate of *S. maltophilia* or *S. liqufaciens* isolates was prepared in 250 mL flasks containing 100 mL of a sterilized nutrient broth medium (agar-free nutrient medium); flasks were inoculated with a loop of each bacteria, which was taken from a 2-day-old culture. Inoculated flasks were incubated on a rotary shaker at 200 rpm for 5 days at 28 ± 2 °C. The culture filtrates of bacteria were obtained through centrifugation for 10 min at 10,000 rpm, then the supernatant was collected and filtrated with a 0.45 µm pore-size syringe filter. Subsequently, the effect of the culture filtrate of *S. maltophilia* and *S. liqufaciens* on the radial growth of *S. cepivora* was tested using PDA plates amended with culture filtrates of the test bacteria. An appropriate volume of the *S. maltophilia* and *S. liqufaciens* culture filtrate was mixed with PDA before pouring into 9 cm-diameter Petri dishes to obtain concentrations of 0, 20, 40, 60, and 80% (v/v). The plates without culture filtrates were used as a negative control, whereas plates with 0, 20, 40, 60, and 80% (v/v) of the recommended dose (25 mL per liter) of the Folicure fungicide were used as a positive control. After solidification, 0.6 cm mycelial discs from active cultures of the pathogenic fungus were placed into the center PDA plates. The plates were incubated at 20 ± 2 °C for 10 days. Each concentration was repeated with six replicate plates. The percentage of a decrease in fungal growth was calculated using Equation (3) as follows:

$$\text{Mycelial inhibition } (\%) = \frac{Mycelial\ growth\ in\ control - Mycelial\ growth\ in\ treatment}{Mycelial\ growth\ in\ control} \times 100 \quad (3)$$

2.4. Field Experiments

Field experiments were conducted in an open field naturally infected with *S. cepivora* during two successive seasons, 2020/2021 and 2021/2022, respectively. The experiments were carried out in a randomized complete design with 12 biological replicates for each treatment or control (negative (mock) and positive (fungicide)). The area of each plot was 4.5 m^2 and consisted of 3 rows; each row was 2 m in length and 75 cm in width. Onion seedlings were transplanted after being dipped in the cell-free cultural filtrates of *S. maltophilia*, *S. liqufaciens*, the nutrient broth (mock), or fungicide treatment. An additional dose of bacterial filtrates (20 mL per plant), as well as controls, were added through a root drench application 14 days post transplanting (dpt). Irrigation, fertilization, and other agricultural practices were conducted as recommended for onion production. Folicure 25% EC (Tebuconazole 25%), a triazole fungicide, was used in the current study as a recommended fungicide (positive control) by dipping onion seedlings for 5 min in 25 mL of fungicide per liter of water just before transplanting. Additionally, according to [39], the grown plants were sprayed with 187.5 mL/100 L of water after 6 and 12 weeks after transplanting.

2.4.1. Disease Assessment

Plants were uprooted after 150 dpt to evaluate the disease incidence and severity according to the 5-degree scale as mentioned above.

2.4.2. Vegetative and Yield Parameters

Plants from each plot were uprooted 150 dpt to calculate the plant height (cm), root length (cm), and number of leaves, as well as the bulb circumference (cm), bulb fresh weight (g), and bulb yield (kg·plot^{-1}).

2.4.3. Total Soluble Phenolic and Flavonoid Compounds

According to [40] and using a Folin–Ciocalteu reagent (FCR), the total soluble phenolic compounds were evaluated. Briefly, phenolic compounds were extracted from 100 mg of fresh onion leaves with 20 mL of 80% methanol for 24 h. After extraction, 1 mL of 10% FCR was added to 0.2 mL of a methanolic extract and then vortexed for 30 s. After 3 min, 0.8 mL of 7.5% sodium carbonates (*w*/*v*) were added to the mixture. Subsequently, the mixture was incubated at room temperature for 30 min, and the absorption was measured at 765 nm. The total soluble phenolic content is expressed as mg of gallic acid equivalents per gram of fresh weight (mg GAE g^{-1} FW). Moreover, according to the methods described by [41], the total soluble flavonoids were evaluated. Briefly, 1 mL of a methanolic extract of onion leaves was mixed with 1 mL of aluminum chloride, $ALCL_3$ (2% in methanol). The mixture was strongly shaken and incubated for 15 min at room temperature, then the absorption was measured at 430 nm. The flavonoid concentration was expressed as mg of Rutin equivalents per gram of fresh weight (mg RE g^{-1} FW).

2.4.4. Enzymatic Activity

Approximately 0.5 g of fresh onion leaves was collected at 1, 2, 3, 4, and 5 days post treatment with culture filtrates of *S. maltophilia* and *S. liqufaciens*. Onion leaves were homogenized in 5 mL of a 50 mM Tris buffer (pH 7.8) containing 1 mM of EDTA-Na2 and 7.5% polyvinylpyrrolidone PVP in a cold mortar and pestle for guaiacol-dependent peroxidases (POX) and polyphenol oxidase (PPO). The leaves' extract was centrifuged at 15,000× *g* for 20 min at 4 °C. Then, the supernatants were used for the enzyme assays.

The POX activity was assessed by measuring the formation of the guaiacol-bound product at 436 nm according to [42]. The reaction mixture contained 2.2 mL of a 100 mM sodium phosphate buffer (pH 6.0), 100 µL of guaiacol, 100 µL of 12 mM H_2O_2, and 10 µL of a crude enzyme extract. The increase in the absorption at 436 nm (A_{436}) was measured as the conjugate was formed using an extinction coefficient of 26.6 mM^{-1} cm^{-1} for the conjugate. Additionally, the PPO activity was determined according to the method described in [43]. The reaction mixture contained 3 mL of a buffered catechol solution (0.01 M), freshly prepared in a 0.1 M phosphate buffer (pH 6.0). The reaction was started by adding 100 µL of a crude enzyme extract. Changes in the absorbance at 495 nm (A_{495}) were recorded every 30 s for 3 min.

2.5. Gas Chromatography–Mass Spectrophotometry (GC-MS) Analysis

2.5.1. Chemical Composition of Bacterial Culture Filtrates

To investigate the chemical composition and to identify active components of culture filtrates, a 48-h-old bacterial culture was centrifugated and the supernatant was collected. The supernatant was concentrated using evaporation at 50 °C in a rotary evaporator. The residue that contained the secondary metabolites and chemical compounds was analyzed using GC–MS after solving in n-hexane [44]. The GC–MS analysis was performed using a Clarus 580/560 S PerkinElmer instrument (PerkinElmer, Inc., Waltham, MA, USA) equipped with a capillary column (30 m × 0.25 mm ID, film thickness of 0.25 μm). Helium was used as the carrier gas at a flow rate of 1 mL.min^{-1} with a solvent delay of 6 min. The initial temperature was 80 °C for 7 min, then it was increased to 140 °C at the rate of 10 °C per minute and held for 1 min, then to 200 °C at the same rate and held for 2 min, and finally increased to 260 °C at the rate of 5 °C per minute and held for 2 min. The source and injector temperatures were 200 °C and 280 °C, respectively. For the analysis, 1 μL was injected with a spilt ratio of 1:20. Detected compounds were tentatively identified by comparing their retention times and mass spectrum with library entries in the NIST 2011 mass spectral database (National Institute of Standards and Technology, Gaithersburg, MA, USA) and the Golm Metabolome database.

2.5.2. Quantification of Indole-3-Acetic Acid (IAA) and Tryptophan Using GC-MS

To better understand the stimulant effect of both bacterial culture filtrates on onion growth, the endogenous levels of tryptophan (the precursor of auxins) and the main auxin, IAA, were investigated. Briefly, both compounds were extracted from 100 mg of ground plant tissues using acidic 80% methanol as described in our previous studies [45–47], then derivatized with methyl chloroformate [48,49] and analyzed using GC-MS running in the selective ion monitoring (SIM) mode [45,47]. Targeted metabolites were analyzed using the same instrument described above with our previously described thermos-program [45–47]. Collected chromatograms were analyzed using TurboMass software (Perkin Elmer, Waltham, MA, USA). Tryptophan and IAA were initially identified by comparing their mass spectra with library entries of the same libraries listed above, then their identification was confirmed by comparing their retention times (RT) and mass spectra to authentic standards.

2.6. *Statistical Analysis*

All experiments were carried out using a completely randomized design with 12 replicates for each treatment. The analysis of variance (ANOVA) statistical model followed by post hoc pairwise comparisons using the Tukey honestly significant difference test were used to compare variances between means of different treatments (HSD; $p \leq 0.05$). Moreover, a simple linear regression (SLR) analysis was carried out to better understand the relationship between the mycelial growth inhibition percentage and the concentrations of culture filtrates of *S. maltophilia* and *S. liquefaciens*.

3. Results

3.1. *Isolation and Identification of S. cepivora*

Seven isolates were isolated from diseased onion plants showing typical symptoms of white rot disease. Although the seven isolates were pathogenic and produced typical white rot on onion plants under greenhouse conditions (Figure 1A), isolate #4 was the most aggressive one and resulted in the highest disease severity (90.69 ± 4.16%) on onion plants. Isolate #4 showed typical morphological characteristics and colony texture with *Sclerotium* sp. when grown on a PDA medium (Figure 1B,C). Briefly, isolate #4 formed white cultures with fluffy to compact mycelia (Figure 1B) that were covered by small round or irregular-shaped dark brown sclerotia (Figure 1C) similar to those observed on naturally infected onion plants. However, sexual structures were absent and were not observed. Collectively, the pathological and morphological characteristics proposed that the isolated fungus was *S. cepivorum*, the causal agent of onion white rot disease. To further confirm the identification, the most aggressive isolate (isolate #4) was selected for a further genetic identification based on the sequence of the internal transcribed spacer (ITS) region (Figure 1C). Briefly, the query sequence showed a high similarity with the large subunit ribosomal RNA gene of *S. cepivora*—strain CBS 321.65 (GenBank Accession No. MH870230.1) (Figure 1D). The new sequence was deposited in the NCBI database and named "*S. cepivora*—Isolate 2023" (GenBank Accession No. OQ392600).

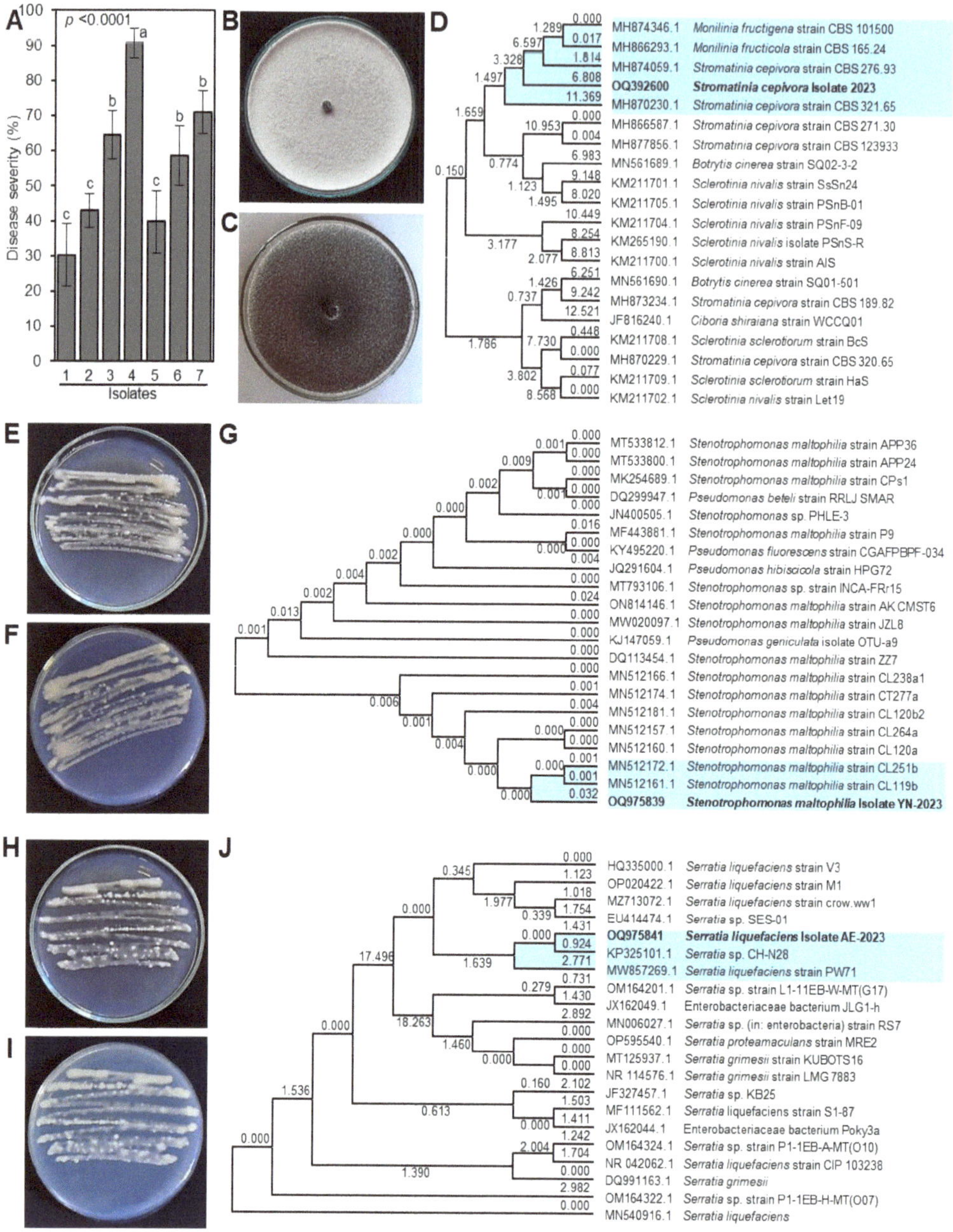

Figure 1. The pathogenicity, morphological characterization, and molecular identification of the isolates of the phytopathogenic fungus *S. cepivora* and the bacterial bioagents *S. maltophilia* and *S. liquefaciens*. (**A**) The disease severity (%) of various isolates of *S. cepivora* on onion plants (cultivar Giza 20) under greenhouse conditions. Bars denote the means ± standard deviations (means ± SD). Different letters signify statistically significant differences among isolates using the Tukey HSD test ($p < 0.05$). (**B**) The growth and morphological characteristics of the phytopathogenic fungus *S. cepivora* on potato dextrose agar (PDA) media after 7 days of incubation at 20 ± 2 °C. (**C**) Sclerotia of *S. cepivora* formed on PDA. (**D**) The evolutionary analysis with the Maximum Likelihood method and Tamura-Nei model using the

ITS-5.8S rDNA sequence of *S. cepivora*—Isolate 2023 (GenBank Accession No. OQ392600) in comparison with 20 reference strains/isolates retrieved from the recent available data in National Center for Biotechnology Information (NCBI) GenBank (https://www.ncbi.nlm.nih.gov/; accessed on 7 February 2023). (**E**,**F**) The growth and morphological characteristics of the biocontrol agent *S. maltophilia* on the nutrient agar medium from the top and the bottom of the Petri dish, respectively, after 7 days of incubation at 28 ± 2 °C. (**G**) The evolutionary analysis with the Maximum Likelihood method and Tamura-Nei model using the 16S rRNA sequence of *S. maltophilia*—Isolate YN-2023 (GenBank Accession No. OQ975839) in comparison with 20 reference strains/isolates retrieved from the recently available data in NCBI GenBank. (**H**,**I**) The growth and morphological characteristics of the biocontrol agent *S. liquefaciens* on the nutrient agar medium from the top and the bottom of the Petri dish, respectively, after 7 days of incubation at 28 ± 2 °C. (**J**) The evolutionary analysis with the Maximum Likelihood method and Tamura-Nei model using the 16S rRNA sequence of *S. liquefaciens*—Isolate AE-2023 (GenBank Accession No. OQ975841) in comparison with 20 reference strains/isolates retrieved from the recently available data in NCBI GenBank. In panels D, G, and J, the query sequence is bolded, and its closest reference strains/isolates are highlighted in light blue.

3.2. Isolation and Identification of Bacterial Bioagent

Two bacterial bioagents were isolated and characterized in this study. The first bacterium was from healthy onion roots. It formed small (approximately 3 mm in diameter), circular, smooth, convex colonies with a ligh–yellow tint on the nutrient agar medium after 24 h post incubation at 37 °C (Figure 1E) and usually produced a yellowish soluble pigment that could be observed from the bottom side of the plate (Figure 1F). A microscopic examination of the isolated bioagent showed that the bacterium is a straight, motile, Gram-negative rod. Moreover, the identification of the isolated bioagent was confirmed based on the sequence of the 16S ribosomal RNA gene. The evolutionary analysis showed that the query sequence showed a high similarity with *S. maltophilia*—strain CL119b and strain CL251b (GenBank Accession No. MN512161.1 and MN512172.1, respectively) (Figure 1G). The new sequence was deposited in the NCBI database and named "*S. maltophilia*—Isolate YN-2023" (GenBank Accession No. OQ975839).

Another bacterial bioagent was endophytic and isolated from healthy bean leaves. It formed circular, smooth, white raised colonies with entire margins when it was grown on a nutrient agar medium (Figure 1H,I). A microscopic examination showed that the bacterium is a straight, motile, Gram-negative rod. Moreover, the evolutionary analysis based on the sequence of the 16S ribosomal RNA gene showed that the query sequence showed as highly similar to *S. liquefaciens*—strain PW71 (GenBank Accession No. MW857269.1) and *Serratia* sp.—CH-N28 (GenBank Accession No. KP325101.1) (Figure 1J). The new sequence was deposited in the NCBI database and named "*S. liquefaciens*—Isolate AE-2023" (GenBank Accession No. OQ975841).

3.3. *S. maltophilia* and *S. Liquefaciens* Inhibited the Mycelial Growth of *S. cepivora*

The antifungal activities of both bioagents (*S. maltophilia* and *S. liquefaciens*) against *S. cepivora* were tested in vitro using dual culture and culture filtrate techniques. In the dual culture plates, both bioagents showed strong fungistatic activity against *S. cepivora* (Figure 2A) and significantly reduced its radial mycelial growth on PDA (Figure 2B). Although the Folicure fungicide (Tebuconazole 25%) had the lowest radial mycelia growth of *S. cepivora* (2.85 ± 0.50 cm), both *S. maltophilia* and *S. liquefaciens* significantly reduced the radial mycelia growth of *S. cepivora* (3.71 ± 0.16 and 4.95 ± 0.32 cm, respectively) compared with the mock control (Figure 2B). In other words, both *S. maltophilia* and *S. liquefaciens* notably inhibited the mycelia growth of *S. cepivora* by 58.74 ± 1.75% and 44.95 ± 3.51%, respectively, compared with the mock control (Figure 2C).

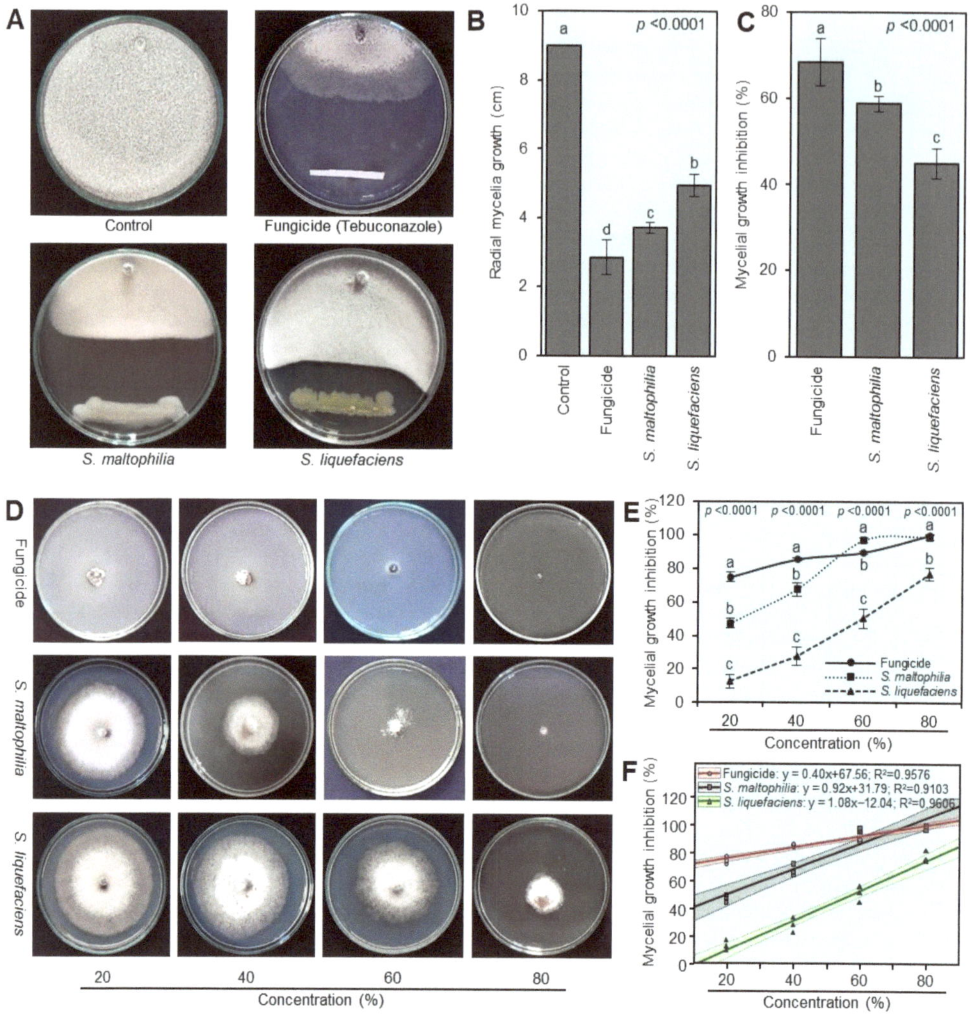

Figure 2. In vitro antifungal activity of the bacterial bioagents *S. maltophilia* and *S. liquefaciens* or their cell-free culture filtrates against the phytopathogenic fungus *S. cepivora*. (**A**) Antifungal activity of *S. maltophilia* and *S. liquefaciens* against *S. cepivora* using the double culture assay. (**B**,**C**) The radial mycelia growth (cm) and mycelial growth inhibition (%) of *S. cepivora* after treatment with *S. maltophilia* and *S. liquefaciens* using the double culture assay. Bars denote the means ± standard deviations (means ± SD). Different letters signify statistically significant differences among isolates using the Tukey HSD test ($p < 0.05$). (**D**) Antifungal activity of different concentrations of cell-free culture filtrates of *S. maltophilia* and *S. liquefaciens* against *S. cepivora*. (**E**) The mycelial growth inhibition (%) of *S. cepivora* after treatment with different concentrations (20, 40, 60, 80% (v/v)) of cell-free culture filtrates of *S. maltophilia* and *S. liquefaciens* or the positive control (Folicure fungicide). Dots denote the means ± standard deviations (means ± SD). Different letters signify statistically significant differences among isolates using the Tukey HSD test ($p < 0.05$). (**F**) The simple linear regression (SLR) analysis of the relationship between the mycelial growth inhibition (%) and different concentrations of culture filtrates of *S. maltophilia*, *S. liquefaciens*, and the Folicure fungicide.

Moreover, the utilization of bioagent culture filtrates demonstrated strong dose-dependent antifungal activity against *S. cepivora* (Figure 2D). Briefly, increasing the percentage of the culture filtrate in PDA media from 20 to 80% significantly increased the inhibition of mycelial growth of *S. cepivora* (Figure 2E). It is worth mentioning that the mycelial growth inhibition (%) due to the utilization of the highest concentration (80%) of the culture filtrate of *S. maltophilia* was similar to the positive control (Folicure fungicide) and significantly inhibited the mycelial growth of *S. cepivora* (99.00 ± 1.73%; Figure 2E). Furthermore, the simple linear regression (SLR) showed that the mycelial growth inhibition percentage was positively correlated with the concentrations of culture filtrates of *S. maltophilia* (y = 0.92x + 31.79; R^2 = 0.9103), *S. liquefaciens* (y = 1.08x − 12.04; R^2 = 0.9606), and the Folicure fungicide (y = 0.40x + 67.56; R^2 = 0.9576) (Figure 2F).

3.4. Culture filtrates of S. maltophilia and S. liquefaciens Reduced the Development of White Rot Disease

In general, the utilization of the culture filtrates of both bioagents (*S. maltophilia* and *S. liquefaciens*) significantly decreased the development of white rot disease on treated onion plants compared with the mock-treated infected plants (control) as expressed by the disease incidence (%) and disease severity (%) during two successive seasons, 2020/2021 and 2021/2022, respectively (Figure 3). Briefly, the white rot disease incidence dropped from 85.58 ± 1.26% in the control plants to 19.08 ± 1.31 and 49.74 ± 2.47% when plants were treated with the culture filtrates of *S. maltophilia* or *S. liquefaciens*, respectively, during the 2020/2021 season (Figure 3A). Likewise, the disease severity declined from 77.89 ± 2.18% to 29.39 ± 2.16 and 43.92 ± 1.57% when onions were treated with the culture filtrates of *S. maltophilia* or *S. liquefaciens*, respectively, during the same season (Figure 3B). The same trend was observed in the second season, 2021/2022 (Figure 3C,D).

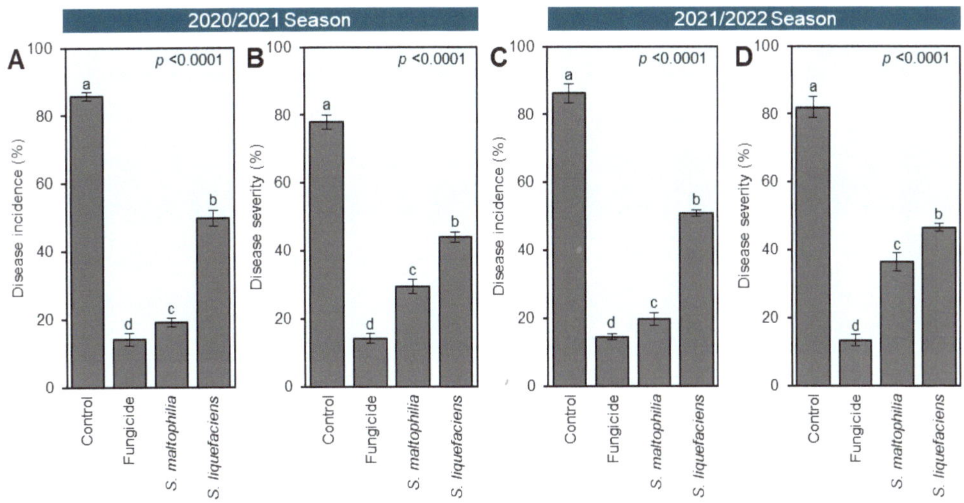

Figure 3. The effect of cell-free culture filtrates of *S. maltophilia* and *S. liquefaciens* on onion white rot disease, caused by *S. cepivora* under field conditions during two successive seasons (2020/2021 and 2021/2022). (**A**,**B**) The disease incidence (%) and severity (%) of onion white rot disease, respectively, during the 2020/2021 season. (**C**,**D**) The disease incidence (%) and severity (%) of onion white rot disease, respectively, during the 2021/2022 season. Bars denote the means ± standard deviations (means ± SD). Different letters signify statistically significant differences among isolates using the Tukey HSD test ($p < 0.05$).

3.5. Cell-Free Culture Filtrates of S. maltophilia and S. liquefaciens Stimulated the Growth of S. cepivora-Infected Onion Plants

During two successive seasons, 2020/2021 and 2021/2022, culture filtrates of *S. maltophilia* and *S. liquefaciens* notably increased the number of leaves per plant, plant height, and fresh weight of the shoot system with no significant differences between both bacterial agents but significantly higher differences compared to the mock- and fungicide-treated controls (Figure 4). However, the application of the culture filtrate of *S. liquefaciens* recorded the longest root length (12.11 ± 0.51 and 12.27 ± 0.96 cm) followed by the culture filtrate of *S. maltophilia* (7.99 ± 0.85 and 7.70 ± 0.74 cm), which were significantly higher than the mock-treated control (2.02 ± 0.60 and 2.22 ± 0.83 cm) during 2020/2021 and 2021/2022, respectively (Figure 4C,D). Collectively, these findings suggest that the root drench application of both bacterial culture filtrates has no phytotoxicity on the treated onion plants, but they even stimulate their growth.

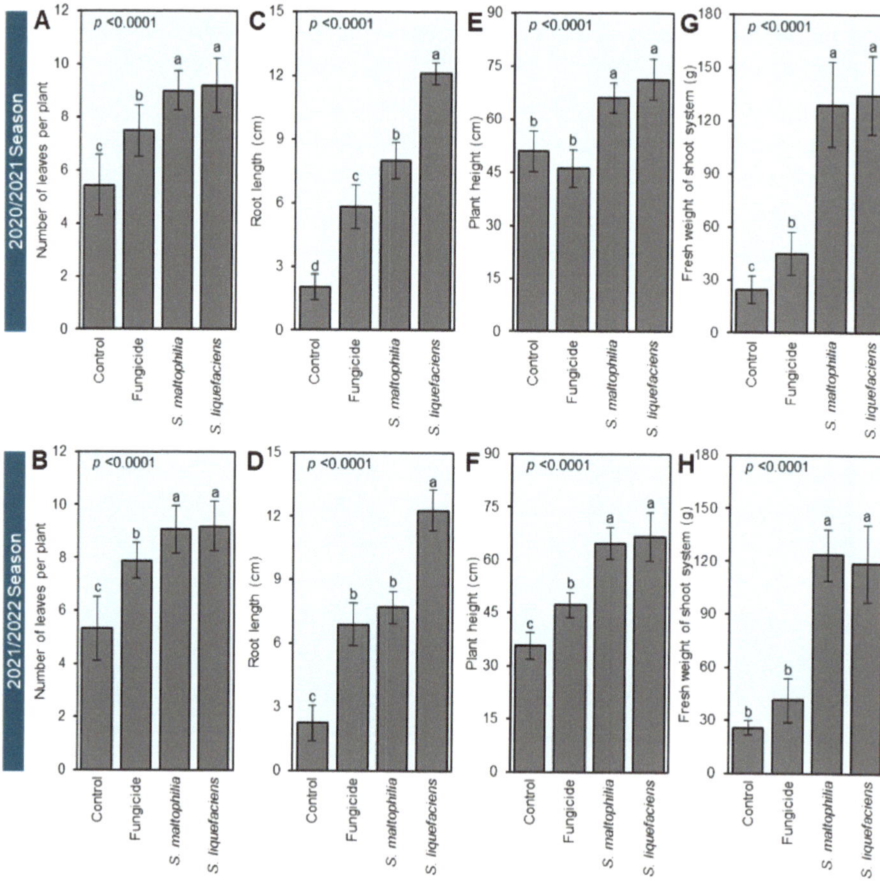

Figure 4. The effect of cell-free culture filtrates of *S. maltophilia* and *S. liquefaciens* on the growth parameters of *S. cepivora*-infested onion plants under field conditions during two successive seasons (2020/2021 and 2021/2022). (**A**,**B**) The number of leaves per plant, (**C**,**D**) root length (cm), (**E**,**F**) plant height (cm), and (**G**,**H**) fresh weight of the shoot system (g) of *S. cepivora*-infested onion plants during 2020/2021 and 2021/2022 seasons, respectively. Bars denote the means ± standard deviations (means ± SD). Different letters signify statistically significant differences among isolates using the Tukey HSD test ($p < 0.05$).

3.6. *Culture Filtrates of S. maltophilia and S. liquefaciens Enhanced the Yield Components of S. cepivora-Infected Onion Plants*

Generally, the root drench application of culture filtrates of both bacterial bioagents remarkably enhanced all yield attributes of *S. cepivora*-infected onion plants including the bulb diameter (Figure 5A,B), bulb fresh weight (Figure 5C,D), and bulb yield per plot (Figure 5E,F) compared with the mock-treated control during two successive seasons, 2020/2021 and 2021/2022, respectively. Although there were no significant differences between the bulb circumference of *S. maltophilia*- and *S. liquefaciens*-treated onions (29.92 ± 1.62 and 30.34 ± 1.28 cm, respectively), the culture filtrate of *S. maltophilia* had the highest bulb fresh weight (171.27 ± 9.58 g) and bulb yield per plot (15.66 ± 1.17 kg·plot^{-1}), which was comparable to the positive control (Folicure fungicide) during the first season. However, there were no significant differences between *S. maltophilia*- and *S. liquefaciens*-treated onion plants in the second season in terms of the bulb fresh weight and bulb yield per plot.

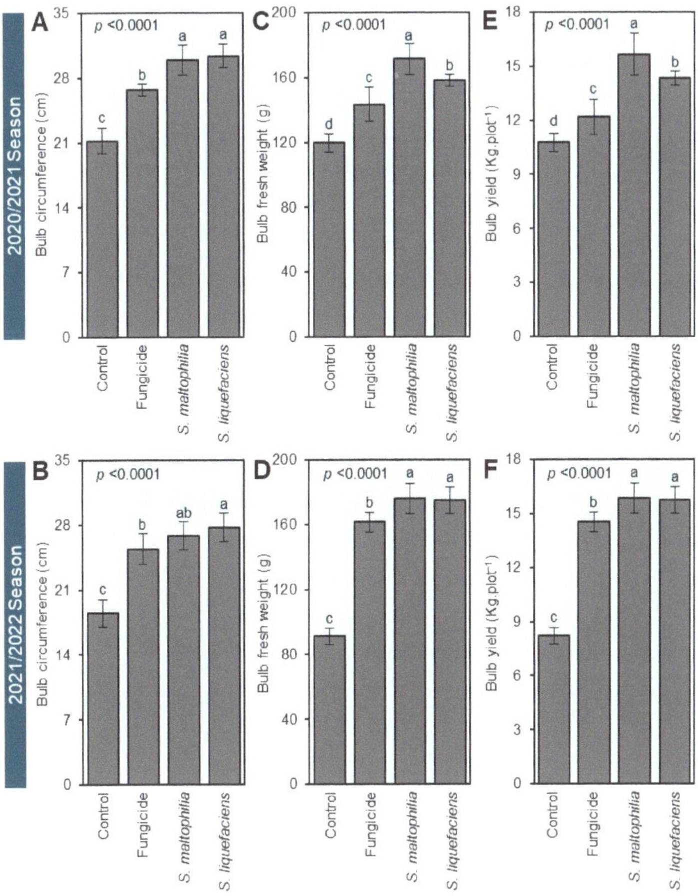

Figure 5. The effect of cell-free culture filtrates of *S. maltophilia* and *S. liquefaciens* on the yield components of *S. cepivora*-infested onion plants under field conditions during two successive seasons (2020/2021 and 2021/2022). (**A**,**B**) The bulb circumference (cm), (**C**,**D**) bulb fresh weight (g), and (**E**,**F**) bulb yield (kg·plot^{-1}) of *S. cepivora*-infested onion plants during 2020/2021 and 2021/2022 seasons, respectively. Bars denote the means ± standard deviations (means ± SD). Different letters signify statistically significant differences among isolates using the Tukey HSD test ($p < 0.05$).

3.7. Cell-Free Culture Filtrates of *S. maltophilia* and *S. liquefaciens* Stimulate the Antioxidant Defense Machinery of *S. cepivora*-Infected Onion Plants

Both enzymatic (peroxidase (POX) and polyphenol oxidase (PPO) activity) and non-enzymatic (total soluble phenolics and flavonoids) antioxidant defense machineries were further investigated to better understand the physiological and biochemical mechanisms of *S. maltophilia* and *S. liquefaciens* (Figure 6). Briefly, the root drench application of culture filtrates of both bacterial bioagents gradually enhanced the enzymatic activity of POX (Figure 6A) and PPO (Figure 6B) until they reached their highest peak at 48 or 72 h post treatment (hpt), then they dropped slowly. Onion plants that were treated with the culture filtrate of *S. maltophilia* recorded their highest peak of POX ($3.42 \pm 0.29 \times 10^{-2}$ μM of tetraguaiacol g^{-1} FW min^{-1}) at 48 hpt but had the highest peak of PPO ($1.43 \pm 0.16 \times 10^{-2}$ arbitrary units) at 72 hpt. On the contrary, onion plants that were treated with the culture filtrate of *S. liquefaciens* had their highest peak of POX ($7.05 \pm 0.57 \times 10^{-2}$ μM of tetraguaiacol g^{-1} FW min^{-1}) at 72 hpt but were the second highest for PPO ($0.94 \pm 0.1 \times 10^{-2}$ arbitrary units) at the same time point, 72 hpt (Figure 6A,B).

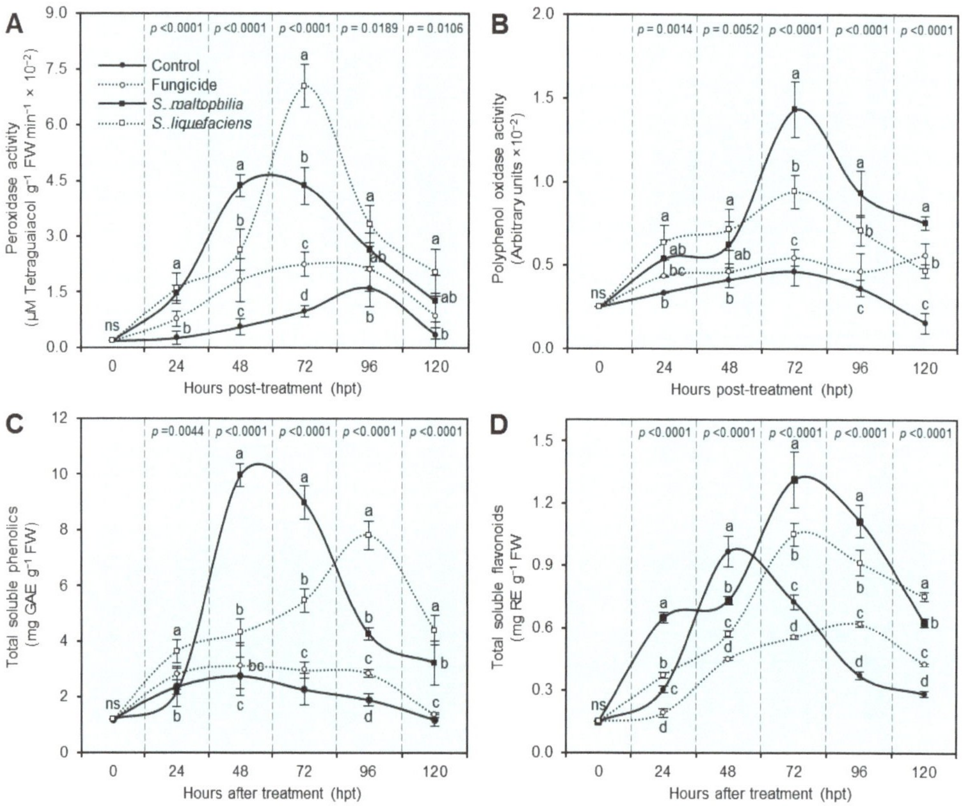

Figure 6. The effect of cell-free culture filtrates of *S. maltophilia* and *S. liquefaciens* on the enzymatic and non-enzymatic antioxidant defense system of *S. cepivora*-infested onion plants under field conditions. (**A**) The peroxidase activity (μM of tetraguaiacol g^{-1} FW min^{-1}), (**B**) polyphenol oxidase activity (arbitrary units), (**C**) total soluble phenolics (mg GAE g^{-1} FW), and (**D**) total soluble flavonoids (mg RE g^{-1} FW). Dots denote the means ± standard deviations (means ± SD). Different letters signify statistically significant differences among isolates using the Tukey HSD test ($p < 0.05$).

Likewise, the non-enzymatic antioxidant defense machinery as expressed by total soluble phenolics (Figure 6C) and flavonoids (Figure 6D) was progressively enhanced due to the root drench application of culture filtrates of both bacterial bioagents until they reached their highest peak then steadily decreased. Briefly, onion plants treated with the culture filtrate of *S. maltophilia* had the highest peak of total soluble phenolics (9.96 ± 0.42 mg GAE g^{-1} FW; Figure 6C) and total soluble flavonoids (1.31 ± 0.14 mg RE g^{-1} FW; Figure 6D) at 48 and 72 hpt, respectively. However, *S. liquefaciens*-treated plants recorded their highest peak of total soluble phenolics (7.82 ± 0.52 mg GAE g^{-1} FW; Figure 6C) later at 96 hpt but reached their highest peak of total soluble flavonoids (1.05 ± 0.06 mg RE g^{-1} FW; Figure 6D) at 72 hpt.

3.8. Culture Filtrates of S. maltophilia *and* S. liquefaciens *and the Endogenous Auxin Content of* S. cepivora*-Infected Onion Plants*

To better understand the physiological and biochemical mechanisms of how culture filtrates of both *S. maltophilia* and *S. liquefaciens* stimulate the growth of *S. cepivora*-infected onion plants, their effect on the endogenous levels of the main auxin, IAA, and its precursor, the amino acid tryptophan, in onion leaves was further investigated. Although the application of the Folicure fungicide did not change the endogenous tryptophan level (600.75 ± 114.52 ng.g^{-1} FW; Figure 7A) and even reduced the IAA content (182.83 ± 11.45 ng.g^{-1} FW; Figure 7B) compared with the non-treated control (669.83 ± 159.15 and 214.42 ± 19.39 ng.g^{-1} FW, respectively), the root drench application of culture filtrates of both bacterial bioagents significantly enhanced the accumulation of IAA and its precursor tryptophan. Onion plants treated with the culture filtrate of *S. maltophilia* had the highest tryptophan (1086.92 ± 90.46 ng.g^{-1} FW) and IAA (420.42 ± 36.86 ng.g^{-1} FW) levels, followed by *S. liquefaciens*-treated plants (904.33 ± 75.84 and 321.42 ± 32.54 ng.g^{-1} FW, respectively).

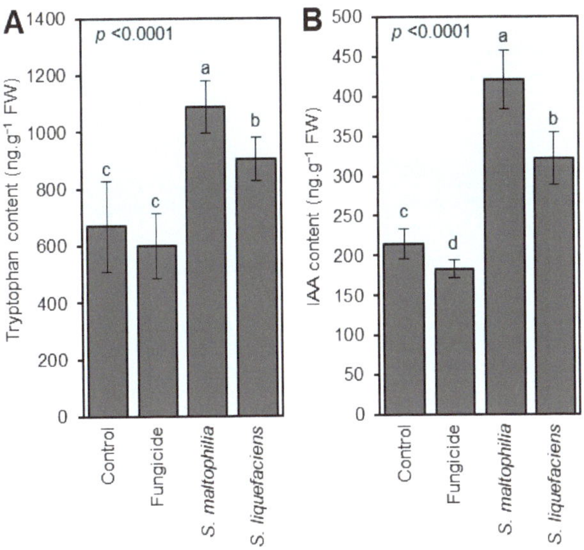

Figure 7. The effect of cell-free culture filtrates of *S. maltophilia* and *S. liquefaciens* on the endogenous levels of the main auxin, IAA, and its precursor, the amino acid tryptophan, in *S. cepivora*-infested onion leaves under field conditions. (**A**) Tryptophan content (ng.g^{-1} FW) and (**B**) IAA content (ng.g^{-1} FW). Bars denote the means ± standard deviations (means ± SD). Different letters signify statistically significant differences among isolates using the Tukey HSD test ($p < 0.05$).

3.9. Chemical Composition of Culture Filtrates of S. maltophilia and S. liquefaciens

In general, 53 compounds were detected and tentatively identified in the culture filtrates of *S. maltophilia* and *S. liquefaciens* (Figure 8). Out of these 53 compounds, 18 compounds were detected only in the culture filtrate of *S. maltophilia*, and 21 compounds were detected specifically in the culture filtrates of *S. liquefaciens*. However, only 14 compounds were detected in both culture filtrates including cyclohexanone, 5-methyl-2-(1-methylethyl), naphthalene, 1,1′-biphenyl, 4-methyl, phenol, 2,4-bis(1,1-dimethylethyl), undecanoic acid, tridecane, benzyl benzoate, hexadecane, palmitic acid, nonadecane, benzoic acid, 3,5-bis(1,1-dimethylethyl)-4-hydroxy, menthol, dodecane, and tetradecane (Figure 8).

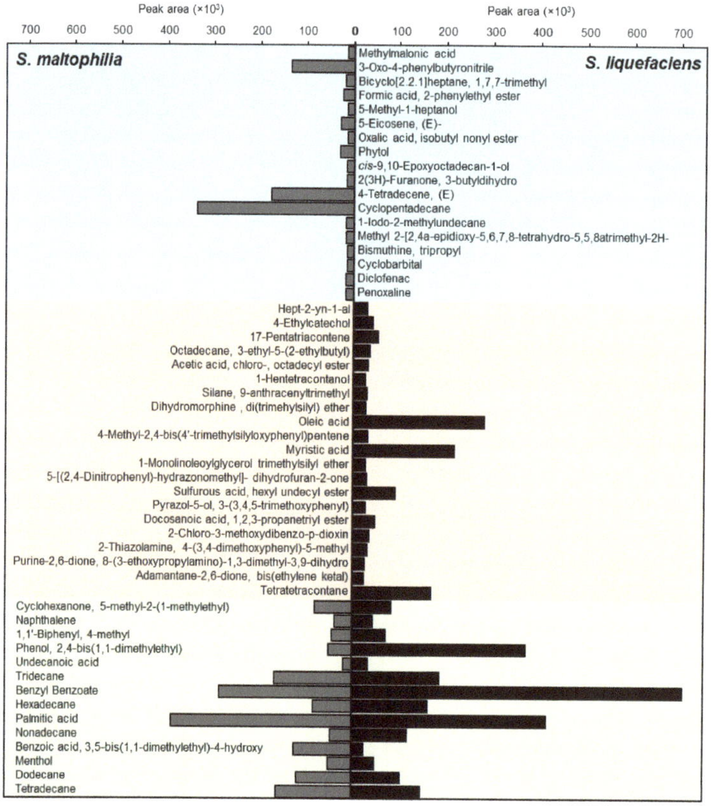

Figure 8. The chemical composition of cell-free culture filtrates of *S. maltophilia* and *S. liquefaciens*. Metabolites were analyzed using GC–MS after solving in n-hexane. Bars represent the peak area of each compound.

4. Discussion

White rot is a serious disease that restricts the commercial production of *Allium* species and causes substantial yield losses worldwide. Although the ascomycetous causal agent, *S. cepivora*, does not produce any asexual spores, it forms a large proportion of infective sclerotia that survive and remain viable in the soil for up to 20 years in the absence of a susceptible host plant [8]. Unfortunately, there has been no effective control mainly for white rot disease until now; however, chemical control using various agrichemicals, particularly fungicides, is the most commonly used strategy to combat this disease. Nevertheless, the extensive use of chemical fungicides has several health and environmental hazards to humans, animals, and non-target organisms. Moreover, pathogens might develop resistance against the repeatedly used fungicides. Due to the efficiency breakdown of chemical fungicides, the search for safer eco-friendly alternative control measures has become a necessity [50]. Biological control might be a promising control strategy against white rot disease, particularly if it is incorporated into integrated pest management programs [13,50,51].

In the current study, *S. maltophilia* and *S. liquefaciens* suppressed the growth of *S. cepivora* and inhibited the development of white rot disease under field conditions. Several studies have previously reported the antagonistic activity of *S. maltophilia* against different soil-borne phytopathogens including bacteria such as *Ralstonia solanacearum*, the causal agent of potato brown rot [20], oomycetes such as *Pythium ultimum* [52,53], fungi such as Rhizoctonia solani [23,54,55], *Fusarium* sp. [54,56], Verticillium dahliae [23], and the sclerotia-producing fungus Sclerotium rolfsii [24]. Likewise, numerous *S. liquefaciens* strains exhibited a strong biocontrol efficacy against soil-borne pathogens including bacteria such as *R. solanacearum* [57] and fungi such as *R. solani* [29], *F. oxysporum* [29,58], and the sclerotia-producing fungus *Sclerotinia sclerotiorum* [29]. However, to the best of our knowledge, this is the first report about the potential application of *S. maltophilia* and *S. liquefaciens* as effective biocontrol agents against *S. cepivora*, the causal agent of onion white rot disease.

Several hypotheses were proposed to explain the antifungal activity of both bioagents. These hypotheses include the production of several extracellular metabolites such as antibiotics, siderophores, quorum-sensing molecules, N-acyl homoserine lactones (AHLs), and cell wall-degrading enzymes [59]. It is worth mentioning that our GC-MS analysis of the culture filtrates of both bacteria showed that both filtrates contain several bioactive compounds with antimicrobial activity such as menthol [60]; benzyl benzoate [61]; oleic, myristic, and palmitic fatty acids [62–64]; tridecane, tetradecane, and hexadecane compounds [65,66]; and phenol,2,4-bis(1,1-diethylethyl) [67]. Taken together, our findings suggest that the antifungal activity of *S. maltophilia* and *S. liquefaciens* might be due to the presence of these compounds in their culture filtrate. However, further studies are required to better understand how the internal mechanisms of menthol operate within a fungal cell.

Another hypothesis that could explain the antifungal activity of *S. maltophilia* and *S. liquefaciens* is due to their ability to produce cell wall-degrading enzymes [59]. It was previously reported that the fungistatic activity of *S. maltophilia* W81 against *Pythium ultimum* is mediated by extracellular proteolytic activity [52] including an inducible extracellular serine protease [53]. Furthermore, the antagonistic activity of *S. maltophilia* against fungal phytopathogens such as Fusarium sp., Rhizoctonia sp., Alternaria sp., and *Bipolaris sorokiniana* is associated with its ability to produce chitinase [54,68].

Additionally, our field experiments showed that both bacterial culture filtrates significantly reduced the disease incidence and disease severity of white rot on treated onion plants compared with the non-treated plants. It is worth noting that none of the bacterial culture filtrates caused any phytotoxicity in treated plants as indicated by a stimulated growth performance (number of leaves per plant, root length, plant height, and fresh weight of the shoot system). However, the biochemical and physiological mechanisms behind these roles are poorly understood. This might be due to the enhancement of enzymatic antioxidants and non-enzymatic antioxidant defense machinery [69]. Peroxidase (POX) and polyphenol oxidase (PPO) are two main components in the enzymatic antioxidant system. POX maintains redox homeostasis via the regulation of H_2O_2 levels [70], whereas PPO is involved in the oxidation of phenolics into highly reactive quinones. Interestingly, our findings proved that the root application of culture filtrates of *S. maltophilia* and *S. liquefaciens* considerably boosted the enzymatic activities of both POX and PPO.

Moreover, the non-enzymatic antioxidant defense machinery relies on phenolics, flavonoids, and lipophilic antioxidants such as carotenoids [71]. Our findings showed that the application of culture filtrates of *S. maltophilia* and *S. liquefaciens* notably enriched the endogenous levels of total soluble phenolics and flavonoids in treated onion leaves. These findings propose that the application of culture filtrates of *S. maltophilia* and *S. liquefaciens* induces multilayered antioxidant defense machinery in *S. cepivora*-infected onion plants to alleviate the risky consequences of reactive oxygen species (ROS) and preserve their homeostasis. Additionally, *Serratia* sp. was previously reported to trigger a plant's defensive mechanisms against miscellaneous phytopathogens via the activation of induced systemic resistance (ISR) after a proper stimulation [59].

Furthermore, culture filtrates of *S. maltophilia* and *S. liquefaciens* might reduce the disease incidence and severity of white rot on treated onion plants by prolonging the sclerotial differentiation period. It was previously reported that menthol delays sclerotial differentiation by up to 145 h [60]. As we mentioned above, the GC-MS analysis showed that both culture filtrates were rich in menthol, which might have prolonged the sclerotial differentiation period and resulted in a reduced disease incidence and severity. Moreover, the culture filtrate of *S. maltophilia* was rich in fatty acids such as oleic and myristic acids. Oleic acid was previously reported to inhibit the growth of numerous phytopathogenic fungi such as *R. solani*, *P. ultimum*, *Pyrenophora avenae*, and *Crinipellis perniciosa* [62]. Likewise, myristic acid negatively affected mycelial growth and spore germination of four phytopathogenic fungi including *A. solani*, *Colletotrichum lagenarium*, *F. oxysporum* f. sp. *Cucumerinum*, and *F. oxysporum* f. sp. *lycopersici* [63]. Although some oxylipins (fatty acid derivatives) such as (±)-cis-12,13-Epoxy-9(Z)-octadecenoic acid, (±)-threo-12,13-Dihydroxy-9(Z)-octadecenoic acid, and (±)-threo-9,10-Dihydroxy-12(Z)-octadecenoic acid showed potent antifungal activity against *S. sclerotiorum* [72], the antifungal activity of fatty acids and their derivatives are poorly studied. Collectively, these findings suggest that extracellular metabolites in culture filtrates of *S. maltophilia* and *S. liquefaciens* inhibit the growth of *S. cepivora* or delay the sclerotial differentiation of its sclerotia, resulting in a reduced disease incidence and disease severity. However, further studies are required to better understand the potential role(s) of extracellular metabolites of *S. maltophilia* and *S. liquefaciens*.

In addition to their protective role against white rot disease, culture filtrates of *S. maltophilia* and *S. liquefaciens* showed strong bio-stimulant properties. Our field experiments showed that both culture filtrates stimulated the growth of *S. cepivora*-infected onion plants compared with the mock-treated control as expressed by more leaves per plant, longer roots, a higher plant height, and a heavier fresh weight of the shoot system. Moreover, both culture filtrates enhanced the yield components of *S. cepivora*-infected onion plants as expressed by a bigger bulb diameter, higher bulb fresh weight, and bulb yield per plot during two successive seasons, 2020/2021 and 2021/2022, respectively. Bio-stimulant and growth promotion properties of *S. maltophilia* and *S. liquefaciens* might be due to their ability to induce the accumulation of auxins and their precursor tryptophan within treated onion plants. Both *S. maltophilia* and *S. liquefaciens* embrace several mechanisms to facilitate plant growth promotion such as enhancing nutrient uptake, producing siderophores, and synthesizing stimulatory phytohormones like IAA [59,73–79]. Both bacteria have complete metabolic pathways linked to their plant growth promotion properties, including the IAA biosynthesis pathway [73–79]. Production of IAA by different strains of S. Maltophilia [74–76], as well as *S. liquefaciens* [77–79], was previously reported. These characteristics highlight the enrichment of the culture filtrates of both bacteria with growth-promoting phytohormone IAA. Moreover, our findings showed that the root drench application of culture filtrates of both bacterial bioagents significantly enhanced the accumulation of IAA and its precursor tryptophan and stimulated the growth of *S. cepivora*-infected onion plants.

5. Conclusions

In conclusion, our findings highlight the potential importance of *S. maltophilia* and *S. liquefaciens* as plant growth-promoting bacteria with potent bio-control properties against *S. cepivora*, the causal agent of onion white rot disease (Figure 9). Our findings showed that these two bacteria and their cell-free culture filtrates notably inhibited the mycelial growth of *S. cepivora*, which might be due to the enriched chemical composition of these filtrates, particularly the high levels of menthol, oleic acid, myristic acid, and other volatile organic compounds. Moreover, culture filtrates of *S. maltophilia* and *S. liquefaciens* significantly reduced the development of white rot disease, which might be due to their antifungal activities and/or due to the activation of both enzymatic (POX and PPO) and non-enzymatic (phenolics and flavonoids) antioxidant defense machineries of *S. cepivora*-infected onion plants. Last but not least, culture filtrates of *S. maltophilia* and *S. liquefaciens* promoted the growth of *S. cepivora*-infected onion plants and enhanced their yield components, which might be a result of a reduced disease severity and/or the induced IAA accumulation. Although further studies are required to better understand the molecular, biochemical, and physiological mechanisms involved in the fungistatic activity of *S. maltophilia* and *S. liquefaciens* or their cell-free culture filtrates against *S. cepivora*, our findings suggest that both bioagents might be eco-friendly alternative control measures to reduce the utilization of chemical fungicides entirely or partially for the safer production of onions in *S. cepivora*-infested soils.

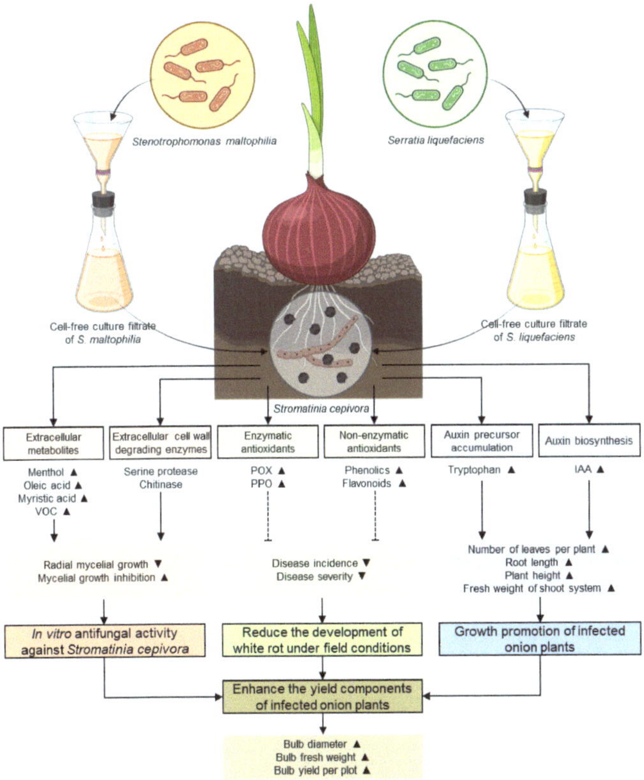

Figure 9. A schematic representation of the potential importance of *S. maltophilia* and *S. liquefaciens* as plant growth-promoting bacteria with potent bio-control properties against *S. cepivora*, the causal agent of onion white rot disease. Briefly, *S. maltophilia* and *S. liquefaciens* and their cell-free culture filtrates have antifungal activities and can inhibit the mycelial growth of *S. cepivora* due to their enriched chemical composition of menthol, oleic acid, myristic acid, and other volatile organic compounds (VOC). As a result, cell-free culture filtrates of both bioagents might reduce the development of white rot disease due to the activation of both enzymatic (POX and PPO) and non-enzymatic (phenolics and flavonoids) antioxidant defense machineries of *S. cepivora*-infected onion plants. Finally, culture filtrates of *S. maltophilia* and *S. liquefaciens* might promote the growth of *S. cepivora*-infected onion plants and enhance their yield components because of a reduced disease severity and/or the induced IAA accumulation. The up-arrow indicates increased levels, whereas down-arrow indicates decreased levels. Solid lines with arrows indicate positive reactions, while dashed lines with whiskers signify negative reactions. For more details, see the main text.

Supplementary Materials: The following supporting information can be downloaded at: https://www.mdpi.com/article/10.3390/horticulturae9070780/s1, Table S1: Sequences from *S. cepivora* and some other fungal species that produce significant alignments with *S. cepivora*—Isolate 2023 using NCBI database.; Table S2: Sequences from *S. maltophilia* and some other bacteria species that produce significant alignments with *S. maltophilia*—Isolate YN-2023 using NCBI database.; Table S3: Sequences from *S. liquefaciens* and some other bacterial species that produce significant alignments with *S. liquefaciens*—Isolate AE-2023 using NCBI database.

Author Contributions: Conceptualization, Y.N., A.A.E. and A.E.-N.; methodology, Y.N. and A.E.-N.; software, Y.N.; validation, H.E.M.O., Y.N., A.A.E., M.H.E.-M. and A.E.-N.; formal analysis, Y.N.; investigation, Y.N. and A.E.-N.; resources, H.E.M.O., A.A.E. and M.H.E.-M.; data curation, Y.N., A.A.E. and A.E.-N.; writing—original draft preparation, Y.N. and A.E.-N.; writing—review and editing, Y.N. and A.A.E.; visualization, Y.N.; supervision, Y.N. and A.A.E.; project administration, Y.N., A.A.E. and A.E.-N.; funding acquisition, H.E.M.O. and M.H.E.-M. All authors have read and agreed to the published version of the manuscript.

Funding: The authors extend their appreciation to the Deanship for Research & Innovation, Ministry of Education in Saudi Arabia for funding this research work through project number: IFP22UQU4320730DSR030.

Data Availability Statement: The datasets generated and/or analyzed during the current study are available from the corresponding author upon reasonable request.

Conflicts of Interest: The authors declare that there is no conflict of interest and that they have no known competing financial interests or personal relationships that could appear to have influenced the work reported in this paper.

References

1. Shultz, S. Onions. *J. Agric. Food Inf.* **2010**, *11*, 8–15.
2. Javadzadeh, A.; Ghorbanihaghjo, A.; Bonyadi, S.; Rashidi, M.R.; Mesgari, M.; Rashtchizadeh, N.; Argani, H. Preventive Effect of Onion Juice on Selenite-Induced Experimental Cataract. *Indian J. Ophthalmol.* **2009**, *57*, 185. [PubMed]
3. FAOSTAT. Food and Agriculture Organization of the United Nations. Available online: https://www.fao.org/faostat/en/#data/QCL (accessed on 2 June 2023).
4. Dutta, R.; Nadig, S.M.; Manjunathagowda, D.C.; Gurav, V.S.; Singh, M. Anthracnose of Onion (*Allium cepa* L.): A Twister Disease. *Pathogens* **2022**, *11*, 884. [PubMed]
5. Darwesh, O.M.; Elshahawy, I.E. Silver Nanoparticles Inactivate Sclerotial Formation in Controlling White Rot Disease in Onion and Garlic Caused by the Soil Borne Fungus *Stromatinia cepivora*. *Eur. J. Plant Pathol.* **2021**, *160*, 917–934.
6. Elshahawy, I.E. Reduced Sclerotial Viability of *Stromatinia cepivora* and Control of White Rot Disease of Onion and Garlic by Means of Soil Bio-Solarization. *Eur. J. Plant Pathol.* **2021**, *160*, 519–540.
7. Valdés-Santiago, L.; Vargas-Bernal, R.; Herrera-Pérez, G.; Colli-Mull, J.G.; Ordaz-Arias, A. Application of Two-Photon Microscopy to Study *Sclerotium cepivorum* Berk Sclerotia Isolated from Naturally Infested Soil and Produced In Vitro. *Curr. Microbiol.* **2021**, *78*, 749–755.
8. Coley-Smith, J.R.; Mitchell, C.M.; Sansford, C.E. Long-term Survival of Sclerotia of *Sclerotium cepivorum* and *Stromatinia gladioli*. *Plant Pathol.* **1990**, *39*, 58–69.
9. Amin, M.; Tadele, S.; Selvaraj, T. White Rot (*Sclerotium cepivorum* Berk)—An Aggressive Pest of Onion and Garlic in Ethiopia: An Overview. *J. Agric. Biotechnol. Sustain. Dev.* **2014**, *6*, 6–15. [CrossRef]
10. Melero-Vara, J.M.; Basallote-Ureba, A.M.P.-L.; Basallote-Ureba, M.J. Comparison of Physical, Chemical and Biological Methods of Controlling Garlic White Rot. *Eur. J. Plant Pathol.* **2000**, *106*, 581–588.
11. Coventry, E.; Noble, R.; Mead, A.; Whipps, J.M. Suppression of Allium White Rot (*Sclerotium cepivorum*) in Different Soils Using Vegetable Wastes. *Eur. J. Plant Pathol.* **2005**, *111*, 101–112.
12. Clarkson, J.P.; Payne, T.; Mead, A.; Whipps, J.M. Selection of Fungal Biological Control Agents of *Sclerotium cepivorum* for Control of White Rot by Sclerotial Degradation in a UK Soil. *Plant Pathol.* **2002**, *516*, 735–745.
13. Shalaby, M.E.; Ghoniem, K.E. Biological and Fungicidal Antagonism of *Sclerotium cepivorum* for Controlling Onion White Rot Disease. *Ann. Microbiol.* **2013**, *63*, 1579–1589. [CrossRef]
14. Coley-Smith, J.R. White Rot Disease of *Allium*: Problems of Soil-Borne Diseases in Microcosm. *Plant Pathol.* **1990**, *39*, 214–222.
15. Zubrod, J.P.; Bundschuh, M.; Arts, G.; Brühl, C.A.; Imfeld, G.; Knäbel, A.; Payraudeau, S.; Rasmussen, J.J.; Rohr, J.; Scharmüller, A.; et al. Fungicides: An Overlooked Pesticide Class? *Environ. Sci. Technol.* **2019**, *53*, 3347–3365. [PubMed]
16. Deresa, E.M.; Diriba, T.F. Phytochemicals as Alternative Fungicides for Controlling Plant Diseases: A Comprehensive Review of Their Efficacy, Commercial Representatives, Advantages, Challenges for Adoption, and Possible Solutions. *Heliyon* **2023**, *9*, e13810. [CrossRef] [PubMed]

17. Chet, I.; Inbar, J. Biological Control of Fungal Pathogens. *Appl. Biochem. Biotechnol.* **1994**, *48*, 37–43. [CrossRef]
18. Sharma, I. Phytopathogenic Fungi and Their Biocontrol Applications. In *Fungi Bio-Prospects in Sustainable Agriculture, Environment and Nano-Technology, Volume 1: Fungal Diversity of Sustainable Agriculture*; Sharma, V.K., Shah, M.P., Parmar, S., Kumar, A., Eds.; Academic Press: Cambridge, MA, USA, 2021; Volume 1, pp. 155–188. ISBN 9780128213940.
19. Aliye, N.; Fininsa, C.; Hiskias, Y. Evaluation of Rhizosphere Bacterial Antagonists for Their Potential to Bioprotect Potato (*Solanum Tuberosum*) against Bacterial Wilt (*Ralstonia solanacearum*). *Biol. Control* **2008**, *47*, 282–288.
20. Messiha, N.A.S.; van Diepeningen, A.D.; Farag, N.S.; Abdallah, S.A.; Janse, J.D.; van Bruggen, A.H.C. *Stenotrophomonas Maltophilia*: A New Potential Biocontrol Agent of *Ralstonia solanacearum*, Causal Agent of Potato Brown Rot. *Eur. J. Plant Pathol.* **2007**, *118*, 211–225. [CrossRef]
21. Debette, J.; Blondeau, R. Presence of *Pseudomonas maltophilia* in the Rhizosphere of Several Cultivated Plants. *Can. J. Microbiol.* **1980**, *26*, 460–463.
22. Garbeva, P.; Van Overbeek, L.S.; Van Vuurde, J.W.L.; Van Elsas, J.D. Analysis of Endophytic Bacterial Communities of Potato by Plating and Denaturing Gradient Gel Electrophoresis (DGGE) of 16S rDNA Based PCR Fragments. *Microb. Ecol.* **2001**, *41*, 369–383. [PubMed]
23. Berg, G.; Marten, P.; Ballin, G. *Stenotrophomonas maltophilia* in the Rhizosphere of Oilseed Rape—Occurrence, Characterization and Interaction with Phytopathogenic Fungi. *Microbiol. Res.* **1996**, *151*, 19–27. [CrossRef]
24. Sultana, F.; Hossain, M.M. Assessing the Potentials of Bacterial Antagonists for Plant Growth Promotion, Nutrient Acquisition, and Biological Control of Southern Blight Disease in Tomato. *PLoS ONE* **2022**, *17*, e0267253.
25. Badri Fariman, A.; Abbasiliasi, S.; Akmar Abdullah, S.N.; Mohd Saud, H.; Wong, M.Y. *Stenotrophomonas maltophilia* Isolate UPMKH2 with the Abilities to Suppress Rice Blast Disease and Increase Yield a Promising Biocontrol Agent. *Physiol. Mol. Plant Pathol.* **2022**, *121*, 101872. [CrossRef]
26. Lodewyckx, C.; Vangronsveld, J.; Porteous, F.; Edward, R.B.; Taghavi, S.; Mezgeay, M.; Van Der Lelie, D. Endophytic Bacteria and Their Potential Applications. *Crit. Rev. Plant Sci.* **2002**, *21*, 583–606.
27. Santoyo, G.; Moreno-hagelsieb, G.; Orozco-mosqueda, C.; Glick, B.R. Plant Growth-Promoting Bacterial Endophytes. *Microbiol. Res.* **2016**, *183*, 92–99. [CrossRef]
28. Liu, Y.; Liu, W.; Liang, Z. Endophytic Bacteria from *Pinellia ternata*, a New Source of Purine Alkaloids and Bacterial Manure. *Pharm. Biol.* **2015**, *53*, 1545–1548.
29. Michail, G.; Reizopoulou, A.; Vagelas, I. Evaluation of The Biocontrol Efficacy of *Serratia proteamaculans* and *S. liquefaciens* Isolated From Bats Guano Pile From a Subterrestrial Cave (Greece). *Agric. Biol. (Sel'skokhozyaistvennaya Biol.)* **2022**, *57*, 566–578. [CrossRef]
30. Tian, S.P.; Bertolini, P. Effects of Low Temperature on Mycelial Growth and Spore Germination of *Botrytis allii* in Culture and on Its Pathogenicity to Stored Garlic Bulbs. *Plant Pathol.* **1995**, *44*, 1008–1015. [CrossRef]
31. Zewide, T.; Fininsa, C.; Sakhuja, P.K. Management of White Rot (*Sclerotium cepivorum*) of Garlic Using Fungicides in Ethiopia. *Crop Prot.* **2007**, *26*, 856–866. [CrossRef]
32. El-Nagar, A.; Elzaawely, A.A.; Xuan, T.D.; Gaber, M.; El-Wakeil, N.; El-Sayed, Y.; Nehela, Y. Metal Complexation of Bis-Chalcone Derivatives Enhances Their Efficacy against Fusarium Wilt Disease, Caused by *Fusarium equiseti*, via Induction of Antioxidant Defense Machinery. *Plants* **2022**, *11*, 2418. [CrossRef]
33. Nehela, Y.; Mazrou, Y.S.A.; Taha, N.A.; Elzaawely, A.A.; Xuan, T.D.; Makhlouf, A.H.; El-Nagar, A. Hydroxylated Cinnamates Enhance Tomato Resilience to *Alternaria alternata*, the Causal Agent of Early Blight Disease, and Stimulate Growth and Yield Traits. *Plants* **2023**, *12*, 1775. [CrossRef] [PubMed]
34. Tamura, K.; Nei, M. Estimation of the Number of Nucleotide Substitutions in the Control Region of Mitochondrial DNA in Humans and Chimpanzees. *Mol. Biol. Evol.* **1993**, *10*, 512–526. [CrossRef] [PubMed]
35. Tamura, K.; Stecher, G.; Kumar, S. MEGA11: Molecular Evolutionary Genetics Analysis Version 11. *Mol. Biol. Evol.* **2021**, *38*, 3022–3027. [CrossRef] [PubMed]
36. Johnson, L.F.; Curl, E.A. Methods for Research on the Ecology of Soil-Borne Plant Pathogens. In *Methods for Research on the Ecology of Soil-Borne Plant Pathogens*; Burgess Publishing Co.: Clayton, NC, USA, 1972.
37. Bulgari, D.; Bozkurt, A.I.; Casati, P.; Çağlayan, K.; Quaglino, F.; Bianco, P.A. Endophytic Bacterial Community Living in Roots of Healthy and '*Candidatus* Phytoplasma Mali'-Infected Apple (*Malus domestica*, Borkh.) Trees. *Antonie Van Leeuwenhoek* **2012**, *102*, 677–687. [CrossRef] [PubMed]
38. Al-Hussini, H.S.; Al-Rawahi, A.Y.; Al-Marhoon, A.A.; Al-Abri, S.A.; Al-Mahmooli, I.H.; Al-Sadi, A.M.; Velazhahan, R. Biological Control of Damping-off of Tomato Caused by *Pythium aphanidermatum* by Using Native Antagonistic Rhizobacteria Isolated from Omani Soil. *J. Plant Pathol.* **2019**, *101*, 315–322.
39. Amin, M.M.; Ahmed, M.F.A. Biological Control of Onion White Rot Disease Using Potential *Bacillus subtilis* Isolates. *Egypt. J. Biol. Pest Control* **2023**, *33*, 27. [CrossRef]
40. Kähkönen, M.P.; Hopia, A.I.; Vuorela, H.J.; Rauha, J.P.; Pihlaja, K.; Kujala, T.S.; Heinonen, M. Antioxidant Activity of Plant Extracts Containing Phenolic Compounds. *J. Agric. Food Chem.* **1999**, *47*, 3954–3962. [CrossRef]
41. Djeridane, A.; Yousfi, M.; Nadjemi, B.; Boutassouna, D.; Stocker, P.; Vidal, N. Antioxidant Activity of Some Algerian Medicinal Plants Extracts Containing Phenolic Compounds. *Food Chem.* **2006**, *97*, 654–660.
42. Harrach, B.D.; Fodor, J.; Pogány, M.; Preuss, J.; Barna, B. Antioxidant, Ethylene and Membrane Leakage Responses to Powdery Mildew Infection of near-Isogenic Barley Lines with Various Types of Resistance. *Eur. J. Plant Pathol.* **2008**, *121*, 21–33. [CrossRef]

43. Malik, C.P.; Singh, M.B. *Plant Enzymology and Histo-Enzymology*; Kalyani Publishers: New Delhi, India, 1980.
44. Hijaz, F.; Nehela, Y.; Killiny, N. Possible Role of Plant Volatiles in Tolerance against Huanglongbing in Citrus. *Plant Signal. Behav.* **2016**, *11*, e1138193. [CrossRef]
45. Nehela, Y.; Hijaz, F.; Elzaawely, A.A.; El-Zahaby, H.M.; Killiny, N. Phytohormone Profiling of the Sweet Orange (*Citrus sinensis* (L.) Osbeck) Leaves and Roots Using GC-MS-Based Method. *J. Plant Physiol.* **2016**, *199*, 12–17. [CrossRef]
46. Killiny, N.; Nehela, Y. Metabolomic Response to Huanglongbing: Role of Carboxylic Compounds in *Citrus sinensis* Response to '*Candidatus* Liberibacter asiaticus' and Its Vector. *Diaphorina citri. Mol. Plant-Microbe Interact.* **2017**, *30*, 666–678. [CrossRef] [PubMed]
47. Nehela, Y.; Hijaz, F.; Elzaawely, A.A.; El-Zahaby, H.M.; Killiny, N. Citrus Phytohormonal Response to *Candidatus* Liberibacter Asiaticus and Its Vector. *Diaphorina citri. Physiol. Mol. Plant Pathol.* **2018**, *102*, 24–35. [CrossRef]
48. Nehela, Y.; Killiny, N. Melatonin Is Involved in Citrus Response to the Pathogen Huanglongbing via Modulation of Phytohormonal Biosynthesis. *Plant Physiol.* **2020**, *184*, 2216–2239. [CrossRef] [PubMed]
49. Nehela, Y.; Taha, N.A.; Elzaawely, A.A.; Xuan, T.D.; Amin, M.A.; Ahmed, M.E.; El-Nagar, A. Benzoic Acid and Its Hydroxylated Derivatives Suppress Early Blight of Tomato (*Alternaria solani*) via the Induction of Salicylic Acid Biosynthesis and Enzymatic and Nonenzymatic Antioxidant Defense Machinery. *J. Fungi* **2021**, *7*, 663. [CrossRef]
50. Hussain, W.A.; Elzaawely, A.A.; El Sheery, N.I.; Ismail, A.A.; El-Zahaby, H.M. Biological Control of Onion White Rot Disease Caused by *Sclerotium cepivorum*. *Environ. Biodivers. Soil Secur.* **2017**, *1*, 101–107. [CrossRef]
51. Rivera-Méndez, W.; Obregón, M.; Morán-Diez, M.E.; Hermosa, R.; Monte, E. *Trichoderma asperellum* Biocontrol Activity and Induction of Systemic Defenses against *Sclerotium cepivorum* in Onion Plants under Tropical Climate Conditions. *Biol. Control* **2020**, *141*, 104145. [CrossRef]
52. Dunne, C.; Crowley, J.J.; Moënne-Loccoz, Y.; Dowling, D.N.; De Bruijn, F.J.; O'Gara, F. Biological Control of *Pythium ultimum* by *Stenotrophomonas maltophilia* W81 Is Mediated by an Extracellular Proteolytic Activity. *Microbiology* **1997**, *143*, 3921–3931. [CrossRef]
53. Dunne, C.; Moenne-Loccoz, Y.; de Bruijn, F.J.; O'Gara, F. Overproduction of an Inducible Extracellular Serine Protease Improves Biological Control of *Pythium ultimum* by *Stenotrophomonas maltophilia* Strain W81. *Microbiology* **2000**, *146*, 2069–2078. [CrossRef]
54. Jankiewicz, U.; Brzezinska, M.S.; Saks, E. Identification and Characterization of a Chitinase of *Stenotrophomonas maltophilia*, a Bacterium That Is Antagonistic towards Fungal Phytopathogens. *J. Biosci. Bioeng.* **2012**, *113*, 30–35. [CrossRef]
55. Giesler, L.J.; Yuen, G.Y. Evaluation of *Stenotrophomonas maltophilia* Strain C3 for Biocontrol of Brown Patch Disease. *Crop Prot.* **1998**, *17*, 509–513. [CrossRef]
56. Dewi, R.R.; Rahmah, S.M.; Taruna, A.; Aini, L.Q.; Fernando, I.; Abadi, A.L.; Syib'li, M.A. The Effectiveness Comparison Between Application of Indigenous Arbuscular Mycorrhizal Fungal Community and *Stenotrophomonas maltophilia* to Suppress Fusarium Wilt Incidence on Local Garlic Plant (Lumbu Hijau). *AGRIVITA J. Agric. Sci.* **2023**, *45*, 131–146. [CrossRef]
57. Antonio Saministrado-Fudalan, P.; Sendaydiego, J.P.; Sinco, A.L. *Serratia liquefaciens* from IMO: Antibacterial and Chitinolytic Potentials. *Int. J. Sci. Basic Appl. Res.* **2017**, *36*, 306–317.
58. Sneh, B.; Agami, O.; Baker, R. Biological Control of Fusarium-Wilt in Carnation with *Serratia liquefaciens* and Hafnia Alvei Isolated from Rhizosphere of Carnation. *J. Phytopathol.* **1985**, *113*, 271–276. [CrossRef]
59. Kshetri, L.; Naseem, F.; Pandey, P. Role of *Serratia* sp. as Biocontrol Agent and Plant Growth Stimulator, with Prospects of Biotic Stress Management in Plant. In *Plant Growth Promoting Rhizobacteria for Sustainable Stress Management: Volume 2: Rhizobacteria in Biotic Stress Management*; Springer: Singapore, 2019; pp. 169–200. [CrossRef]
60. Lucini, E.I.; Zunino, M.P.; López, M.L.; Zygadlo, J.A. Effect of Monoterpenes on Lipid Composition and Sclerotial Development of *Sclerotium cepivorum* Berk. *J. Phytopathol.* **2006**, *154*, 441–446. [CrossRef]
61. Diastuti, H.; Chasani, M. Suwandri Antibacterial Activity of Benzyl Benzoate and Crotepoxide from *Kaempferia rotunda* L. Rhizome. *Indones. J. Chem.* **2020**, *20*, 9–15. [CrossRef]
62. Walters, D.; Raynor, L.; Mitchell, A.; Walker, R.; Walker, K. Antifungal Activities of Four Fatty Acids against Plant Pathogenic Fungi. *Mycopathologia* **2004**, *157*, 87–90. [CrossRef]
63. Liu, S.; Ruan, W.; Li, J.; Xu, H.; Wang, J.; Gao, Y.; Wang, J. Biological Control of Phytopathogenic Fungi by Fatty Acids. *Mycopathologia* **2008**, *166*, 93–102. [CrossRef]
64. Casillas-Vargas, G.; Ocasio-Malavé, C.; Medina, S.; Morales-Guzmán, C.; Del Valle, R.G.; Carballeira, N.M.; Sanabria-Ríos, D.J. Antibacterial Fatty Acids: An Update of Possible Mechanisms of Action and Implications in the Development of the next-Generation of Antibacterial Agents. *Prog. Lipid Res.* **2021**, *82*, 101093. [CrossRef]
65. Yogeswari, S.; Ramalakshmi, S.; Neelavathy, R.; Muthumary, J. Identification and Comparative Studies of Different Volatile Fractions from Monochaetia Kansensis by GCMS. *Glob. J. Pharmacol.* **2012**, *6*, 65–71.
66. Faridha Begum, I.; Mohankumar, R.; Jeevan, M.; Ramani, K. GC–MS Analysis of Bio-Active Molecules Derived from *Paracoccus pantotrophus* FMR19 and the Antimicrobial Activity Against Bacterial Pathogens and MDROs. *Indian J. Microbiol.* **2016**, *56*, 426–432. [CrossRef]
67. Wagay, N.A.; Mohiuddin, Y.G.; Khan, N.A. Phytochemical Evaluation and Identification of Bioactive Phytochemical Evaluation and Identification of Bioactive Compounds in Camel Thorn Alhagi Pseudalhagi (M. Bieb.). *Int. J. Adv. Res. Sci. Eng.* **2018**, *7*, 300–311.

68. Zhang, Z.; Yuen, G.Y. The Role of Chitinase Production by *Stenotrophomonas maltophilia* Strain C3 in Biological Control of Bipolaris Sorokiniana. *Phytopathology* **2007**, *90*, 384–389. [CrossRef]
69. Sharma, P.; Jha, A.B.; Dubey, R.S.; Pessarakli, M. Reactive Oxygen Species, Oxidative Damage, and Antioxidative Defense Mechanism in Plants under Stressful Conditions. *J. Bot.* **2012**, *2012*, 217037. [CrossRef]
70. Rajput, V.D.; Harish; Singh, R.K.; Verma, K.K.; Sharma, L.; Quiroz-Figueroa, F.R.; Meena, M.; Gour, V.S.; Minkina, T.; Sushkova, S.; et al. Recent Developments in Enzymatic Antioxidant Defence Mechanism in Plants with Special Reference to Abiotic Stress. *Biology* **2021**, *10*, 267. [CrossRef] [PubMed]
71. Racchi, M.L. Antioxidant Defenses in Plants with Attention to *Prunus* and *Citrus* spp. *Antioxidants* **2013**, *2*, 340–369. [PubMed]
72. Granér, G.; Hamberg, M.; Meijer, J. Screening of Oxylipins for Control of Oilseed Rape (*Brassica napus*) Fungal Pathogens. *Phytochemistry* **2003**, *63*, 89–95. [CrossRef]
73. Desai, M.D.; Suthar, K.P.; Singh, D.; Parekh, V.; Khunt, M.D.; Haidar, A. Molecular Evidence of Phytohormonal Modulatory Effect of *Serratia liquefaciens* on Rice. *Appl. Biol. Res.* **2020**, *22*, 126. [CrossRef]
74. Hassan, T.U.; Bano, A. Comparative Effects of Wild Type *Stenotrophomonas maltophilia* and Its Indole Acetic Acid-Deficient Mutants on Wheat. *Plant Biol.* **2016**, *18*, 835–841. [CrossRef] [PubMed]
75. Ambawade, M.S.; Pathade, G.R. Production of Indole Acetic Acid (IAA) by *Stenotrophomonas maltophilia* BE25 Isolated from Roots of Banana (*Musa* spp). *Int. J. Sci. Res.* **2013**, *4*, 2644–2650.
76. Singh, R.P.; Jha, P.N. The PGPR *Stenotrophomonas maltophilia* SBP-9 Augments Resistance against Biotic and Abiotic Stress in Wheat Plants. *Front. Microbiol.* **2017**, *8*, 1945. [CrossRef] [PubMed]
77. Caneschi, W.L.; Sanchez, A.B.; Felestrino, É.B.; de Carvalho Lemes, C.G.; Cordeiro, I.F.; Fonseca, N.P.; Villa, M.M.; Vieira, I.T.; Moraes, L.Â.G.; de Almeida Barbosa Assis, R.; et al. *Serratia liquefaciens* FG3 Isolated from a Metallophyte Plant Sheds Light on the Evolution and Mechanisms of Adaptive Traits in Extreme Environments. *Sci. Rep.* **2019**, *9*, 18006. [CrossRef] [PubMed]
78. Cheng, C.; Nie, Z.W.; He, L.Y.; Sheng, X.F. Rice-Derived Facultative Endophytic *Serratia liquefaciens* F2 Decreases Rice Grain Arsenic Accumulation in Arsenic-Polluted Soil. *Environ. Pollut.* **2020**, *259*, 113832. [CrossRef] [PubMed]
79. Han, H.; Wang, Q.; He, L.Y.; Sheng, X.F. Increased Biomass and Reduced Rapeseed Cd Accumulation of Oilseed Rape in the Presence of Cd-Immobilizing and Polyamine-Producing Bacteria. *J. Hazard. Mater.* **2018**, *353*, 280–289. [CrossRef] [PubMed]

horticulturae

MDPI

Review

A Review on Biocontrol Agents as Sustainable Approach for Crop Disease Management: Applications, Production, and Future Perspectives

Anshika Tyagi [1,†], Tensangmu Lama Tamang [1,†], Hamdy Kashtoh [1,†], Rakeeb Ahmad Mir [2], Zahoor Ahmad Mir [3], Subaya Manzoor [4], Nazia Manzar [5], Gousia Gani [6], Shailesh Kumar Vishwakarma [5], Mohammed A. Almalki [7,*] and Sajad Ali [1,*]

1 Department of Biotechnology, Yeungnam University, Gyeongsan 38541, Republic of Korea; hamdy_kashtoh@ynu.ac.kr (H.K.)
2 Department of Biotechnology, School of Life Sciences, Central University of Kashmir, Ganderbal 191201, India
3 Department of Plant Science and Agriculture, University of Manitoba, Winnipeg, MB R2M0TB1, Canada; zahoorbio@gmail.com
4 Division of Plant Pathology, Faculty of Agriculture SKUAST-K, Wadura, Sopore 193201, India
5 Plant Pathology Lab, ICAR-National Bureau of Agriculturally Important Microorganisms, Maunath Bhanjan 275103, India
6 Division of Basic Science and Humanities, Faculty of Agriculture-SKUAST-K, Wadura, Sopore 193201, India
7 Department of Biological Sciences, College of Science, King Faisal University, Al-Ahsa, Al Hofuf 31982, Saudi Arabia
* Correspondence: malmalki@kfu.edu.sa (M.A.A.); sajadmicro@yu.ac.kr (S.A.)
† These authors contributed equally to this work.

Abstract: Horticultural crops are vulnerable to diverse microbial infections, which have a detrimental impact on their growth, fruit quality, and productivity. Currently, chemical pesticides are widely employed to manage diseases in horticultural crops, but they have negative effects on the environment, human health, soil physiochemical properties, and biodiversity. Additionally, the use of pesticides has facilitated the development and spread of resistant pathovars, which have emerged as a serious concern in contemporary agriculture. Nonetheless, the adverse consequences of chemical pesticides on the environment and public health have worried scientists greatly in recent years, which has led to a switch to the use of biocontrol agents such as bacteria, fungi, and insects to control plant pathogens. Biocontrol agents (BCAs) form an integral part of organic farming, which is regarded as the future of sustainable agriculture. Hence, harnessing the potential of BCAs is an important viable strategy to control microbial disease in horticultural crops in a way that is also ecofriendly and can improve the soil health. Here, we discuss the role of the biological control of microbial diseases in crops. We also discuss different microbial-based BCAs such as fungal, bacterial, and viral and their role in disease management. Next, we discuss the factors that affect the performance of the BCAs under field conditions. This review also highlights the genetic engineering of BCAs to enhance their biocontrol efficiency and other growth traits. Finally, we highlight the challenges and opportunities of biocontrol-based disease management in horticulture crops and future research directions to boost their efficacy and applications.

Keywords: horticulture; biocontrol agents; diseases; pesticides; production; metabolic engineering

1. Introduction

Horticultural crops are an important source of income with high nutritional, medicinal, and industrial significance [1]. Farmers' adaptation to horticultural farming has significantly increased in recent times because of the huge demand, substantial income, and technological advancement. Agriculture and economic diversification can be accelerated via horticulture farming [1]. However, the production of horticulture crops is affected by both biotic and abiotic stresses [2]. Among them, phytopathogens pose serious threats to horticultural crops across the globe. For instance, fungal pathogens such as *Alternaria* spp., *Phytophthora infestans*, *Septoria lycopersici*, *Fusarium oxysporum* f. sp. *lycopersici*, *Botrytis cinerea*, and *Verticillium dahliae* have been found to significantly affect the quality and yield production of horticulturally important crops [3–7]. In addition, a large number of seed-born fungal pathogens viz., *Macrophomina phaseolina*, *Colletotrichum*, *Fusarium*, *Ascochyta pinodes*, and *Sclerotinia sclerotiorum* have been reported to impair crop growth and yield [6–10]. On the other hand, many bacterial diseases like canker, soft rot, leaf spot/spot, wilt, blight, speck, and brown spot affect the fruit quality and crop productivity [11,12]. Moreover, different types of viruses cause severe diseases in horticultural crops by hampering their physiological processes and hindering growth, leading to higher yield losses [13,14].

Currently, the majority of the farmers use pesticides as a standard method of managing disease in horticultural crops, which has huge implications on human health and the environment. The excessive application of pesticides has polluted water bodies, leading to a growing apprehension regarding the spread of toxic chemicals to humans [15]. Despite the use of antifungals, on a global scale, growers lose about 10–23% of crops due to infestations of fungal pathogens [16]. The situation is further worsened by an additional 20% crop loss during the post-harvest process [16]. On the other hand, for the management of bacterial disease, antibiotics are used, which have also become a major concern due the emergence of antibiotic resistant pathovars. For instance, strAB streptomycin-resistance genes have been found in *Pseudomonas syringae*, *Erwinia amylovora*, and *Xanthomonas campestris* [17]. Previous study has shown that bacterial plant pathogens have developed resistance against a wide range of antibiotics such as streptomycin and tetracycline [17]. Therefore, there is a need to find a viable alternative to address this problem and manage bacterial pathogens in horticulture. Although transgenic technology has also been adapted for improving disease resistance in horticulture crops, apart from several benefits such as developing resistant plant varieties, transgenic plants raise serious concerns regarding health risks and environmental concerns [18].

BCAs can play a vital role in disease management and increase food production in a more sustainable way [19]. However, researchers should also thoroughly investigate the range of BCAs to reduce the use of synthetic chemicals that are damaging the balance in ecosystems as well as address pathogen resistance to BCAs. The use of pesticides has expanded during the last decade due to their contribution to improving agricultural productivity by controlling the prevalence and persistence of plant pathogens [20]. Previous studies support the potential benefits of pesticides to control plant pathogens [21,22]. Eventually, there is a strong anticipation that scientific endeavors will strongly need to mitigate the use of hazardous pesticides to contain the peril they pose to human well-being and disturbance in the ecosystem. All of the above-mentioned strategies and their potential constraints have forced scientists to rely on natural methods of controlling plant pathogens. One such approach is the use of BCAs, which have a huge number of benefits and the least constraints. In addition to the use of disease-resistant cultivars, the intervention of BCAs has played a pivotal role in integrated pest management strategies to reduce the use of chemical-based pesticides. Specific BCAs systematically penetrate the target pathogen without affecting the crop plant [23]. Additionally, the public perception of using BCAs as natural and environmentally friendly makes them more convenient for agricultural use. The BCAs confer disease management by enhancing the immunity of crop plants, secreting antimicrobial compounds that inhibit pathogen prevalence and growth [23]. The importance of BCAs and the principle of synergically assessing their efficiency, efficacy, durability, precision, and environmental safety further widen the scope for their use in our agricultural systems. The current review article is a compendium to address the contribution of BCAs in controlling plant pathogens and also highlights the potential benefits of BCAs in comparison to chemical pesticides.

2. Biocontrol as a Viable Approach for Controlling Disease in Horticulture Crops

BCAs have emerged as a viable tool not only for controlling pathogens and pests, but also in improving plant growth and soil health, which make them an important tool in maintaining the one-health principle [24]. However, several enhancements such as scaling up production, optimizing formulation techniques and delivery methods, increasing cost-effectiveness, and meeting regulatory requirements are necessary before they can be successfully implemented in the field [25]. The global scientific community is at pace with exploring BCAs in terms of understanding the mechanisms behind their strategies to control pathogens [26]. BCAs include bacteria, viruses, fungi, insects, mites, nematodes, yeasts, and protozoa, which control pathogens through a wide range of biological mechanisms [27]. Upon interaction with the target pathogen, BCAs modulate their genetic machinery to produce arsenals of compounds to circumvent the pathogen dominance and disease prevalence [28,29]. Bacterial species such as *Bacillus thuringiensis, Bacillus subtilis, Bacillus amyloliquefaciens, Bacillus licheniformis, Bacillus velezensis, Bacillus pumilus,* and *Bacillus mojavensis* are used as BCAs against plant fungal pathogens [30–32]. The biocontrol mechanism mediated by these bacterial species includes the production of antibiotics, siderophore production, and the production of antimicrobial metabolites including lytic enzymes, and most importantly, inducing systemic resistance mechanisms against specific pathogens [33]. In addition, BCAs also trigger the production of growth phytohormones such as cytokinins (CK), indole-3-acetic acid (IAA), and gibberellins (GA) as well as the defense hormones salicylic acid (SA), abscisic acid (ABA), and jasmonic acid (JA) to improve growth and adaptive responses during stress conditions [34,35]. Moreover, some BCAs also produce metabolic intermediates such as ACC deaminase, the enzyme that degrades 1-aminocyclopropane-1-carboxylic acid (ACC) to regulate the ethylene concentration under biotic stress conditions [35]. Consequently, these reports suggest that BCAs operate multiple mechanisms to control pathogens in addition to improving the growth and metabolism in crop plants [36]. Furthermore, we show how BCAs use a multifaceted mode of action for controlling plant pathogens or pests in Figure 1.

Many antimicrobial compounds (AMCs) largely belonging to secondary metabolites are known to be secreted by microbes such as bacteria, fungi, and actinomycetes [37]. The classes of antimicrobial agents produced by bacteria include non-ribosomal lipopeptides and peptides (NRPs), ribosomal peptides (RPs), and polyketides (PKs) [38]. A large class of NRPs belongs to cyclic lipopeptides (LPs), which exert their action against a wide range of fungal and bacterial pathogens. This class of antibacterial agents mediates their action by attaching to a membrane to cause perforation, resulting in the leakage of ions, which is followed by depolarization of the membrane, where, once entering the cells, it inhibits the replication, transcription, and translation of bacterial pathogens [39]. Similarly, several types of LPs exhibit antifungal properties and exert their action by disrupting cell walls by synthesizing chitin and (1–3)-β-D-glucan synthases, which in turn disrupts osmotic regulation and the morphological architecture of fungal pathogens [40]. In addition, the *Bacillus* species contains three families of LPs like iturin, fengycin, and surfactin, which possess both antifungal as well as antibacterial properties [29,41]. Hussain et al. [42] recently reported that the LPs iturin A and bacillomycin F produced by the *Bacillus siamensis* Sh420 strain inhibited the growth of *Fusarium graminearum*. Similarly, previous studies have shown that iturin A and bacillomycin F from the *B. siamensis* JFL15 strain showed strong antifungal activity against different plant fungal pathogens such as *Colletotrichum nymphaeae, Rhizoctonia solani,* and *Magnaporthe grisea* [43]. The β-1,3-glucanase and chitinase secreted by *B. siamensis* QN2MO-1 contribute to the antifungal activities against *Fusarium* wilt in tomato plants, and also promotes growth and improves fruit quality [44]. LPs have a great potential as biocontrol elements synthesized by bacterial species to enhance agricultural productivity. For instance, the bacteriocins belonging to RP have been reported to exhibit broad-spectrum antimicrobial activities against closely related bacterial pathogens by disrupting the cell wall and damaging cytoplasmic membranes [38,45]. Several *Bacillus* species produce some important bacteriocins like amisin, amylolysin, ericin, entianin, thuricin, subtilin, subtilosin A, and subtilosin B, which are used as potential BCAs against a wide range of bacterial pathogens [38,46]. Bacteriocins like leuconocin S and leucocin A-UAL are produced by Gram-positive and Gram-negative bacterial species, even though their lysis mechanisms completely differ. The most widely accepted mechanism includes the absorption of bacteriocins by cell surfaces, followed by the inhibition of cell wall synthesis, enhancing the permeabilization of the cell membrane and the inhibition of RNases and DNases [45,47].

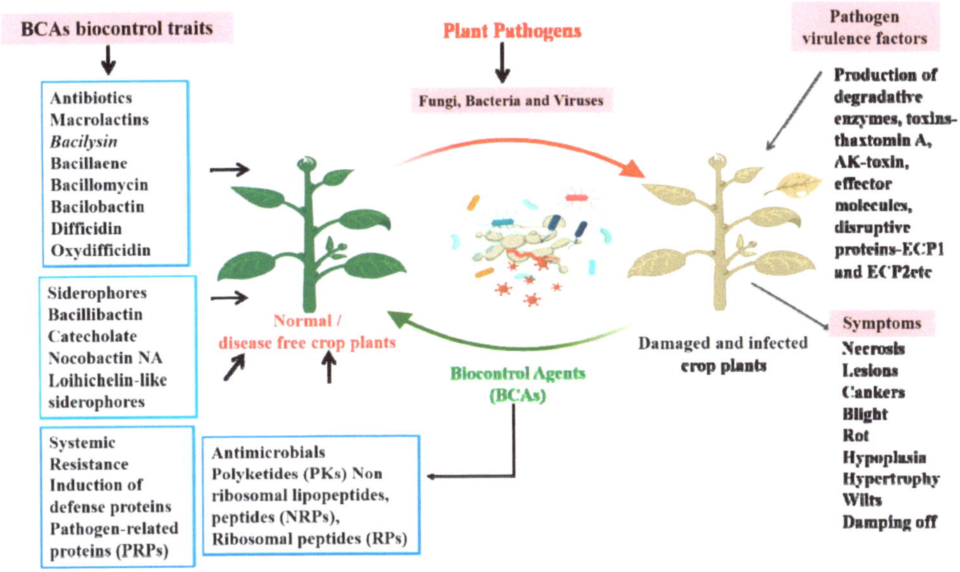

Figure 1. Schematic diagram depicting the defense mechanism operated by BCAs against plant pathogens. The production of bioactive molecules by BCAs mediates a diverse range of defense responses against pathogens. They produce antibiotics, siderophores, antimicrobial metabolites, and induce systemic resistance mechanisms against specific pathogens. This defense response consequently improves plant growth and regulates the defense mechanisms against a wide range of pathogens.

Polyketides and other peptides are another class of antimicrobial agents produced by BCAs that possess antifungal and antibacterial activities [46,48]. For instance, oxydifficidin and difficidin, produced by *Bacillus methylotrophicus* DR-08, exert antibacterial properties against *R. solanacearum*, the causative agent of bacterial wilt in tomato plants [49]. Additionally, macrolactin and bacillomycin D, isolated from *B. amyloliquefaciens* NJN-6, were found to possess antifungal properties against *F. oxysporum* and *R. solanacearum* in banana [50]. The polyketide viz. bacillaene, difficidin, and macrolactin, produced by *B. amyloliquefaciens* DSBA-11, inhibited the growth of the *Ralstonia pseudosolanacearum* pathogen in tomato plants [51]. Chen et al. [52] reported that the bacterial strain *B. velezensis* SDTB038 possessed antifungal properties against *Fusarium* crown and root rot in tomatoes by synthesizing metabolites like bacillaene, bactin, bacilysin, difficidin, fengycin, macrolactin H, and surfactin. A wide range of fungal species have previously been reported to be used as microbial formulations to contain the pathogenesis of pathogenic fungi *Botrytis*, *Fusarium*, *Sclerotinia*, and *Pythium* [53]. The biocontrol of these fungal species relies on mechanisms such as mycoparasitism, the synthesis of antimicrobial metabolites, and competition for nutrients and space [54,55].

The *Trichoderma* species-based biocontrol mechanism relies on antibiosis, mycoparasitism, competition for resources and space, and inducing defense systems. The mechanism of antibiosis relies on the production of primary and secondary metabolites that are critical in controlling the growth and division of pathogens. Reports suggest that *Trichoderma* spp., releases around 390 non-volatile metabolites, some of which include caffeic acid, cathequin, ferulic acid, gliotoxin, gliovirin, heptelidic acid, nematolin, sepedonin, 3,4,15-scirpenetriol, and viridian [56]. For instance, *Trichoderma* spp. helps in nutrient and water uptake, enhances the metabolism of plants, and induces defense signaling pathways by establishing a positive interaction with crop plants [57]. Strains of the fungal species *Trichoderma harzianum* have been characterized and identified as potential candidates to control *F. oxysporum* f. sp. *lactucae* (Fol) in baby lettuce (*Lactuca sativa* L. var *acephala*) [58].

3. Factors Affecting the BCAs Activities and Strategies to Overcome Limitations of BCA Effectiveness

Despite huge efforts, the impact of biocontrol has remained reasonably insignificant in managing to control diseases in crops in comparison to its synthetic counterparts [23]. Most importantly, BCAs face greater challenges when transforming their roles from experimental conditions to crop fields exposed to various environmental factors [59]. For instance, the type of plant species and genotype have a significant role in the effectiveness of BCAs [60]. Umer et al. [61] reported that the production of antimicrobials from BCAs as well as their colonization and induction of systematic resistance is dependent on the type of plant and varies from species to species. Consequently, for a targeted application of BCAs, scientists must uncover the complexities of host plants, BCAs, and pathogens to avoid less effectiveness and failure [62]. The pathogen also greatly influences the biocontrol potential of specific types of BCAs. Reports suggest that each pathogen interacts with the host plant differently in terms of pathogenesis and prevalence due to the ecological fitness and genetic variability in specific pathogens [63,64]. Additionally, a large number of pests have developed resistance to BCA antimicrobials. For example, the resistance of several pests such as *Diabrotica virgifera*, *Busseola fusca*, *Helicoverpa zea*, *Spodoptera frugiperda*, and *Pectinophora gossypiella* has been reported toward Bt toxins [59]. The nature of BCAs and the class of biocontrol agents they release is another factor that is critical for efficacy as potential BCAs. Consequently, to understand the nature of BCAs, it is critical to investigate the distribution pattern of BCAs in the rhizosphere [63,64]. Furthermore, the mode of action of BCAs, pathogen selection, and the ability to resist harsh climatic conditions play defining factors in the success of BCAs in disease management [65].

Environmental conditions such as biotic and abiotic factors play a pivotal role in the ability of BCAs to kill specific pathogens [59]. It is very important to select a specific type of BCA that has higher stability and efficacy in a wide range of environmental conditions like heat stress, drought stress, salinity stress, type of soil, and competition with other biotic species [66]. Understanding the factors that influence the BCAs and their efficacy is critical before their application in crop fields. Currently, the efficacy, viability, and susceptibility of the BCAs to stressful environmental conditions is one of the major problems that has limited its expansion [23]. Considering this limitation, the efficacy of biocontrol BCAs can be increased by preparing small liquid formulations. For instance, to control the devastating impact of *Sclerotium rolfsii*, Sutthisa et al. [67] prepared 5% trehalose-*Trichoderma asperellum* MSU007 as a liquid formulation. Another strategy to enhance the efficacy of BCAs is to combine them with fungicides in order to weaken and stress the pathogens to render them more susceptible to subsequent attacks by BCAs [68]. For instance, the combination of Fusarium sp. isolates with a benzimidazole fungicide improved the effectiveness of Fusarium sp. in controlling the Fusarium wilt of carnation or cyclamen [69]. Consequently, by employing a combination of BCAs and synthetic chemicals, the duration of effectiveness will be increased, and therefore, the use of fungicides or antibacterial chemicals will be reduced, leading to the sustainable agricultural production of crops. The combinatorial approach of using nanoparticles and BCAs is another versatile method for enhancing the efficacy of BCAs. For example, Vehapi et al. [70] used a formulation based on nanoparticles (BpNPs), polyvinyl alcohol (PVA), and sodium alginate (SA) in combination with *B. pumilus* as plant protection against *Colletotrichum gloeosporioides* infestation to inhibit postharvest fungal decay disease in crop plants. The huge number of inclusive studies demonstrates the potential of BCAs to enhance agricultural productivity at a global scale [71]. However, the affectivity of BCAs is determined by the number of crops per farm, the size of the farms, and the cultivation strategies [72,73]. Our existing agricultural practices must be redesigned to favor biocontrol applications. Apart from the use of BCAs, monitoring measures are also important to make decisions regarding the time of application, preventive measures, and the use of pathogen-resistant varieties are crucial to decrease the use of synthetic chemicals for pest control. Additional considerations include the choice of cropping system and specific environmental conditions that help avoid pest pressure and disease progression.

4. Fungal Species as BCAs: Mechanisms and Applications

Fungal species are one of the important groups of BCAs that are commonly used to control pests and pathogens mainly due to their target specificity, quick generation time, and relatively high reproduction rate [74]. Many genera of fungi have been utilized extensively as efficient BCAs against fungal phytopathogens such as *Alternaria* [75] *Penicillium* [76], *Aspergillus* [77], *Fusarium* [78], *Rhizoctonia* [79], *Colletotrichum* [80], and other devastating pests such as insects and nematodes [81]. *Trichoderma* spp. is the most well-known fungal antagonist being studied and employed as a microbial fungicide. Their integration promotes sustainable agricultural practices and aids in the efficient management of plant diseases [82].

Numerous fungi, particularly *Trichoderma*, have been found to have broad-spectrum antagonistic activity against a variety of phytopathogens [75–80]. *Trichoderma* is a common genus of fungus organisms that live in soil and function as mycoparasites, saprotrophs, and plant symbionts [82]. Filamentous fungi of this genus have been studied in detail and used as BCAs against agricultural plant diseases [83]. In order to provide effective plant disease management, a great deal of research has been conducted in the last 10 years on the direct and indirect control potential of BCAs against phytopathogens. In 1794, *Trichoderma* was initially isolated from soil and decaying organic materials [84]. Nowadays, *Trichoderma* is the source of more than 60% of effective biofungicides used worldwide [85]. Several antibacterial secondary metabolites including viridiofungin, trichokonin, and lysozyme have been identified from *T. harzianum*. These compounds have shown an antibacterial activity against bacterial plant pathogens such as *Clavibacter michiganensis* and *Erwinia amylovora* [86]. *R. solanacearum*, a causing agent of bacterial wilt disease in tomato, chilli, and other crops, is responsible for enormous agricultural losses globally [87]. Many approaches have been studied for the management of *R. solanacearum*. The crude extract of *T. harzianum* could inhibit the growth of *R. solanacearum* when in vitro and in planta experiments were conducted on tomato plants. Disruption of bacterial cells was confirmed by scanning electron microscopy [88]. Applications of the *T. asperellum* isolates T4 and T8 in the field delayed the onset of wilt, reduced the incidence of *R. solanacearum*-caused illness, and improved tomato plant growth and yield [89].

The fungal strain BCP, also called *Acremonium strictum*, has the ability to hinder *B. cinerea* growth by severely lysing the host hyphae [90]. Using light microscopy, the lysis of *B. cinerea* mycelia was verified [90]. *Fusarium* wilt was found to be effectively controlled by the Fo47 strain of *F. oxysporum*, which was recognized as nonpathogenic. It was also discovered that Fo47 could raise the bioactivity of chitinase, β-1,3-glucanase, and β-1,4-glucosidase in tomato plants, indicating that Fo47 could cause resistance in tomatoes [91]. Upon applying a mixture of *T. harzianum* and rice straw (RST) and oil palm (EFB) compost to okra plants, the plants exhibited resilience against Choanephora wet rot disease. Up to 85.04% less illness severity was observed when *T. harzianum* was suspended in water [92]. Jaihan et al. [93] used the extracts of entomopathogenic fungus *Ophiocordyceps sobolifera* against *Colletotrichum* spp. in chili plants and found that it significantly inhibited the mycelial growth and conidial germination of all the tested *Colletotrichum* spp. under in vitro conditions. A previous study showed that the coexpression of *β-1,3-* and *β-1,6*-glucanase genes in *T. virens* significantly inhibited pathogens such as *R. solani*, *Rhizopus oryzae*, and

Pythium ultimum (Oomycota, Chromista) [94]. In a field study, commercial formulations of *T. virens* (Soilgard) and *T. harzianum* (Rootshield) were used to prevent *Fusarium* wilt in tomato plants, where it was found that these formulations prevented 62 and 68% of the disease, respectively [95]. *Fusarium* wilt of potatoes (*F. oxysporum*) was better controlled by *T. harzianum* in a greenhouse setting compared to a field setting, but a higher yield was obtained in the field [96]. Previous study has reported that gliotoxin, an important secondary metabolite of *Trichoderma virens* T23, suppressed *Sclerotium rolfsii*, which causes destructive soilborne disease in many plants [97]. Isolates of *T. harzianum*, *T. viride*, and *T. spirale* showed different inhibitory effects against the mycelial growth of *R. solani* and *F. oxysporum* f. sp. phaseoli of bean in laboratory, greenhouse, and field conditions. *T. harzianum* formulation-treated plots had a yield comparable to the healthy control. Boureghda and Bouznad [98] reported that three strains of *T. atroviride* and *T. harzianum* (Th. 16) were the most effective isolates in protecting chickpea seedlings against Fusarium wilt among several isolates of *T. atroviride, T. harzianum*, and *T. longibrachiatum*. They also observed that an increase in the vegetative growth of the tested plants was correlated with a decrease in the severity of the disease. Tsai et al. [99] reported that among five *Trichoderma* strains isolated from *Anoectochilus rhizospheres*, a conidial formulation of a strain of *T. asperellum* mixed with carboxymethyl cellulose (CoCMC) had an excellent ability in controlling the stem rot disease (*F. oxysporum*) of Taiwan Anoectochilus and could completely protect the tested plants against this disease for 9 weeks. Schubert et al. [100] examined the potential of *Trichoderma* spp. as a wound treatment for controlling wood decay fungi in urban trees and reported that *T. atroviride* (T-15603.1) could be successfully applied as a biological wound treatment against wood decay fungi. Another previous study reported that *Penicillium citrinum* and *Aspergillus terreus* showed biocontrol activity against the *Sclerotium rolfsii* pathogen by inducing jasmonic acid and salicylic accumulation in sunflower plants [101]. Among the fungal BCAs, Trichoderma is the most widely used BCA across the globe. However, there is need to explore novel fungal BCAs with growth promoting traits as well as effective biocontrol activity that will boost agricultural productivity.

Mechanism of Action

Fungal BCAs can employ various mechanisms such as parasitism, competition for nutrients and space, prevention of pathogen colonization in specific host tissues, antibiosis, and the induction of plant resistance against diseases to target pathogens, thereby inhibiting their growth, sporulation, and spread within infected plants [45,55]. Multiple approaches have been developed to investigate the activity mechanisms of selected BCAs [102]. Interestingly, some BCAs initially thought to hinder pathogen development through mycoparasitism or antibiotic production were later found to induce systemic resistance in plants against pathogens [82]. Besides protecting plants from diseases, certain fungal BCAs have the ability to increase plant growth, leading to increased biomass and yield. Hyperparasitism is a fungal propensity characterized by direct antagonistic interactions, whereby other microorganisms eliminate a pathogen [59]. Sometimes, a fungal species can be parasitic on different species of fungi; this tendency is called mycoparasitism [103]. When cucumbers in greenhouse conditions were treated with conidial suspensions of the hyperparasite *Ampelomyces quisqualis*, they were found to be effective against *Sphaerotheca fuliginea*, a causative agent of powdery mildew of cucumber. *A. quisqualis* parasitized *S. fuliginea* extensively in studies with commercial cucumber crops [104]. Previous studies have reported that *Trichoderma* spp. are useful in the biocontrol of *R. solani*, the main fungus that causes damping-off and root rot in many crops [105,106]. Due to their hyperparasitic nature, *Trichoderma* spp. are among the most studied fungi as BCAs because they effectively protect crops against a wide range of plant pathogens. On the other hand, antagonistic relationships, or antibiosis, are biological interactions between two or more organisms that are harmful to one or more of them. Antagonistic fungi, for instance, produce antimicrobial compounds to inhibit the growth of pathogenic fungi in their vicinity [107]. Fungi produce one or more antimicrobial compounds and other secondary metabolites with antibiotic activity [108]. The most studied fungal species for antibiosis is *Trichoderma* [109], and *Gliocladium* spp. [110] and *T. virens* (syn. *Gliocladium virens*) secrete gliotoxin and gliovirin [111,112]. A previous study reported that the inoculation of strawberry plants with *Trichoderma* spp. controlled *B. cinerea* infection as well as improved growth [113].

Competition occurs when two organisms vie for nutrients or space. Certain species of fungi and yeasts can impede phytopathogens through competitive mechanisms, diminishing nutrient availability. Fungus can inhibit the pathogens through the suppression of the spore germination rates and decrease the germ tube development of pathogens [114]. Endophytic fungi can reside inside the plant tissues; this can be either cellular interspace or intracellular space [115]. Competition for essential nutrients often leads to pathogen starvation, a common cause of microorganism mortality, consequently facilitating the biological control of fungal phytopathogens [116]. *Trichoderma* spp. produce various secondary metabolites including nonribosomal peptides, polyketides, peptaibols, pyrones, siderophores, and volatile and non-volatile terpenes, which possess pharmaceutical and biotechnological significance [117]. The symbiotic association of *Trichoderma* with plant root systems enhances mineral and water uptake while conferring resistance against pathogenic organisms. Under iron-deficient conditions, many fungi secrete siderophores, iron-binding ligands facilitating the mobilization of environmental iron [118]. These siderophores enhance the rhizosphere competence of *T. asperellum* strain T34, rendering it effective as a BCA against *F. oxysporum* f. sp. *lycopersici* (Fol) on tomato plants [100]. For instance, the efficient biocontrol of *Pythium* and *F. oxysporum* mediated by *Trichoderma* was found to be associated with the bioavailability of iron [119].

Trichoderma spp. has gained significant attention in the field of agriculture due to its potential as a BCA against various phytopathogens. Its effectiveness lies not only in its direct antagonistic action against pathogens, but also in its ability to trigger various regulatory mechanisms in plants, leading to enhanced disease resistance [82]. The *T. asperelloides* PSU-P1 strain was able to inhibit the growth of *Stagonosporopsis cucurbitacearum* by expressing PR genes, chitinase, and the glucanase enzyme. *S. cucurbitacearum* is responsible for disease in muskmelon. Scanning electron microscopy (SEM) analysis confirmed that *T. asperelloides* PSU-P1 mediated the destruction of *S. cucurbitacearum* hyphae [120]. The biocontrol of *Sorghum anthracnose*, caused by *Colletotrichum graminicola*, with *T. asperellum* and *T. harzianum* was demonstrated in one study. In this study, seeds bio-primed using the T3 (*T. asperellum*) isolate resulted in the highest increased activities of the antioxidant enzymes, with superoxide dismutase showing a 36.63% increase, peroxidase exhibiting a 43.59% increase, and polyphenol oxidase showing a 40.96% increase at 48 h post pathogen inoculation. Following the 15th day of pathogen inoculation, lignification was increased in the sorghum roots in most treatments, indicating a reinforcement of the defense mechanism. The field trials with T3 were found to be most effective [80]. Many other species of *Trichoderma* have been reported for their prominent plant growth promotion and biocontrol properties. The noticeable species of *Trichoderma* are *T. atroviride* [121], *T. harzianum* [122], *T. koningii* [123], *T. hebeiense* [124], *T. longibrachiatum* [125], *T. polysporum* [126], *T. reesei* [127], *T. virens* [128], *T. viride* [129]. *T. koningii*, and *B. megaterium*, either alone or in combination have been tested for controlling root-knot nematodes and *F. oxysporum* in potato plants [130].

Induced resistance (IR) stands out as a crucial mechanism of biocontrol in plants and is effective against both soilborne and foliar pathogens, as highlighted by Hossainet al. [131]. This resistance mechanism hampers pathogen growth and spread by triggering the secretion of defense-related enzymes like chitinases, proteases, and peroxidases [132]. The fungi *Penicillium* and *Trichoderma* have been shown to induced systemic resistance (ISR against pathogens. Previous study has reported that the inoculation of plant growth-promoting fungi (PGPF) in tobacco plants not only increased their growth, but also induced immunity against potato virus Y (PVY) infection. In comparison to non-inoculated plants, the treated plants exhibited better growth, reduced virus concentrations, and activated genes associated with defense, indicating the efficacy of PGPF in boosting plant protection [133].

Salicylic acid (SA), produced by the *T. harzianum* T39 strain, has been shown to induce resistance against *B. cinerea* in bean plants [134]. Applying a BCA directly to a diseased plant part exhibited ISR, while using dead cells of the BCA showcased local induced resistance (IR). For instance, the use of dead cells of T39 can hinder powdery mildew infection in cucumber and *B. cinerea* infection in tobacco, pepper, and beans. Additionally, inoculating the cucumber seedlings' leaves, and roots with *T. harzianum* led to increased peroxidase and chitinase activity [135]. Root endophytic fungal isolates KB2S2-15 (*ectomycorrhiza*) and KA2S1-42 (Pleosporales) showed the complete suppression of *P. infestans* sporangia germination, with a slightly lower inhibition with KB1S1-4. Pre-treating leaflets with a 5% extract from these endophytes resulted in the total suppression of *P. infestans* mycelial development and late blight symptoms. This suggests that these biocontrol candidates could be used to control late blight disease [136]. Different fungal BCAs for disease management in sustainable agriculture are shown in Table 1.

Table 1. List of fungal BCAs for disease management in plants.

Fungal Species	Mode of Action	Disease/Host	Reference
Ampelomyces quisqualis	Hyperparasitism	Effective against powdery mildew in grapes	[137]
Trichoderma harzianum	Induction of systemic resistance in pepper plants along with capsidol accumulation	Effective against blight and fruit rot of peppers (*Phytophthora capsica*)	[138–140]
Aureobasidium pullulans	Antagonism	Effective against gray mold of strawberry	[141–144]
Pichia guilliermondii	Antagonism	Manage blue mold (*Penicillium italicum*) on citrus fruits	[145–147]
Cladosporium cladosporioides *Cladosporium pseudocladosporioides*	Antagonism	Manage white rust disease in chrysanthemum	[148]
Candida famata	Antagonism	Manage *Penicillium digitatum* in orange fruits	[145,149]
T. harzianum	Antagonism	Fusarium crown and root rot (FCRR) of tomato	[150]
T. atroviride T95	Mycoparasitism	*Botryosphaeria berengeriana* f. sp. *piricola*	[151]
T. harzianum (Th.J.89-2), *T. viride* (*TV.J.92-1* and ITCC-1433), *T. auxeoviride*	Antagonism	Dry root rot chilli (*R. solani*)	[152]
T. koningii	Mycoparasitism	*Sclerotinia sclerotiorum*	[153]
Trichoderma longibrachiatum	Antagonism	*Sclerotium cepivorum*	[154]
Trichoderma asperellum	Antagonism	Fusarium root and stem rot disease Apple replant disease	[155,156]
T. viride	Mycoparasitism	*F. oxysporum* f. sp. *adzuki* and *Pythium arrhenomanes*	[157]
T. viride	Antagonism	*Phytophthora infestans*	[158]
T. viride	Antagonism	*Alternaria alternata*, *Alternaria solani*, *Colletotrichum* spp., *Fusarium solani*, *Fusarium equiseti*, and *F. oxysporum*	[159]
T. virens	Mycoparasitism	*Pythium ultimum*	[160]
T. virens	Mycoparasitism	*Phytophthora erythroseptica*	[161]
Aureobasidium pullulan	Cell wall-degrading enzymes, synthesis of antifungal compounds, and mycoparasitism	Potential to control *Penicillium* expansum, *B. cinerea*, *Aspergillus niger*, and *Rhizopus stolonifer* in grapes	[162,163]
C. gloeosporioides and *R. vinctus*		Panama wilt of banana	[164]

Table 1. *Cont.*

Fungal Species	Mode of Action	Disease/Host	Reference
Chaetomium globosum	Mycofungicide compound production	Seed rot and damping-off of several seed- and soilborne plant pathogens like *P. infestans* (late blight of potato), *Pythium ultimum* (*Pythium damping-off* of sugar beet), *Alternaria raphani*, and *A. brassicola*	[165–167]
Paecilomyces	Antagonism	Apple rot, apple collar rot, and crown rot (*Phytophthora cactorum*)	[168,169]
Aspergillius versicolor lm6–50	Antagonism	Powdery scab in potato (*Spongospora subterranea*)	[170]

In the past two decades, there has been significant progress in utilizing fungi as biological control agents (BCAs), with several commercially available BCA products already on the market. The future expansion of fungal BCAs hinges on the successful development of resting spores and robust mycelia. However, BCAs alone may not suffice to control all types of plant diseases across diverse conditions. While the mechanisms of action for some BCAs are becoming clearer, further research and development are necessary to better understand their behavior. The genetic transformation of fungi holds promise for enhancing BCA performance under varying environmental conditions, however, potential risks related to their application into the environment require thorough investigation to establish acceptable implementation guidelines.

5. Bacteria as BCAs for Plant Disease Management

Beneficial bacteria such as *Bacillus, Bradyrhizobium, Paenibacillus, Pseudomonas, Acinetobacter, Azotobacter, Azospirillum, Rhizobium,* and *Streptomyces* are known to promote not only plant growth, but also act as portent BCAs [171,172] that use different strategies to combat pathogen attack. For example, they can secrete diverse antimicrobial compounds such as antibiotics, lipopeptides, biosurfactants, microbial volatile compounds, bacteriocins, and cell-wall degrading enzymes against different bacterial and fungal pathogens [29,172–174]. On the one hand, they can trigger the host immune system by acting as elicitors or priming agents. Additionally, BCAs may potentially disrupt the pathogens' quorum sensing (QS) system by preventing or enzymatically breaking down the formation of signal molecules that initiate infections [174,175]. When it comes to nutrition acquisition, highly competitive bacterial BCA can outcompete the pathogen by using low-molecular-weight siderophores that have an affinity for ferrous iron to colonize and persist in the infected site [175]. The first use of bacteria as a BCA for plant protection was *Bacillus thuringiensis* [176]. Similarly, after the discovery of penicillin as an antibiotic to control bacterial pathogens, *Staphylococcus* was the first to revolutionize the scientific community [177]. The primary bacteria utilized as BCAs in horticulture crops are strains of *Bacillus, Pseudomonas*, and *Serratia* [173]. The significant contribution of microbial pathogens in controlling major diseases in horticulture crops is widely recognized. The bacteria release secondary metabolites to inhibit the growth of fungal pathogens and serve as potential biopesticides. *Bacillus* species are the most commonly used biopesticides and they also enhance plant defense by activating the induced systemic response (ISR) (Figure 2). Ongena et al. [178] reported that bean and tomato plants with a high concentration of *Bacillus* in their rhizosphere exhibited enhanced disease resistance against *B. cinerea*.

Biological control agents show great promise in managing diseases and reducing the spread of pathogens at the site of infection. BCAs are integrated into the soil to improve the diversity of beneficial microorganisms in the vicinity of the rhizosphere. It restricts pathogen growth through various processes including parasitism and the production of lytic enzymes. Some bacteria emit volatile organic compounds such as acetoin, 2,3-butadial, and 3-hydroxy-2butanone that stimulate defense enzymes and restrict pathogen growth [179]. Similarly, the VOCs produced by *P. fluorescens* WR-1 have restricted the growth of *R. solanacearum* in tomato plants [180]. Applying beneficial microorganisms

(BCAs) to root exudate enhances plant defense by activating the induced systemic resistance (ISR) pathway and promoting the expression of defense-related genes. The root colonization by BCAs is an important feature in the delivery of secondary metabolites and cell-wall degrading enzymes. Additionally, the application of *Pseudomonas* strain *PCL1751* to tomato plants improves their resistance to the root rot pathogen *F. oxysporum* [181]. In another study conducted by Schuhegger et al. [182], it was found that *Serratia liquefaciens* and *Pseudomonas putida* have the ability to stimulate induced systemic resistance (ISR) in tomato plants, providing defense against *Alternaria alternata*. Induced systemic resistance is a crucial aspect of disease resistance as it facilitates the activation of defense hormones and ultimately results in the production of resistance genes. In Table 2, we summarize the list of potential bacterial BCAs that have been used against different pathogens in sustainable agriculture.

Table 2. Role of bacterial BCAs to control pathogens in different crop systems.

S. No.	Bacterial Strain	Pathogen against	Disease	Crop	Dose/CFU (g^{-1}/mL^{-1})	Mode of Action	Reference
1.	*Bacillus altitudinis* GS-16	*Colletotrichum gloeosporioides*	Anthracnose disease	Tea	1×10^8 cfu/mL.	Damage cell permeability and integrity	[183]
2.	*Bacillus thuringiensis*	*B. cinerea*	Grey mold	Bean, potato, and rapeseed	10^6 cfu g^{-1}	Inhibition mycelium growth	[184]
3.	*Bacillus subtilis* GYUN-2311	*Colletotrichum* spp.	Anthracnose diseases	Apple and hot pepper	10^8 cfu/mL	Inhibition mycelium growth	[185]
4.	*Bacillus velezensis* S4	*Magnaporthe oryzae*	Rice blight	Rice	2×10^7 cfu·mL^{-1}	Inhibits fungal and oomycete hyphae growth and alters appressoria	[186]
5.	*Pseudomonas aeruginosa* 91	*Fusarium oxysporum* f. sp. *cubense*	Banana Fusarium wilt	Banana	~10^7	Interacts directly with host cells via flagella, pili and lipoproteins	[187]
6.	*Pseudomonas aeruginosa* BRp3	*Xanthomonas oryzae* pv. *oryzae*	Bacterial leaf blight	Rice	10^9 cfu mL^{-1}	Induction of defense related enzymes, production of 4-hydroxy-2-alkylquinolines (HAQs) along with siderophores	[188]
7.	*Pseudomonas aeruginosa* CQ-40	*Botrytis cinerea*	Gray mold	Tomato	2.5×10^8 cfu mL^{-1}	Type III secretion system	[189]
8.	*Pseudomonas strain IALR1619*	*Pythium ultimum*	Damping-off	Lettuce and Cucumber	7.47×10^8 cfu	Inhibit mycelial growth	[190]
9.	*Bacillus velezensis*	*Ralstonia solanacearum*	Bacterial wilt	Tomato	1×10^8 cfu/mL	Downregulation of genes associated with spore germination and growth	[191]
10.	*Bacillus subtilis XZ18-3*	*Rhizoctonia cerealis*	Foliar blight	Wheat	$3.7 \pm 0.2 \times 10^{11}$ cfu/g	Inhibit *Rhizoctonia cerealis* mycelial growth and induce mycelial swelling and rupture.	[192]

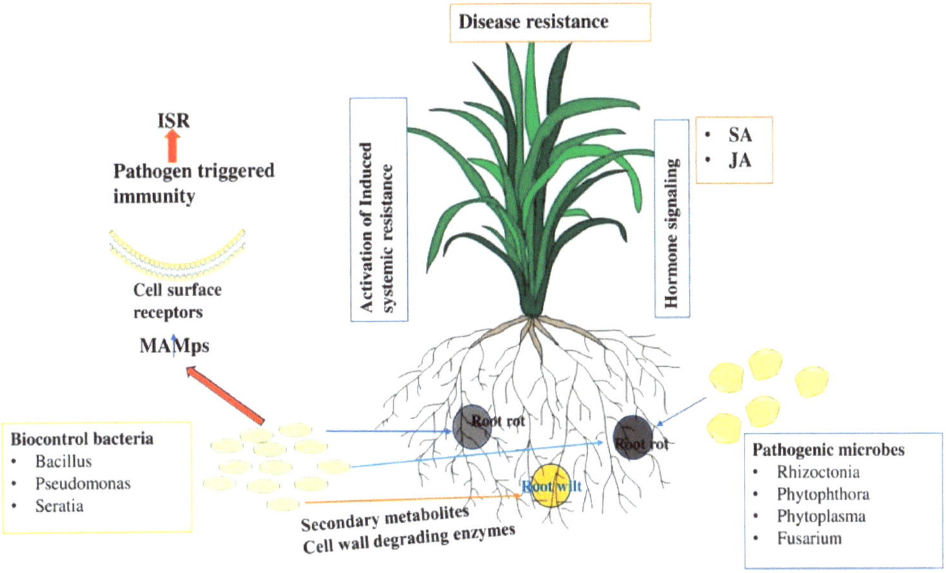

Figure 2. Schematic diagram showing the role of bacterial biocontrol agents in plant disease resistance. The induction of defense hormones triggers the production of SA and jasmonic acid (JA), which in turn confers resistance to fungal infections. The role of secondary metabolites secreted by beneficial microbes leads to the activation of induced systemic resistance (ISR). They can also produce diverse antimicrobial compounds such as antibiotics, lipopeptides, biosurfactants, microbial volatile compounds, bacteriocins, and cell-wall degrading enzymes to inhibit pathogen growth and disease progression.

6. Viruses as Biocontrol Agents

Viruses are important BCAs that act against a wide range of microbial pathogens and pests [193]. Viruses often carry a negative reputation due to their association with diseases and pandemics. However, it is important to note that the majority of viral species are harmless or even helpful, playing crucial roles in ecosystems [194]. As the most abundant entities on Earth, viruses are believed to have a significant ecological impact in maintaining the balance of organism populations. All three domains of life are susceptible to viral infection; however, many endocellular bacteria such as phytoplasmas are conspicuously immune to viral infection. Since viruses are frequently regarded as the ultimate parasites and do not usually cause detectable harm to their hosts, they are appealing candidates for BCAs [195]. A century ago, Mallmann and Hemstreet [196] initially proposed the use of bacteriophages as BCAs against bacteria. Currently, researchers are interested in investigating phages as BCAs for bacterial disease control, primarily because of the fast development of antibiotic resistance in bacterial pathogens that has resulted in serious agricultural loss [197]. Interestingly, phages are viruses that only infect bacteria; they do not directly harm plants or mammals. On the other hand, the utilization of the mycovirus *Cryphonectria hypovirus* 1 (CHV1), which was found to induce hypovirulence in the ascomycetous fungus *Cryphonectria parasitica*, was the first instance of a virus being a successful BCA in fungi [193]. Since then, CHV1 has been used to impede the spread of chestnut blight in forests and orchards [198]. The first instances of virus biocontrol in insects were the applications of baculoviruses to manage populations of lepidopteran insects [199]. Viruses can also be employed to manage viral disease in plants via cross protection. Here, attenuated plant virus strains are used as a BCA in order to safeguard crops from pathogenic strains of the same or similar viral species. The basis for this crop protection technique is the cross protection/interference phenomena, which was initially studied in tobacco plants infected with a yellow strain of the tobacco mosaic virus (TMV). Figure 3, shows viral BCAs employ different strategies to safeguard plants from pathogen attacks.

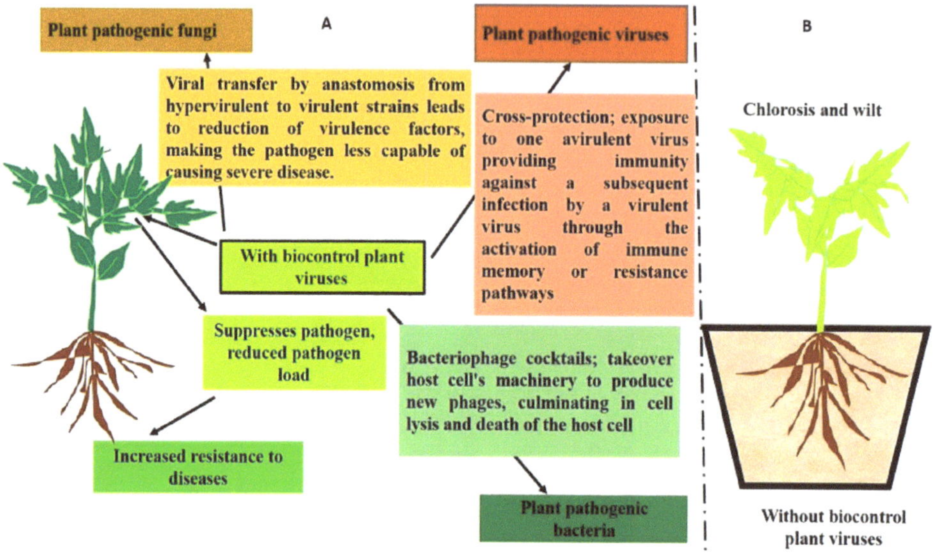

Figure 3. Schematic diagram showing the role of viruses as BCAs in controlling plant pathogens and pests. (**A**) Depicts plants with viral biocontrol agents that protect them from pathogens by using diverse strategies such as activating immune signaling pathways, infecting pathogens to alter their virulence, triggering pathogen cell lysis, and cross infection by activating memory resistance. (**B**) Depicts plants without viral BCAs being challenged by pathogens, which leads to disease development and ultimately host death.

6.1. Biocontrol Potential of Mycoviruses against Fungal Plant Pathogens

Mycoviruses are obligate parasites found in the majority of economically significant plant-pathogenic fungi. Latent infections are often linked to mycovirus–fungus interactions; however, mycovirus infection can potentially be advantageous or deleterious to the fungal host. They have a variety of genome types, but the most common ones are double-stranded RNA (dsRNA) or positive- and negative-sense single-stranded RNA (ssRNA+/−) genomes. Single-stranded DNA (ssDNA) genomes are far less common and are conspicuously absent from dsDNA viruses [200,201]. These have a mono- or multi-segmented genome, naked or encapsidated [202]. Mycovirus-infected hypovirulent *Cryphonectria parasitica* was effectively used in the past to control chestnut blight, which led to an increased investigation into mycoviruses and their potential for the biocontrol of pathogenic fungus in plants across different pathosystems [203]. Based on artificial transection methods, it was found that mycoviruses have broad host ranges and induce hypovirulence. For instance, species that are phylogenetically close to *C. parasitica* such as *C. radicalis*, *C. havanensis*, *C. cubensis*, and *Endothia gyrosa* were artificially infected using the synthetic transcript of CHV1-EP713, which showed that CHV1 was not only able to multiply, but also induced hypovirulence as well as other phenotypic changes in all of these fungal strains [204,205]. Previous reports have shown that CHV1 can also reduce the virulence and growth of important pant pathogens such as *Valsa ceratosperma* and *Phomopsis* G-type, highlighting their potential as BCAs [206]. The infection of *C. parasitica*, *Glomerella cingulate*, and *V. ceratosperma* with mycovirus *Rosellinia necatrix partitivirus 1* (RnPV1) also inhibited growth and reduced their virulence [207]. On the other hand, Xiao et al. [208] used purified virions of mycovirus *Sclerotinia sclerotiorum partitivirus 1* (SsPV1) to infect *B. cinerea*, which also showed growth alteration, stable persistent infection, and the induction of reduced virulence. Lee et al. [209]

used protoplast fusion to transmit mycovirus FgV1-DK21 to *F. graminearum*, *F. asiaticum*, *F. oxysporum* f. sp. *lycopercisi*, and *C. parasitica*. All of these hosts allowed FgV1-DK21 to proliferate, even the distantly related *C. parasitica*, where it resulted in growth changes and decreased pathogenicity, which were different from those seen in CHV1 infections. These studies further support the notion that mycoviruses are potential BCAs and could be applied to the control of a wide range of plant fungal pathogens.

6.2. Phage-Based Biobactericides

Plant pathogenic bacteria also pose a serious threat to crop productivity. Currently, many bacterial pathogens have developed antibiotic resistance, which is a major concern to the environment and human health. For example, *Xanthomonas campestris*, *Erwinia amylovora*, and *Pseudomonas syringae* have been found to harbor antibiotic resistance genes (strAB) that confer resistance against streptomycin [210]. To overcome this problem, harnessing phage-based BCAs is a viable strategy for controlling bacterial pathogens in sustainable agriculture. Bacteriophages, often known as phages, are viruses that cause bacterial hosts to degrade by selectively infecting and multiplying within them as antimicrobial agents. In order to control bacterial disease and increase crop yields, phage therapy is a viable, effective strategy against resistant bacterial plant pathogens [211]. The U.S. Environmental Protection Agency (EPA) has approved four distinct phages from Omnilytics as active ingredients for bactericides since 2005. The first cocktail, known as AgriPhage, guards tomato and pepper plants in both greenhouses and open fields from bacterial spot and speck by including bacteriophages against *X. campestris* pv. *vesicatoria* and *P. syringae*. Similarly, Omnilytics has created concoctions to combat fire blight on pear and apple trees as well as bacterial canker disease on citrus and tomatoes [195]. Unfortunately, the literature lacks specific information regarding the phage composition and anticipated reductions in symptom development of the current cocktails; however, preliminary formulations of the *X. campestris* pv. *vesicatoria* phage cocktail were found to reduce tomato plant bacterial spot disease severity by an average of 17% [212]. Five greenhouse trials found that a cocktail against *X. citri* subsp. *citri* comprising of at least some of the Omnilytics phages might lower citrus canker disease severity by an average of 59% [213]. A&P Inphatec, a subsidiary of Otsuka Pharmaceutical, developed XylPhi-PD, a phage product that the EPA licensed in 2021 to guard grapevines in California against Pierce's disease, which is brought on by *Xylella fastidiosa* [194]. In a four-site field study conducted in California, the business revealed that the constant injection of XylPhi-PD in the xylem reduced disease incidence by 57%. The Hungarian government has given Enviroinvest permission to distribute the phage cocktail Erwiphage Plus domestically as an anti-*E. amylovora* bactericide. This medicine is modified every year to prevent the development of resistance, and it can only be used under very specific circumstances during the flowering season. Each year, emergency authorization is granted for as few as 120 days, spanning from mid-March to mid-July [214]. An engineered cocktail of four lytic phages has been reported as an effective BCA of *X. fastidiosa* and its associated infections including PD, olive-quick decline syndrome, and oleander, coffee, and almond leaf scorch [215,216]. Ahmad et al. [217] showed that the filamentous phage XacF1 caused loss of virulence, reduced EPS production, restricted motility, and delayed growth rate in *Xanthomonas axonopodis* pv. *citri*, the causative agent of citrus canker disease, which further highlights its biocontrol potential against bacterial pathogens. On the other hand, phages such as *vRsoP-WF2*, *vRsoP-WR2*, and *vRsoP-WM2* were also identified as effective BCAs against *R. solanacearum* [218]. Two particular phage species, *E. amylovora Siphoviridae* phage (PhiEaH1) [214], and PhiEaH2 [219], were isolated locally and effectively decreased the incidence of fire blight cases in field studies.

6.3. Cross-Protection

Cross-protection is a strategy that pre-immunizes plants against mild virus strains in order to fight viral infections [195]. Plants are shielded against secondary infection by other viral strains by cross protection. McKinney was the first to demonstrate the cross-protection effect in tobacco plants chronically infected with a tobacco mosaic virus (TMV) light green isolate, which decreased the onset of yellow symptoms caused by a TMV yellow mosaic isolate [220]. There has been some success using it to manage viral illnesses such as cross-protecting tomatoes against mild strains of the tobacco mosaic virus, papaya against minor strains of the papaya ring spot virus, and citrus against moderate strains of the citrus tristeza virus [221]. On the other hand, yellow symptoms were not suppressed by the TMV moderate dark green isolate. Many viruses such as the potato virus X, potato leafroll virus, and citrus tristeza virus (CTV) have been shown to exhibit cross protection [222]. "Plant vaccines" are another term for attenuated isolates. Sap from appropriate leaves can be made to proliferate an attenuated strain in order to provide a real plant vaccination. Next, approximately ten day old seedlings are mechanically given this sap along with carborundum in their cotyledons [223]. An attenuated virus cannot infect another viral species that is closely related to it once it has infected a whole plant. As a result, attenuated viruses are crucial agents of biological control. Salaman [224] initially looked into the cross-protective effects of plant viruses in 1933 and discovered that a moderate isolate of potato virus X (PVX) may prevent a severe PVX isolate from becoming infected. Holmes subsequently created a mild isolate of TMV by heating a virulent strain and observed that in plants infected with the mild isolate, the symptoms brought on by infection with the virulent isolate were reduced [225]. For field studies involving over 2000 citrus rootstocks, a mild isolate of CTV was created in 1951 [226]. The findings suggested that citrus trees protected against more severe CTV isolates were infected with milder isolates. Posnette and Todd [227] employed a similar study approach when they tested a mild isolate of cacao virus 1A in an African area where swollen shoot disease was common. In that study, only 35 out of 416 trees infected with the mild isolate exhibited severe symptoms of infection compared to 273 out of 387 uninfected trees that displayed severe symptoms. This outcome was interpreted as proof that inoculating cacao trees with a moderate viral isolate may successfully protect them in the real world. As a result, attenuated viruses have been the subject of both fundamental and applied research for many years. In addition, cross-protection against the pepino mosaic virus has been recently created and widely used in Europe [228]. These strains of the virus also received separate authorization from the European Food Safety Agency in 2015. Since then, cross-protection has been used in open-field settings to guard against significant viral diseases caused by the papaya ringspot virus in papaya [229] and the citrus tristeza virus (CTV) in citrus species [229]. In South Africa, Peru, and the U.S., cross protection has been proven to be an effective management strategy for the citrus stem pitting disease caused by CTV [230].

7. Biocontrol Agents' Production and Formulations

7.1. Production of BCAs

A critical factor that must be considered when selecting a BCA for commercial development is the availability of a cost-effective manufacturing and stabilization technology that yields an optimally effective form of the antagonist [23]. More studies on the practical aspects of mass production and formulation need to be undertaken to make new biocontrol products stable, effective, safer, and more cost effective [231]. One essential stage in creating effective BCAs for market use is the production of BCAs. The basic objective of industrial scale-up processes, which involve either liquid or solid phase fermentation, is to produce a high amount of biomass at the lowest feasible cost [231]. This is typical procedure to create bacteria and yeasts by liquid fermentation in a continuously stirred tank in order to attain a high aeration. However, a lot of molds are mostly made in a solid state. Regardless of the mass production method, the main objective is to have the highest yield at the lowest cost [232]. The practical efficacy of a BCA greatly depends on the quality of the inoculant, itself a function of the production and formulation processes [233]. The two main processes in production optimization are determining the composition of the medium and enhancing the growing conditions. To manage the BCA production operation and attain an excessive number of cells, create secondary metabolites, or both, it is vital to comprehend their mode of action. Erlenmeyer flasks are typically used in a low-volume laboratory setting for the first phase, which involves evaluating a huge number of substances and concentrations. After that, 2 to 5 L laboratory bioreactors are used to properly calibrate the growth settings. Ultimately, 100–300 L pilot plant bioreactors will be used for the initial large-scale manufacturing, and it will not be hard to scale up the operation to commercial requirements if the pilot plant results are satisfactory. In specifically made plastic bags (VALMIC®) encompassing turba:vermiculite (1:1 *w*/*w*), the generation of *Penicillium frequentans* strain Pf909, a BCA in the management of brown rot in stone fruit, was developed [234]. By applying potato extract, V-8 juice, molasses, and wheat fiber, the BCA *Trichoderma* spp. was grown in both liquid form [235] and on solid media containing various cereals such as sorghum and millet [236]. As an example, the formation of the conidia of the antagonist *Ampelomyces quisqualis* has been shown in a variety of liquid media including a sugar-based medium enhanced with shrimp shell powder [237] and potato dextrose broth adjusted with 2.5% glycerol [238]. In order to guarantee a high, strong, and productive microbial population, careful research is required to enhance the growth conditions (temperature, pH, agitation, aeration, initial inoculum, and process duration) after the optimal growth medium for each microorganism has been determined [239]. A valuable product can be recovered after fermentation by focusing on acquiring cells (or spores), the supernatant containing the released metabolites, or even both, depending on the BCA. One instance of BCA that is recommended is the recovery and formulation of *B. amyloliquefaciens* [240], as both forms are involved in its mode of action to manage diseases. This includes both the vegetative and endospore cells as well as the produced metabolites. Microorganisms are typically present in low concentrations and are combined with other molecules from which they must be removed through a number of steps. These processes include centrifugation, flocculation, and filtering (pressure, rotary vacuum drum, and flocculation) [241].

7.2. BCA Formulation

When developing a formulation, a number of factors need to be considered including the cost of production and acquisition and compatibility with agricultural machinery [242]. As the link between fermentation and field application, the formulation is sometimes seen as the main obstacle to the commercialization of BCAs [243,244]. Both liquid and solid BCA formulations come in a wide range of varieties. Solid formulations can increase the BCA stability, although liquid formulations are typically simple and less expensive to produce. A formulated product is primarily made up of, or may contain: (1) the active ingredient, which is synthesized microorganism cells and secondary metabolites, (2) carriers, which are often inert materials that carry and deliver the active ingredient in the target site, and (3) adjuvants, which are substances that protect the active ingredient from high temperatures, desiccation, ultraviolet radiation, and other environmental stresses while also facilitating the product's spread and dispersal in the intended environment. The microbe must be compatible with all adjuvants employed in the formulation process and not negatively affected by any of them [245].

7.2.1. Liquid Formulation

In order to maximize the product's viability and improve the BCA's adhesion, surfactant, and dispersion capabilities, liquid formulations combine whole cultures or cell suspensions with additives like stabilizers, colorants, and extra nutrients [246]. Compared to formulations based on solids, these formulations are less expensive and simpler to process. Particles can be suspended in a variety of liquids such as oil or water, dispersants, surfactants, suspender components, or a carrier liquid in some liquid compositions. When the processed culture is combined with emulsifiers, surfactants, and/or mineral or vegetable oils that facilitate their subsequent dispersion in water, the mixture is referred to as oil-based. The oils that are utilized must not be harmful to humans, plants, microorganisms, and animals. Since oils provide a defensive effect that lengthens the shelf life of microorganisms, it is generally thought that oil-based formulations are appropriate for foliar applications in dry ambient circumstances [247]. The development of the filamentous fungus *T. harzianum*, an efficient opponent to prevent *Botrytis* rot in apples, is an illustration of this system and was conducted by emulsion [248]. The yeast *Hanseniaspora guilliermondii* isolate YBB3 was developed as a liquid formulation based on glycerol to manage *Aspergillus* rot in grapes. This formulation proved to be more effective than solid formulations that mixed the biomass produced with talc/kaolin powder alone or modified with yeast extract, sucrose, and sodium alginate [249].

7.2.2. Solid Formulation

Although dried products have short viability rates due to both the thermal and dehydration stresses that occur during the drying operation, they are nevertheless a practical solution because of their low production costs and ease of storage and transportation [250]. As a result, every isolate requires a technique that is experimentally tailored to it. The following methods of dehydration are efficient: fluidized bed drying, spray drying, and freeze-drying. Vacuum desiccation, often known as freeze-drying or lyophilization, is a technique for dehydrating labile goods. In short, the sampled liquid needs to be chilled until the solutes crystallize, then the non-crystallizing solutes create an amorphous matrix, and the freezable solution water finally turns into ice. Next, the water is removed from the amorphous matrix by sublimating the ice in vacuum-controlled conditions. Ultimately, the product's moisture content is disported [251]. This method is frequently employed for the conservation of microorganisms, particularly those kept in microbiology collections. According to Prakash et al. [252], it keeps microorganisms viable for over twenty years and does not require specific low-temperature settings for continued management. By eliminating the moisture from the liquid droplets, a product can be transformed from a liquid to a solid state in the form of powder through spray-drying, also known as atomization. The scattered droplets (10–200 m), which are mostly created by pressure nozzles, pneumatic type atomizers, or rotary wheel/disc atomizers, are combined with low-humidity hot air (150–170 °C) inside the chamber. Subsequently, rapid mass transfer and related heat reactions cause the moisture to swiftly escape as vapor from the suspended droplets. Until the required particle properties are attained, the droplets continue to dry inside the drying chamber. Ultimately, external apparatus like cyclones and/or bag-filter houses are used to separate the dried particles from the drying air and collect them afterward [253]. The only BCAs that can be created with this method are those that can withstand high temperatures such as those that can produce thermoresistant endospores. *B. amyloliquefaciens* strain CPA-8, created to suppress *Monilinia* spp. in nectarines and peaches, has shown some promising effects [253]. Powders, granules, and spheres are dried, granulated, and coated using fluidized bed technology [254]. Its foundation is the strong interactions that occur between particles. Via a perforated plate, heated air that has been filtered and perhaps dehumidified (between 35 and 45 °C) enters the product, counteracting gravity and fluidizing the material that has been previously sucked or placed into the granulator. However, in order to use this product, it must first be extruded and then chopped into short rods or "pellets". Filters retain small particles that are transported by the air flow to the cylindrical expansion chamber, allowing them to return to the conical product chamber. As the process proceeds, this cycle is mostly controlled by changes in the product's temperature and the fluidization air flow [254]. Because this technology uses relatively mild temperatures, takes little time, and is inexpensive, it is especially well-suited for BCAs that are sensitive to heat. Numerous BCAs including *Epicoccum nigrum* and *Penicillium oxalicum* have been successfully conserved using this technique [255].

The process of encapsulation involves the microorganism being placed inside a specific matrix or capsule that may be biodegradable if it is made of polyacrylamides, guar gum, gum arabic, sodium alginate, chitosan, or other biocomposites [256]. Vemmer et al. [227,257] demonstrated that these matrices are effective carriers because they enclose living cells and shield them from harsh environmental elements like UV rays and rain. Additionally, González et al. [258] reported that these microcapsules are easily manipulable, can be kept long-term in room temperature storage, and are excellent for BCAs that require a delayed release such those placed in soil. The alginate encapsulation of *Gliocladium virens* (Soil Gard) [259] and the encapsulations of *B. thuringiensis* in shell-core hydrocapsules, which have a liquid center surrounded by a polymer membrane [260], are two examples of this type of formulation.

8. Role of Genetic Engineering for Improving BCAs Traits

The use of BCAs for disease and pest management has significantly increased in recent times because of the negative impact of chemical pesticides on soil fertility, the environment, and human health. However, there are many challenges while using BCAs such as their environmental stability, efficacy, pest or disease suppression ability, host range, target selection, reducing the ecological risk, and improving the mass production or storage [261,262]. Recent developments in genomics and biotechnology have facilitated improvements in the above traits in BCAs and also aids in the identification of novel BCAs, their characterization, and the genetic by-products [261–263]. The availability of microbial genome data has opened new dimensions for the discovery of a large array of BCAs [263]. The integrated approaches of omics technologies will provide a strong foundation for in-depth insights regarding the interactions between BCAs and their hosts. Through the application of omics approaches such as the genomics, transcriptomics, and proteomics of BCAs, pathogens, and crop plants, a specific set of interactome could reveal the key mechanisms and factors influencing the efficacy and applicability of an organism as a BCA [264,265]. The efficacy of BCAs can be increased using genetic engineering. For instance, in *Pseudomonas fluorescens* F113, the loss of the *sadB*, *wspR*, and *kinB* genes results in increased motility and improved root colonization, demonstrating good biocontrol activity against *Phytophthora cactorum* in strawberry plants and *Fusarium oxysporum* f. sp. Radicis-lycopersici in tomato plants [266]. Conversely, significant mycosubtilin production resulting from the overexpression of the mycosubtilin gene utilizing the *Staphylococcus aureus* constitutive promoter in *Bacillus subtilis* strain ATCC 6633 enhances *Pythium aphaniderm* suppression [267]. Introduction of the chitinase gene in *Bacillus thuringiensis 3023* from *Serratia marcescens* showed broad biocontrol activity against various pests [268]. Genetically modified *Pseudomonas fluorescens* strain BL915 with high antipathogenic substances (APS) like pyrrolnitrin showed strong biocontrol activity against soilborne and seedling pathogens [269]. Similarly, the overexpression of *Cry* genes in *B. thuringiensis* showed strong insecticide activity against susceptible and Bt-resistant insects in tobacco plants [270]. Genetically-modified *Pseudomonas* spp., through the introduction a translational enhancer from Bacillus confers strong biocontrol activity. Genome shuffling has been used in *Bacillus subtilis*, *Streptomyces melanosporofaciens*, and *Streptomyces bikiniensis* for the improvement in biocontrol traits against *Fusarium oxysporum* f. sp. *melonis*, *Phytophthora infestans*, *Fusarium oxysporum* f. sp. *Cucumerinum* [271], *Streptomyces melanosporofaciens* [272], and *Streptomyces bikiniensis* [273]. Recently, the *Sarocladium oryzae* mutant strain was found to efficiently control *Rhizoctonia solani* and *Sclerotinia sclerotiorum* pathogens [274]. Transformation of the *Serratia marcescens* chitinase gene into *Pseudomonas fluorescens* improved the biocontrol efficiency against *R. solani* on bean seedlings [275]. Similarly, the introduction of pyrrolnitrin coding genes from *Pseudomonas protegens Pf-5* to *P. synxantha 2–79* showed strong biocontrol activity against wheat pathogens. It has been demonstrated that *T. virens* cotton strains containing two copies of the ech42 gene exhibited increased antagonistic activity against *R. solani* [276]. Sun et al. [277] genetically modified the *Clonostachys rosea* strain, which showed strong biocontrol activity against *Sclerotinia* rot of soybean. To increase its effectiveness against soilborne infections, a genetically engineered strain of *Pseudomonas putida* was developed with *phz* or *phl* biosynthesis gene loci, which produced constitutively phenazine-1-carboxylate (PCA) or 2,4-diacetylphloroglucinol (DAPG) [278]. Another study found that once the trichodermin gene (tri5-trichodiene synthase) was cloned and overexpressed in *T. brevicompactum*, it produced higher trichodermin, which exhibited strong biocontrol activity against *Aspergillus fumigatus* and *Fusarium* spp. Ma et al., 2003. A previous study showed that *T. virens* Δtvk1 mutants showed improved biocontrol efficacy against *R. solani* along with the overproduction of lytic enzymes and the increased expression of genes linked to mycoparasitism [279]. These studies further support the notion that genetic engineering is the key to developing long-term effective BCAs. Although genetic engineering has been a major driver for BCA development, it has some possible health risks and lateral gene flow to non-target animals, which might pose threats to the environment. In this regard, genome editing is an alternative for improving the traits of BCAs more precisely and effectively without endangering the environment. The use of CRISPR and Cas genes has expedited the creation of versatile and economical genomic engineering toolkits that rely on the programmable targeting of CRISPR–Cas technologies [280]. Genome editing using CRISPR/Cas has been used for improving diverse biocontrol traits in *Trichoderma* species [281]. In future, CRISPR/Cas based gene editing for the improvement of biocontrol traits is the most promising tool for developing future climate resilient, efficient, and ecofriendly BCAs in sustainable agriculture.

An understanding of the biosynthetic pathway of antimicrobial metabolites in BCAs is important for improving their biocontrol efficiency. For example, the antibiosis action of BCAs is often achieved by the synthesis of low-molecular-weight chemicals that either directly or indirectly impede the development of pathogens. These encompass a wide range of chemical groups including polyketides [282], terpenes [283], and peptides [284,285]. Thus, metabolic engineering is needed to increase the secondary metabolite production of BCAs, which may potentially result in the synthesis of new antimicrobial compounds. Metabolic engineering holds great promise for the development of future BCAs, offering innovative strategies to enhance their efficacy, specificity, and environmental sustainability. Modern techniques such as plasmid engineering, RNAi interference, and CRISPR-based genome editing utilized to modify the *T. hazrium* strain for strain enhancement seems promising (Figure 4).

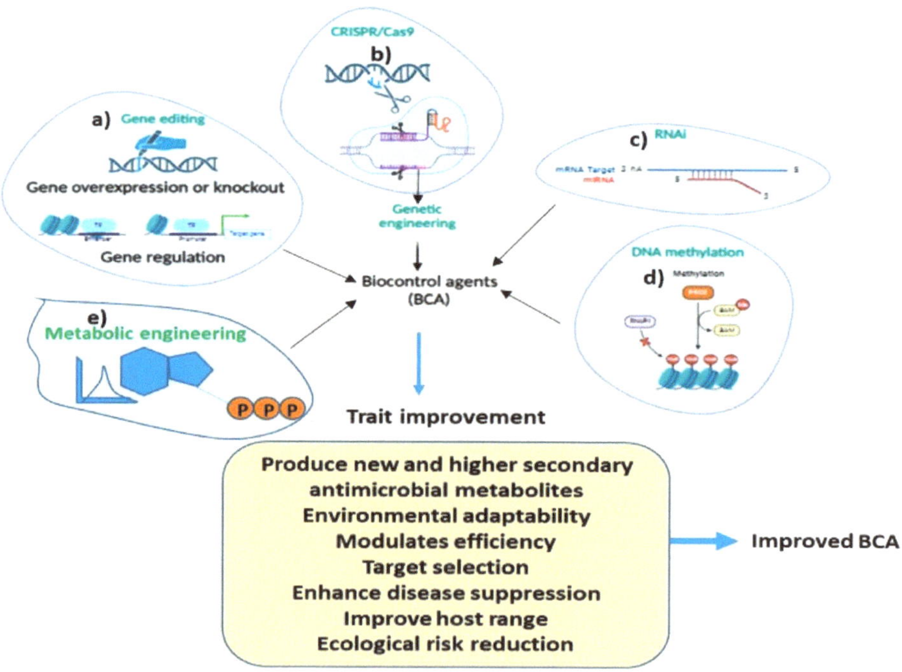

Figure 4. Schematic diagram highlighting the role of different molecular methods for improving the biocontrol traits in BCAs for disease and pest management. (a) gene editing, (b) CRISPR/Cas9, (c) RNAI, (d) DNA methylation and (e) Metabolic engineering.

9. Conclusions

In sustainable agriculture, BCAs are currently being examined as alternatives to synthetic pesticides to manage plant pathogens and pests, mainly due to their safety and ecofriendly nature. These include naturally occurring microorganisms such as fungus, bacteria, and viruses, which are commercially produced as biopesticides and have a variety of agricultural applications in biological and integrated pest management programs. BCAs use diverse mechanisms such as parasitism, antibiosis, or competition and induce the plant defense system to manage diseases or pests. They also play a vital role in improving plant growth and soil fertility. However, the identification and development of a novel class of biocontrol agents possessing increased potency, high productivity in fermenters, extended shelf life, room temperature storage capacity, and excellent compatibility with other control strategies are required for the sustainable management of plant diseases. On the other hand, BCAs must be evaluated for any negative impacts on crops, consumers, and the environment before they can be approved and commercially produced. To further clear the regulatory road and guarantee biosafety, it is also necessary to comprehend how BCAs affect the natural microbiome, which includes the soil and plant microbiomes. Before they may be effectively used in the field, a number of modifications must be made including improving the formulation and delivery strategies, scaling up manufacturing, adhering to regulatory standards, and enhancing cost-effectiveness. Further investigations on certain underdeveloped areas such as the creation of next generation BCAs and the application of biotechnology in conjunction with "omics" approaches to enhance biocontrol efficacy are also necessary for the future development of successful biological control strategies. In this regard, genetic engineering using CRISPR genome editing have the potential to decrease undesirable traits in BCAs and introduce new, desirable traits, which can act against a broad spectrum of pathogens and pests in different crop systems. In climate change scenarios, biological control agents may lose their efficacy if there is a higher fluctuation in temperature, humidity, frequency of rain, and other weather conditions. Therefore, it is important to develop climate resilient biocontrol agents in sustainable agriculture.

Author Contributions: Conceptualization, M.A.A. and S.A.; Methodology, A.T., T.L.T., H.K., R.A.M., Z.A.M., S.M., N.M., G.G., S.K.V., M.A.A. and S.A.; Software, A.T., T.L.T., H.K., R.A.M., Z.A.M., S.M., N.M., G.G., S.K.V., M.A.A. and S.A.; Validation, A.T., T.L.T., H.K., R.A.M., Z.A.M., S.M., N.M., G.G., S.K.V., M.A.A. and S.A.; Formal analysis, A.T., T.L.T., H.K., R.A.M., Z.A.M., S.M., N.M., G.G., S.K.V., M.A.A. and S.A.; Investigation, M.A.A. and S.A.; Resources, A.T., T.L.T., H.K., R.A.M., Z.A.M., S.M., N.M., G.G., S.K.V., M.A.A. and S.A.; Data curation, A.T., T.L.T., H.K., R.A.M., Z.A.M., S.M., N.M., G.G., S.K.V., M.A.A. and S.A.; Writing—original draft preparation, A.T., T.L.T., H.K., R.A.M., Z.A.M., S.M., N.M., G.G., S.K.V., M.A.A. and S.A.; Writing—review and editing, A.T., T.L.T., H.K.; R.A.M., Z.A.M., S.M., N.M.; G.G., S.K.V., M.A.A. and S.A.; Visualization, A.T., T.L.T., H.K., R.A.M., Z.A.M., S.M., N.M., G.G., S.K.V., M.A.A. and S.A.; Supervision, M.A.A. and S.A.; Project administration, M.A.A. and S.A.; Funding acquisition, M.A.A. and S.A. All authors have read and agreed to the published version of the manuscript.

Funding: This research received no external funding.

Data Availability Statement: Not applicable.

Conflicts of Interest: The authors declare no conflicts of interest.

References

1. Mall, M.; Kumar, R.; Akhtar, M.Q. Horticultural crops and abiotic stress challenges. In *Stress Tolerance in Horticultural Crops*; Woodhead Publishing: Cambridge, UK, 2021; pp. 1–19.
2. Xu, J.; Zhang, N.; Wang, K.; Xian, Q.; Dong, J.; Chen, X. Exploring new strategies in diseases resistance of horticultural crops. *Front. Sustain. Food Syst.* **2022**, *6*, 1021350. [CrossRef]
3. Ali, S.; Tyagi, A.; Rajarammohan, S.; Mir, Z.A.; Bae, H. Revisiting Alternaria-host interactions: New insights on its pathogenesis, defense mechanisms and control strategies. *Sci. Hortic.* **2023**, *322*, 112424. [CrossRef]
4. Chaerani, R.; Voorrips, R.E. Tomato early blight (*Alternaria solani*): The pathogen, genetics, and breeding for resistance. *J. Gen. Plant Pathol.* **2006**, *72*, 335–347. [CrossRef]

5. Utkhede, R.S.; Mathur, S. Preventive and curative biological treatments for control of Botrytis cinerea stem canker of greenhouse tomatoes. *BioControl* **2006**, *51*, 363–373. [CrossRef]
6. Charoenporn, C.; Kanokmedhakul, S.; Lin, F.C.; Poeaim, S.; Soytong, K. Evaluation of bio-agent formulations to control Fusarium wilt of tomato. *Afr. J. Biotechnol.* **2010**, *9*, 5836–5844.
7. Panno, S.; Davino, S.; Caruso, A.G.; Bertacca, S.; Crnogorac, A.; Mandić, A.; Noris, E.; Matić, S. A review of the most common and economically important diseases that undermine the cultivation of tomato crop in the mediterranean basin. *Agronomy* **2021**, *11*, 2188. [CrossRef]
8. Khodadadi, F.; González, J.B.; Martin, P.L.; Giroux, E.; Bilodeau, G.J.; Peter, K.A.; Doyle, V.P.; Aćimović, S.G. Identification and characterization of Colletotrichum species causing apple bitter rot in New York and description of *C. noveboracense* sp. nov. *Sci. Rep.* **2020**, *10*, 11043. [CrossRef]
9. Marquez, N.; Giachero, M.L.; Declerck, S.; Ducasse, D.A. *Macrophomina phaseolina*: General Characteristics of Pathogenicity and Methods of Control. *Front. Plant Sci.* **2021**, *12*, 634397. [CrossRef]
10. Tripathi, A.N.; Singh, B.P. Evaluation of seed health of seeds of different vegetable crops. *Veg. News Letter.* **2022**, *9*, 7.
11. Sundin, G.W.; Wang, N. Antibiotic Resistance in Plant-Pathogenic Bacteria. *Annu. Rev. Phytopathol.* **2018**, *25*, 161–180. [CrossRef]
12. Sánchez-Hernández, E.; González-García, V.; Martín-Gil, J.; Lorenzo-Vidal, B.; Palacio-Bielsa, A.; Martín-Ramos, P. Phytochemical Screening and Antibacterial Activity of *Taxus baccata* L. against *Pectobacterium* spp. and *Dickeya chrysanthemi*. *Horticulturae* **2023**, *9*, 201. [CrossRef]
13. Hull, R. Symptoms and Host Range. In *Plant Virology*; Elsevier: Amsterdam, The Netherlands, 2014; pp. 145–198.
14. Mehetre, G.T.; Leo, V.V.; Singh, G.; Sorokan, A.; Maksimov, I.; Yadav, M.K.; Upadhyaya, K.; Hashem, A.; Alsaleh, A.N.; Dawoud, T.M.; et al. Current Developments and Challenges in Plant Viral Diagnostics: A Systematic Review. *Viruses* **2021**, *13*, 412. [CrossRef] [PubMed]
15. Syafrudin, M.; Kristanti, R.A.; Yuniarto, A.; Hadibarata, T.; Rhee, J.; Al-Onazi, W.A.; Algarni, T.S.; Almarri, A.H.; Al-Mohaimeed, A.M. Pesticides in drinking water—A review. *Int. J. Environ. Res. Public Health* **2021**, *18*, 468. [CrossRef] [PubMed]
16. Stukenbrock, E.; Gurr, S. Address the growing urgency of fungal disease in crops. *Nature* **2023**, *617*, 31–34. [CrossRef] [PubMed]
17. Verhaegen, M.; Bergot, T.; Liebana, E.; Stancanelli, G.; Streissl, F.; Mingeot-Leclercq, M.P.; Mahillon, J.; Bragard, C. On the use of antibiotics to control plant pathogenic bacteria: A genetic and genomic perspective. *Front. Microbiol.* **2023**, *14*, 1221478. [CrossRef] [PubMed]
18. Jagtap, A.D.; Sarkale, P.S.; Patil, P.A. Transgenic Plants and Animals in Agriculture: Assessing the Risks and Benefits. *Nat. Camp.* **2024**, *28*, 95–102.
19. van Dijk, M.; Morley, T.; Rau, M.L.; Saghai, Y. A meta-analysis of projected global food demand and population at risk of hunger for the period 2010–2050. *Nat. Food* **2021**, *2*, 494–501. [CrossRef] [PubMed]
20. Mahmood, I.; Imadi, S.R.; Shazadi, K.; Gul, A.; Hakeem, K.R. Effects of pesticides on environment. In *Plant Soil and Microbes*; Springer: Berlin/Heidelberg, Germany, 2016; pp. 253–269.
21. Sandy, Y.A.; Zahro, F.A.; Rizky, D.R.; Fajarwati, S.K.; Effendi, M. Knowledge Level of Farmers regarding the Use of Pesticide for Pest and Disease Control. *J. Agrinika J. Agroteknologi Agribisnis* **2024**, *8*, 12–22.
22. Dalavayi, H.M.; Bala, S.; Choudhury, D. Eco-friendly plant based on botanical pesticides. *Plant Arch.* **2021**, *21*, 2197–2204.
23. He, D.C.; He, M.H.; Amalin, D.M.; Liu, W.; Alvindia, D.G.; Zhan, J. Biological Control of Plant Diseases: An Evolutionary and Eco-Economic Consideration. *Pathogens* **2021**, *10*, 1311. [CrossRef]
24. Lee, J.; Kim, S.; Jung, H.; Koo, B.K.; Han, J.A.; Lee, H.S. Exploiting bacterial genera as biocontrol agents: Mechanisms, interactions and applications in sustainable agriculture. *J. Plant Biol.* **2023**, *66*, 485–498. [CrossRef]
25. Theresa, M.; Radhakrishnan, E.K. Microbial biocontrol formulations for commercial applications. In *Microbiome Stimulants for Crops*; Woodhead Publishing: Cambridge, UK, 2021; pp. 179–192.
26. Galli, M.; Feldmann, F.; Vogler, U.K.; Kogel, K.-H. Can biocontrol be the game-changer in integrated pest management? A review of definitions, methods and strategies. *J. Plant Dis. Prot.* **2024**, *131*, 265–291. [CrossRef]
27. Stenberg, J.A.; Sundh, I.; Becher, P.G.; Björkman, C.; Dubey, M.; Egan, P.A.; Friberg, H.; Gil, J.F.; Jensen, D.F.; Jonsson, M.; et al. When is it biological control? A framework of definitions, mechanisms, and classifications. *J. Pest Sci.* **2021**, *94*, 665–676. [CrossRef]
28. Funck Jensen, D.; Dubey, M.; Jensen, B.; Karlsson, M. Clonostachys rosea for the control of plant diseases. In *Microbial Bioprotectants for Plant Disease Management*; BDS Publishing: Cambridge, UK, 2021; pp. 429–471.
29. Köhl, J.; Kolnaar, R.; Ravensberg, J. Mode of action of microbial biological control agents against plant diseases: Relevance beyond efficacy. *Front. Plant Sci.* **2019**, *10*, 845. [CrossRef] [PubMed]
30. Xu, Z.; Shao, J.; Li, B.; Yan, X.; Shen, Q.; Zhang, R. Contribution of bacillomycin D in Bacillus amyloliquefaciens SQR9 to antifungal activity and biofilm formation. *Appl. Environ. Microbiol.* **2013**, *79*, 808–815. [CrossRef] [PubMed]
31. Kumar, P.; Kamle, M.; Borah, R.; Mahato, D.K.; Sharma, B. Bacillus thuringiensis as microbial biopesticide: Uses and application for sustainable agriculture. *Egypt. J. Biol. Pest Control* **2021**, *31*, 95. [CrossRef]
32. Salazar, B.; Ortiz, A.; Keswani, C.; Minkina, T.; Mandzhieva, S.; Singh, S.P.; Rekadwad, B.; Borriss, R.; Jain, A.; Singh, H.B.; et al. *Bacillus* spp. as bio-factories for antifungal secondary metabolites: Innovation beyond whole organism formulations. *Microb. Ecol.* **2023**, *86*, 1–24. [CrossRef]

33. Zhang, N.; Wang, Z.; Shao, J.; Xu, Z.; Liu, Y.; Xun, W.; Miao, Y.; Shen, Q.; Zhang, R. Biocontrol mechanisms of Bacillus: Improving the efficiency of green agriculture. *Microb. Biotechnol.* **2023**, *16*, 2250–2263. [CrossRef]
34. Hashem, A.; Tabassum, B.; Abd_Allah, E.F. Bacillus subtilis: A plant-growth-promoting rhizobacterium that also impacts biotic stress. *Saudi J. Biol. Sci.* **2019**, *26*, 1291–1297. [CrossRef]
35. Beneduzi, A.; Ambrosini, A.; Passaglia, L.M. Plant growth-promoting Rhizobacteria (PGPR): Their potential as antagonists and biocontrol agents. *Genet. Mol. Biol.* **2012**, *35*, 1044–1051. [CrossRef]
36. Karačić, V.; Miljaković, D.; Marinković, J.; Ignjatov, M.; Milošević, D.; Tamindžić, G.; Ivanović, M. Bacillus Species: Excellent Biocontrol Agents against Tomato Diseases. *Microorganisms* **2024**, *12*, 457. [CrossRef] [PubMed]
37. Berdi, J. Bioactive microbial metabolites. *J. Antibiot.* **2005**, *58*, 1–26. [CrossRef] [PubMed]
38. Caulier, S.; Nannan, C.; Gillis, A.; Licciardi, F.; Bragard, C.; Mahillon, J. Overview of the antimicrobial compounds produced by members of the Bacillus subtilis group. *Front. Microbiol.* **2019**, *10*, 302. [CrossRef] [PubMed]
39. Schneider, T.; Müller, A.; Miess, H.; Gross, H. Cyclic lipopeptides as antibacterial agents-Potent antibiotic activity mediated by intriguing mode of actions. *Int. J. Med. Microbiol.* **2014**, *304*, 37–43. [CrossRef] [PubMed]
40. Helmy, N.M.; Parang, K. Cyclic peptides with antifungal properties derived from bacteria, fungi, plants, and synthetic sources. *Pharmaceuticals* **2023**, *16*, 892. [CrossRef] [PubMed]
41. Meena, K.R.; Kanwar, S.S. Lipopeptides as the antifungal and antibacterial agents: Applications in food safety and therapeutics. *BioMed Res. Int.* **2015**, *2015*, 473050. [CrossRef] [PubMed]
42. Hussain, S.; Tai, B.; Ali, M.; Jahan, I.; Sakina, S.; Wang, G.; Zhang, X.; Yin, Y.; Xing, F. Antifungal potential of lipopeptides produced by the *Bacillus siamensis* Sh420 strain against Fusarium graminearum. *Microbiol. Spectr.* **2024**, *12*, 4008–4023. [CrossRef] [PubMed]
43. Xu, B.H.; Lu, Y.Q.; Ye, Z.W.; Zheng, Q.W.; Wei, T.; Lin, J.F.; Guo, L.Q. Genomics. guided discovery and structure identification of cyclic lipopeptides from the Bacillus siamensis JFL15. *PLoS ONE* **2018**, *13*, e0202893. [CrossRef]
44. Zhang, M.; Li, X.; Pan, Y.; Qi, D.; Zhou, D.; Chen, Y.; Feng, J.; Wei, Y.; Zhao, Y.; Li, K.; et al. Biocontrol mechanism of *Bacillus siamensis* sp. QN2MO.1 against tomato fusarium wilt disease during fruit postharvest and planting. *Microbiol. Res.* **2024**, *283*, 127694. [CrossRef]
45. Epparti, P.; Eligar, S.M.; Sattur, A.; Kumar, B.S.G.; Halami, P.M. Characterization of dual bacteriocins producing *Bacillus subtilis* SC3.7 isolated from fermented food. *Food Sci. Technol.* **2022**, *154*, 112854. [CrossRef]
46. Fira, D.; Dimkić, I.; Berić, T.; Lozo, J.; Stanković, S. Biological control of plant pathogens by Bacillus species. *J. Biotechnol.* **2018**, *285*, 44–55. [CrossRef] [PubMed]
47. Gillor, O.; Benjamin, C.K.; Margaret, A.R. Colicins and Microcins: The next generation antimicrobials. *Adv. Appl. Microbiol.* **2004**, *54*, 129–146. [PubMed]
48. Miljaković, D.; Marinković, J.; Balešević-Tubić, S. The significance of *Bacillus* spp. in disease suppression and growth promotion of field and vegetable crops. *Microorganisms* **2020**, *8*, 1037. [CrossRef] [PubMed]
49. Im, S.M.; Yu, N.H.; Joen, H.W.; Kim, S.O.; Park, H.W.; Park, A.R.; Kim, J.C. Biological control of tomato bacterial wilt by oxydifficidin and difficidin-producing Bacillus methylotrophicus DR-08. *Pestic. Biochem. Physiol.* **2020**, *163*, 130–137. [CrossRef] [PubMed]
50. Yuan, J.; Li, B.; Zhang, N.; Waseem, R.; Shen, Q.; Huang, Q. Production of bacillomycin- and macrolactin-type antibiotics by *Bacillus amyloliquefaciens* NJN-6 for suppressing soilborne plant pathogens. *J. Agric. Food Chem.* **2012**, *60*, 2976–2981. [CrossRef] [PubMed]
51. Singh, D.; Devappa, V.; Yadav, D.K. Suppression of tomato bacterial wilt incited by Ralstonia pseudosolanacearum using polyketide antibiotic-producing *Bacillus* spp. isolated from rhizospheric soil. *Agriculture* **2022**, *12*, 2009. [CrossRef]
52. Chen, Q.; Qiu, Y.; Yuan, Y.; Wang, K.; Wang, H. Biocontrol activity and action mechanism of *Bacillus velezensis* strain SDTB038 against Fusarium crown and root rot of tomato. *Front. Microbiol.* **2022**, *13*, 994716. [CrossRef] [PubMed]
53. Guzmán-Guzmán, P.; Kumar, A.; de Los Santos-Villalobos, S.; Parra-Cota, F.I.; Orozco-Mosqueda, M.D.C.; Fadiji, A.E.; Hyder, S.; Babalola, O.O.; Santoyo, G. *Trichoderma* Species: Our Best Fungal Allies in the Biocontrol of Plant Diseases-A Review. *Plants* **2023**, *12*, 432. [CrossRef]
54. Elad, Y. Biological control of foliar pathogens by means of *Trichoderma harzianum* and potential modes of action. *Crop Prot.* **2000**, *19*, 709–714. [CrossRef]
55. Howell, C.R. Mechanisms employed by Trichoderma species in the biological control of plant diseases: The history and evolution of current concepts. *Plant Dis.* **2003**, *87*, 4–10. [CrossRef]
56. Guo, Q.; Shi, L.; Wang, X.; Li, D.; Yin, Z.; Zhang, J.; Ding, G.; Chen, L. Structures and biological activities of secondary metabolites from the Trichoderma genus (covering 2018–2022). *J. Agric. Food Chem.* **2023**, *71*, 13612–13632. [CrossRef]
57. Asad, S.A. Mechanisms of action and biocontrol potential of *Trichoderma* against fungal plant diseases—A review. *Ecol. Complex.* **2022**, *49*, 100978. [CrossRef]
58. Manganiello, G.; Nicastro, N.; Ortenzi, L.; Pallottino, F.; Costa, C.; Pane, C. Trichoderma Biocontrol Performances against Baby-Lettuce *Fusarium* Wilt Surveyed by Hyperspectral Imaging-Based Machine Learning and Infrared Thermography. *Agriculture* **2024**, *14*, 307. [CrossRef]
59. Tabashnik, B.E.; Brévault, T.; Carrière, Y. Insect resistance to Bt crops: Lessons from the first billion acres. *Nat. Biotechnol.* **2013**, *31*, 510–521. [CrossRef]

60. Timper, P. Conserving and Enhancing Biological Control of Nematodes. *J. Nematol.* **2014**, *46*, 75–89.
61. Umer, M.; Mubeen, M.; Iftikhar, Y.; Shad, M.A.; Usman, H.M.; Sohail, M.A.; Atiq, M.N.; Abbas, A.; Ateeq, M. Role of Rhizobacteria on Plants Growth and Biological Control of Plant Diseases: A Review. *Plant Prot.* **2021**, *5*, 59–73.
62. Matsumoto, A.; Takahashi, Y. Endophytic Actinomycetes: Promising Source of Novel Bioactive Compounds. *J. Antibiot.* **2017**, *70*, 514–519. [CrossRef]
63. Farzand, A.; Moosa, A.; Zubair, M.; Khan, A.R.; Massawe, V.C.; Tahir, H.A.S.; Sheikh, T.M.M.; Ayaz, M.; Gao, X. Suppression of *Sclerotinia sclerotiorum* by the Induction of Systemic Resistance and Regulation of Antioxidant Pathways in Tomato Using Fengycin Produced by *Bacillus amyloliquefaciens* FZB42. *Biomolecules* **2019**, *9*, 613. [CrossRef] [PubMed]
64. Ezrari, S.; Mhidra, O.; Radouane, N.; Tahiri, A.; Polizzi, G.; Lazraq, A.; Lahlali, R. Potential Role of Rhizobacteria Isolated from Citrus Rhizosphere for Biological Control of Citrus Dry Root Rot. *Plants* **2021**, *10*, 872. [CrossRef]
65. Ayaz, M.; Ali, Q.; Farzand, A.; Khan, A.R.; Ling, H.; Gao, X. Nematicidal Volatiles from *Bacillus atrophaeus* GBSC56 Promote Growth and Stimulate Induced Systemic Resistance in Tomato against *Meloidogyne incognita*. *Int. J. Mol. Sci.* **2021**, *22*, 5049. [CrossRef]
66. Zubair, M.; Hanif, A.; Farzand, A.; Sheikh, T.M.M.; Khan, A.R.; Suleman, M.; Ayaz, M.; Gao, X. Genetic Screening and Expression Analysis of Psychrophilic *Bacillus* spp. Reveal Their Potential to Alleviate Cold Stress and Modulate Phytohormones in Wheat. *Microorganisms* **2019**, *7*, 337. [CrossRef]
67. Sutthisa, W.; Hompana, W.; Yutthasin, R. Enhancing Biocontrol Potential: Development and Efficacy Assessment of a Liquid Formulation of *Trichoderma Asperellum* MSU007 against Sclerotium Rrolfsii. *Trends Sci.* **2024**, *21*, 7550. [CrossRef]
68. Hjeljord, L.; Tronsmo, A. Trichoderma and Gliocladium in biological control: An overview. In *Trichoderma & Gliocladium—Enzymes, Biological Control and Commercial Applications*; Harman, G.E., Kubicek, C.P., Eds.; Taylor & Francis Ltd.: London, UK, 1998; pp. 131–151.
69. Minuto, A.; Migheli, Q.; Garibaldi, A. Evaluation of antagonistic strains of *Fusarium* spp. in the biological and integrated control of *Fusarium* wilt of cyclamen. *Crop Prot.* **1995**, *14*, 221–226. [CrossRef]
70. Vehapi, M.; İnan, B.; Kayacan-Cakmakoglu, S.; Sagdic, O.; Özçimen, D.B. Preparation of *Bacillus pumilus* loaded electrosprayed nanoparticles as a plant protective against postharvest fungal decay. *Eur. J. Plant Pathol.* **2024**, *168*, 121–136. [CrossRef]
71. Goulet, F.; Aulagnier, A.; Fouilleux, E. Moving beyond pesticides: Exploring alternatives for a changing food system. *Environ. Sci. Policy* **2023**, *147*, 177–187. [CrossRef]
72. Galluzzo, N. How does eliminating the use of pesticides affect technical efficiency in Italian farms? *Bulg. J. Agric. Sci.* **2023**, *29*, 14–23.
73. Palmisano, T. Narratives and practices of pesticide removal in the Andean valleys of Chile and Argentina. *Environ. Sci. Policy* **2023**, *139*, 149–156. [CrossRef]
74. Thambugala, K.M.; Daranagama, D.A.; Phillips, A.J.L.; Kannangara, S.D.; Promputtha, I. Fungi vs. Fungi in Biocontrol: An Overview of Fungal Antagonists Applied Against Fungal Plant Pathogens. *Front. Cell. Infect. Microbiol.* **2020**, *10*, 604923. [CrossRef]
75. Sánchez-Cruz, R.; Mehta, R.; Atriztán-Hernández, K.; Martínez-Villamil, O.; del Rayo Sánchez-Carbente, M.; Sánchez-Reyes, A.; Folch-Mallol, J.L. Effects on *Capsicum annuum* plants colonized with *Trichoderma atroviride* P. karst strains genetically modified in Taswo1, a gene coding for a protein with Expansin-like activity. *Plants* **2021**, *10*, 1919. [CrossRef]
76. Liu, P.; Luo, L.; Long, C.A. Characterization of competition for nutrients in the biocontrol of *Penicillium italicum* by *Kloeckera apiculata*. *Biol. Control* **2013**, *67*, 157–162. [CrossRef]
77. Ren, X.; Branà, M.T.; Haidukowski, M.; Gallo, A.; Zhang, Q.; Logrieco, A.F.; Altomare, C. Potential of *Trichoderma* spp. for biocontrol of aflatoxin-producing *Aspergillus flavus*. *Toxins* **2022**, *14*, 86. [CrossRef] [PubMed]
78. Matarese, F.; Sarrocco, S.; Gruber, S.; Seidl-Seiboth, V.; Vannacci, G. Biocontrol of *Fusarium* head blight: Interactions between *Trichoderma* and mycotoxigenic Fusarium. *Microbiology* **2012**, *158*, 98–106. [CrossRef] [PubMed]
79. Abbas, A.; Mubeen, M.; Zheng, H.; Sohail, M.A.; Shakeel, Q.; Solanki, M.K.; Zhou, L. *Trichoderma* spp. genes involved in the biocontrol activity against *Rhizoctonia solani*. *Front. Microbiol.* **2022**, *13*, 884469. [CrossRef] [PubMed]
80. Manzar, N.; Singh, Y.; Kashyap, A.S.; Sahu, P.K.; Rajawat, M.V.S.; Bhowmik, A.; Saxena, A.K. Biocontrol potential of native *Trichoderma* spp. against anthracnose of great millet (*Sorghum bicolor* L.) from Tarai and hill regions of India. *Biol. Control* **2021**, *152*, 104474. [CrossRef]
81. Quesada-Moraga, E.; Garrido-Jurado, I.; Yousef-Yousef, M.; González-Mas, N. Multitrophic Interactions of Entomopathogenic Fungi. *Biol. Control* **2022**, *67*, 457–472.
82. Manzar, N.; Kashyap, A.S.; Goutam, R.S.; Rajawat, M.V.S.; Sharma, P.K.; Sharma, S.K.; Singh, H.V. *Trichoderma*: Advent of Versatile Biocontrol Agent, Its Secrets and Insights into Mechanism of Biocontrol Potential. *Sustainability* **2022**, *14*, 12786. [CrossRef]
83. Ghorbanpour, M.; Omidvari, M.; Abbaszadeh-Dahaji, P.; Omidvar, R.; Kariman, K. Mechanisms underlying the protective effects of beneficial fungi against plant diseases. *Biol. Control* **2018**, *117*, 147–157. [CrossRef]
84. Persoon, C.H. Disposita methodical fungorum. *Romers. Neues. Mag. Bot.* **1794**, *1*, 81–128.
85. Abbey, J.A.; Percival, D.; Abbey, L.; Asiedu, S.K.; Prithiviraj, B.; Schilder, A. Biofungicides as alternative to synthetic fungicide control of grey mould (Botrytis cinerea)–prospects and challenges. *Biocontrol Sci. Technol.* **2019**, *29*, 207–228. [CrossRef]
86. El-Hasan, A.; Walker, F.; Schone, J.; Buchenauer, H. Detection of viridiofungin A and another antifungal metabolites excreted by *Trichoderma harzianum* active against different plant pathogens. *Eur. J. Plant Pathol.* **2009**, *124*, 457–470. [CrossRef]

87. Kashyap, A.S.; Manzar, N.; Rajawat, M.V.S.; Kesharwani, A.K.; Singh, R.P.; Dubey, S.C.; Pattanayak, D.; Dhar, S.; Lal, S.K.; Singh, D. Screening and Biocontrol Potential of *Rhizobacteria* Native to Gangetic Plains and Hilly Regions to Induce Systemic Resistance and Promote Plant Growth in Chilli against Bacterial Wilt Disease. *Plants* **2021**, *10*, 2125. [CrossRef] [PubMed]
88. del Carmen, H.; Rodríguez, M.; Evans, H.C.; de Abreu, L.M.; de Macedo, D.M.; Ndacnou, M.K.; Bekele, K.B.; Barreto, R.W. New species and records of *Trichoderma* isolated as mycoparasites and endophytes from cultivated and wild coffee in Africa. *Sci. Rep.* **2021**, *11*, 5671. [CrossRef]
89. Sundheim, L. Control of Cucumber Powdery Mildew by the Hyperparasite *Ampelomyces quisqualis* and Fungicides. *Plant Pathol.* **1982**, *31*, 209–214. [CrossRef]
90. Gurung, M. Evaluation of *Trichoderma asperellum* as Biocontrol Agent for Phytophthora Foot and Root Rot of Citrus. Master's Thesis, Texas A&M University, Kingsville, TX, USA, 2018.
91. Contina, J.B. *Biological Control of Fusarium solani, Rhizoctonia solani and the Pale Cyst Nematode Globodera pallida with Trichoderma harzianum ThzID1*; University of Idaho: Pocatello, ID, USA, 2016.
92. Di Francesco, A.; Martini, C.; Mari, M. Biological control of postharvest diseases by microbial antagonists: How many mechanisms of action? *Eur. J. Plant Pathol.* **2016**, *145*, 711–717. [CrossRef]
93. Jaihan, P.; Sangdee, K.; Sangdee, A. Disease suppressive activity of extracts from entomopathogenic fungus Ophiocordyceps sobolifera against chili anthracnose fungi *Colletotrichum* spp. in a pot experiment. *J. Gen. Plant Pathol.* **2018**, *84*, 237–242. [CrossRef]
94. Abo-Elyousr, K.A.; Abdel-Hafez, S.I.; Abdel-Rahim, I.R. Isolation of *Trichoderma* and evaluation of their antagonistic potential against Alternaria porri. *J. Phytopathol.* **2014**, *162*, 567–574. [CrossRef]
95. Kalimutu, P.K.; Mahardika, I.B.K.; Sagung, P.R.A.A.A. Antagonism Test of *Trichoderma atroviride* and *Gliocladium* sp. Bali Local Isolates As a Disease Control of Blendok Disease (*Botryodiplodia theobromae*) in Grapefruit (*Citrus grandis* L. Osbeck). *Sustain. Environ. Agric. Sci.* **2020**, *4*, 102–110. [CrossRef]
96. Sherkhane, P.D.; Bansal, R.; Banerjee, K.; Chatterjee, S.; Oulkar, D.; Jain, P.; Mukherjee, P.K. Genomics-Driven Discovery of the Gliovirin Biosynthesis Gene Cluster in the Plant Beneficial Fungus *Trichoderma Virens*. *ChemSelect* **2017**, *2*, 3347–3352. [CrossRef]
97. Hua, L.; Zeng, H.; He, L.; Jiang, Q.; Ye, P.; Liu, Y.; Zhang, M. Gliotoxin Is an Important Secondary Metabolite Involved in Suppression of *Sclerotium rolfsii* of *Trichoderma virens* T23. *Phytopathology* **2021**, *111*, 1720–1725. [CrossRef]
98. Boureghda, H.; Bouznad, Z. Biological control of Fusarium wilt of chickpea using isolates of *Trichoderma atroviride*, T. harzianum and *T. longibrachiatum*. *Acta Phytopathol. Entomol. Hung.* **2009**, *44*, 25–38. [CrossRef]
99. Tsai, C.C.; Tzeng, D.S.; Hsieh, S.P.Y. Biological control of Fusarium stem rot of *Anoectochilus formosanus* Hayata by Trichoderma asperellum TA strain. *Plant Pathol. Bull.* **2008**, *17*, 243–254.
100. Schubert, M.; Fink, S.; Schwarze, F.W.M.R. Field experiments to evaluate the application of Trichoderma strain (T.15603.1) for biological control of wood decay fungi in trees. Part II. *Arboric. J.* **2008**, *31*, 249–268. [CrossRef]
101. Waqas, M.; Khana, A.L.; Hamayuna, M.; Shahzad, R.; Kang, S.M.; Kim, J.G.; Lee, I.J. Endophytic Fungi Promote Plant Growth and Mitigate the Adverse Effects of Stem Rot: An Example of *Penicillium citrinum* and *Aspergillus terreus*. *J. Plant Interact.* **2015**, *10*, 280–287. [CrossRef]
102. Gressel, J. Four pillars are required to support a successful biocontrol fungus. *Pest Manag. Sci.* **2024**, *80*, 5–39. [CrossRef] [PubMed]
103. Martínez-Álvarez, P.; Fernández-González, R.A.; Sanz-Ros, A.V.; Pando, V.; Diez, J.J. Two fungal endophytes reduce the severity of pitch canker disease in *Pinus radiata* seedlings. *Biol. Control* **2016**, *94*, 1–10. [CrossRef]
104. El-Saadony, M.T.; Saad, A.M.; Soliman, S.M.; Salem, H.M.; Ahmed, A.I.; Mahmood, M.; AbuQamar, S.F. Plant Growth-Promoting Microorganisms as Biocontrol Agents of Plant Diseases: Mechanisms, Challenges and Future Perspectives. *Front. Plant Sci.* **2022**, *13*, 923880. [CrossRef]
105. Kumari, N.; Srividhya, S. Secondary Metabolites and Lytic Tool Box of Trichoderma and Their Role in Plant Health. In *Molecular Aspects of Plant Beneficial Microbes in Agriculture*; Academic Press: Cambridge, MA, USA, 2020; pp. 305–320.
106. Segarra, G.; Casanova, E.; Avilés, M.; Trillas, I. *Trichoderma asperellum* Strain T34 Controls Fusarium Wilt Disease in Tomato Plants in Soilless Culture through Competition for Iron. *Microb. Ecol.* **2010**, *59*, 141–149. [CrossRef] [PubMed]
107. Vinale, F.; Nigro, M.; Sivasithamparam, K.; Flematti, G.; Ghisalberti, E.L.; Ruocco, M.; Lorito, M. Harzianic Acid: A Novel Siderophore from *Trichoderma harzianum*. *FEMS Microbiol. Lett.* **2013**, *347*, 123–129. [CrossRef]
108. Yan, L.; Khan, R.A.A. Biological Control of Bacterial Wilt in Tomato through the Metabolites Produced by the Biocontrol Fungus, *Trichoderma harzianum*. Egypt. *J. Biol. Pest Control* **2021**, *31*, 1–9.
109. Konappa, N.; Krishnamurthy, S.; Siddaiah, C.N.; Ramachandrappa, N.S.; Chowdappa, S. Evaluation of biological efficacy of Trichoderma asperellum against tomato bacterial wilt caused by Ralstonia solanacearum. *Egypt J. Biol. Pest Control* **2018**, *28*, 63. [CrossRef]
110. Choi, G.J.; Kim, J.C.; Jang, K.S.; Cho, K.Y.; Kim, H.T. Mycoparasitism of Acremonium strictum BCP on *Botrytis cinerea*, the Gray Mold Pathogen. *J. Microbiol. Biotechnol.* **2008**, *18*, 167–170. [PubMed]
111. Fuchs, J.G.; Moënne-Loccoz, Y.; Défago, G. Nonpathogenic Fusarium oxysporum strain Fo47 induces resistance to Fusarium wilt in tomato. *Plant Dis.* **1997**, *81*, 492–496. [CrossRef] [PubMed]
112. Siddiqui, Y.; Sariah, M.; Ismail, M.R.; Ali, A. Trichoderma.fortified compost extracts for the control of choanephora wet rot in okra production. *Crop Prot.* **2008**, *27*, 385–390. [CrossRef]
113. Kowalska, J. Effects of Trichoderma asperellum [T1] on Botrytis cinerea [Pers.: Fr.], growth and yield of organic strawberry. Acta Scientiarum Polonorum. *Hortorum Cultus* **2011**, *10*, 107–114.

114. Djonovic, S.; Vargas, W.A.; Kolomiets, M.V.; Horndeski, M.; Wiest, A.; Kenerley, C.M. A proteinaceous elicitor Sm1 from the beneficial fungus Trichoderma virens is required for induced systemic resistance in maize. *Plant Physiol.* **2007**, *145*, 875–889. [CrossRef] [PubMed]
115. Barakat, R.M.; Al-Masri, M.I. Effect of Trichoderma harzianum in combination with fungicides in controlling gray mould disease (Botrytis cinerea) of strawberry. *Am. J. Plant Sci.* **2017**, *8*, 651–665. [CrossRef]
116. Sallam, N.; Eraky, A.M.I.; Sallam, A. Effect of *Trichoderma* spp. on Fusarium wilt disease of tomato. *Mol. Biol. Rep.* **2019**, *46*, 4463–4470. [CrossRef] [PubMed]
117. Díaz-Urbano, M.; Goicoechea, N.; Velasco, P.; Poveda, J. Development of agricultural bio-inoculants based on mycorrhizal fungi and endophytic filamentous fungi: Co-inoculants for improve plant-physiological responses in sustainable agriculture. *Biol. Control* **2023**, *182*, 105223. [CrossRef]
118. Zhao, L.; Wang, F.; Zhang, Y.; Zhang, J. Involvement of Trichoderma asperellum strain T6 in regulating iron acquisition in plants. *J. Basic Microbiol.* **2014**, *54*, S115–S124. [CrossRef]
119. Půža, V.; Tarasco, E. Interactions between entomopathogenic fungi and entomopathogenic nematodes. *Microorganisms* **2023**, *11*, 163. [CrossRef]
120. Intana, W.; Wonglom, P.; Suwannarach, N.; Sunpapao, A. *Trichoderma asperelloides* PSU-P1 Induced Expression of Pathogenesis-Related Protein Genes against Gummy Stem Blight of Muskmelon (Cucumis melo) in Field Evaluation. *J. Fungi* **2022**, *8*, 156. [CrossRef] [PubMed]
121. Gezgin, Y.; Maral Gül, D.; Sözer Şenşatar, S.; Kara, C.U.; Sargın, S.; Sukan, F.V.; Eltem, R. Evaluation of *Trichoderma atroviride* and *Trichoderma citrinoviride* growth profiles and their potentials as biocontrol agent and biofertilizer. *Turk. J. Biochem.* **2020**, *45*, 163–175. [CrossRef]
122. Zhang, F.; Ge, H.; Zhang, F.; Guo, N.; Wang, Y.; Chen, L.; Li, C. Biocontrol potential of *Trichoderma harzianum* isolate T-aloe against *Sclerotinia sclerotiorum* in soybean. *Plant Physiol. Biochem.* **2016**, *100*, 64–74. [CrossRef] [PubMed]
123. Khan, I.H.; Javaid, A. In Vitro Biocontrol Potential of *Trichoderma pseudokoningii* against *Macrophomina phaseolina*. *Int. J. Agric. Biol.* **2020**, *24*, 730–736.
124. Chen, K.; Zhuang, W.Y. Three new soil-inhabiting species of *Trichoderma* in the *Stromaticum* clade with test of their antagonism to pathogens. *Curr. Microbiol.* **2017**, *74*, 1049–1060. [CrossRef]
125. Degani, O.; Rabinovitz, O.; Becher, P.; Gordani, A.; Chen, A. *Trichoderma longibrachiatum* and *Trichoderma asperellum* confer growth promotion and protection against late wilt disease in the field. *J. Fungi* **2021**, *7*, 444. [CrossRef] [PubMed]
126. Zhu, H.; Ma, Y.; Guo, Q.; Xu, B. Biological weed control using *Trichoderma polysporum* strain HZ-31. *Crop Prot.* **2020**, *134*, 105161. [CrossRef]
127. Hinterdobler, W.; Li, G.; Spiegel, K.; Basyouni-Khamis, S.; Gorfer, M.; Schmoll, M. *Trichoderma reesei* Isolated from Austrian Soil with High Potential for Biotechnological Application. *Front. Microbiol.* **2021**, *12*, 552301. [CrossRef] [PubMed]
128. Hyder, S.; Inam-ul-Haq, M.; Bibi, S.; Humayun, A.; Ghuffar, S.; Iqbal, S. Novel Potential of *Trichoderma* spp. as Biocontrol Agent. *J. Entomol. Zool. Stud.* **2017**, *5*, 214–222.
129. Kumar, P.; Misra, A.K.; Modi, D.R.; Gupta, V.K. Biocontrol Potential of *Trichoderma* Species against Mango Malformation Pathogens. *Arch. Phytopathol. Plant Prot.* **2012**, *45*, 1237–1245. [CrossRef]
130. El-Shennawy, M.Z.; Khalifa, E.Z.; Ammar, M.M.; Mousa, E.M.; Hafez, S.L. Biological Control of the Disease Complex on Potato Caused by Root-Knot Nematode and *Fusarium* Wilt Fungus. *Nematol. Mediterr.* **2012**, *40*, 169–172.
131. Hossain, M.M.; Sultana, F.; Islam, S. Plant Growth-Promoting Fungi (PGPF): Phytostimulation and Induced Systemic Resistance. In *Plant-Microbe Interactions in Agro-Ecological Perspectives: Volume 2: Microbial Interactions and Agro-Ecological Impacts*; Springer: Singapore, 2017; pp. 135–191.
132. Patel, Z.M.; Mahapatra, R.; Jampala, S.S.M. Role of Fungal Elicitors in Plant Defense Mechanism. In *Molecular Aspects of Plant Beneficial Microbes in Agriculture*; Academic Press: Cambridge, MA, USA, 2020; pp. 143–158.
133. Elsharkawy, M.M. Suppression of Potato Virus Y Infection in Tobacco by Plant Growth Promoting Fungi. *Egypt. J. Biol. Pest Control* **2016**, *26*, 695–700.
134. De Meyer, G.; Bigirimana, J.; Elad, Y.; Hofte, M. Induced systemic resistance in Trichoderma harzianum T39 biocontrol of Botrytis cinerea. *Eur. J. Plant Pathol.* **1998**, *104*, 279–286. [CrossRef]
135. Yedidia, I.; Benhamou, N.; Chet, I. Induction of Defense Responses in Cucumber Plants (*Cucumis sativus* L.) by the Biocontrol Agent *Trichoderma harzianum*. *Appl. Environ. Microbiol.* **1999**, *65*, 1061–1070. [CrossRef] [PubMed]
136. El-Hasan, A.; Ngatia, G.; Link, T.I.; Voegele, R.T. Isolation, Identification, and Biocontrol Potential of Root Fungal Endophytes Associated with Solanaceous Plants against Potato Late Blight (*Phytophthora infestans*). *Plants* **2022**, *11*, 1605. [CrossRef] [PubMed]
137. Malviya, D.; Thosar, R.; Kokare, N.; Pawar, S.; Singh, U.B.; Saha, S.; Rai, J.P.; Singh, H.V.; Somkuwar, R.G.; Saxena, A.K. A comparative analysis of microbe-based technologies developed at ICAR-NBAIM against *Erysiphe necator* causing powdery mildew disease in grapes (*Vitis vinifera* L.). *Front. Microbiol.* **2022**, *13*, 871901. [CrossRef] [PubMed]
138. Ahmed, A.S.; Sánchez, C.P.; Candela, M.E. Evaluation of Induction of Systemic Resistance in Pepper Plants (*Capsicum annuum*) to *Phytophthora capsici* using *Trichoderma harzianum* and its Relation with Capsidiol Accumulation. *Eur. J. Plant Pathol.* **2000**, *106*, 817–824. [CrossRef]

139. Santos, M.; Diánez, F.; Sánchez-Montesinos, B.; Huertas, V.; Moreno-Gavira, A.; Esteban García, B.; Garrido-Cárdenas, J.A.; Gea, F.J. Biocontrol of Diseases Caused by *Phytophthora capsici* and *P. parasitica* in Pepper Plants. *J. Fungi* **2023**, *9*, 360. [CrossRef] [PubMed]
140. Segarra, G.; Avilés, M.; Casanova, E.; Borrero, A.; Trillas, I. Effectiveness of biological control of *Phytophthora capsici* in pepper by *Trichoderma asperellum* strain T34. *Phytopathol. Mediterr.* **2013**, *52*, 77–83.
141. Adikaram, N.K.; Joyce, D.C.; Terryc, L.A. Biocontrol activity and induced resistance as a possible mode of action for *Aureobasidium pullulans* against grey mould of strawberry fruit. *Australas. Plant Pathol.* **2002**, *31*, 223–229. [CrossRef]
142. Iqbal, M.; Andreasson, E.; Stenberg, J.A. Biological control of strawberry diseases by Aureobasidium pullulans and sugar beet extract under field conditions. *J. Plant Pathol.* **2023**, *105*, 933–941. [CrossRef]
143. Ippolito, A.; Nigro, F.; Romanazzi, G.; Campanella, V. Field application of Aureobasidium pullulans against Botrytis storage rot of strawberry. In *Non Conventional Methods for the Control of Post-Harvest Disease and Microbiological Spoilage*; Workshop Proceedings; COST: Brussels, Belgium, 1997; pp. 127–133.
144. Iqbal, M.; Jamshaid, M.; Zahid, M.A.; Andreasson, E.; Vetukuri, R.R.; Stenberg, J.A. Biological control of strawberry crown rot, root rot and grey mould by the beneficial fungus Aureobasidium pullulans. *BioControl* **2021**, *66*, 535–545. [CrossRef]
145. Arras, G. Mode of action of an isolate of Candida famata in biological control of *Penicillium digitatum* in orange fruits. *Postharvest Biol. Technol.* **1996**, *8*, 191–198. [CrossRef]
146. Droby, S.; Hofstein, R.; Wilson, C.L.; Wisniewski, M.; Fridlender, B.; Cohen, L.; Weiss, B.; Daus, A.; Timar, D.; Chalutz, E. Pilot Testing of Pichia guilliermondii: A Biocontrol Agent of Postharvest Diseases of Citrus Fruit. *Biol. Control* **1993**, *3*, 47–52. [CrossRef]
147. Lahlali, R.; Hamadi, Y.; El Guilli, M.; Jijakli, M.H. Efficacy assessment of Pichia guilliermondii strain Z1, a new biocontrol agent, against citrus blue mould in Morocco under the influence of temperature and relative humidity. *Biol. Control* **2011**, *56*, 217–224. [CrossRef]
148. Torres, D.E.; Rojas-Martínez, R.I.; Zavaleta-Mejía, E.; Guevara-Fefer, P.; Márquez-Guzmán, G.J.; Pérez-Martínez, C. *Cladosporium cladosporioides* and *Cladosporium pseudocladosporioides* as potential new fungal antagonists of Puccinia horiana Henn., the causal agent of chrysanthemum white rust. *PLoS ONE* **2017**, *12*, 0170782. [CrossRef] [PubMed]
149. Demirci, F. Effects of Pseudomonas fluorescens and Candida famata on blue mould of citrus caused by Penicillium italicum. *Aust. J. Crop Sci.* **2011**, *5*, 341–346.
150. Bandara, A.Y.; Kang, S. Trichoderma application methods differentially affect the tomato growth, rhizomicrobiome, and rhizosphere soil suppressiveness against *Fusarium oxysporum*. *Front. Microbiol.* **2024**, *15*, 1366690. [CrossRef] [PubMed]
151. Sivan, A.; Ucko, O.; Chet, I. Biological control of Fusarium crown rot of tomato by *Trichoderma harzianum* under field conditions. *Plant Dis.* **1987**, *71*, 587–592. [CrossRef]
152. Kexiang, G.; Xiaoguang, L.; Yonghong, L.; Tianbo, Z.; Shuliang, W. Potential of *Trichoderma harzianum* and *T. atroviride* to control *Botryosphaeria berengeriana* f. sp. piricola, the cause of apple ring rot. *J. Phytopathol.* **2002**, *150*, 271–276. [CrossRef]
153. Bunker, R.N.; Kusum Mathur, K.M. Antagonism of local biocontrol agents to *Rhizoctonia solani* inciting dry root rot of chilli. *J. Mycol. Plant Pathol.* **2001**, *31*, 50–53.
154. Trutmann, P.; Keane, P.J. Trichoderma koningii as a biological control agent for *Sclerotinia sclerotiorum* in Southern Australia. *Soil Biol. Biochem.* **1990**, *22*, 43–50. [CrossRef]
155. Camacho-Luna, V.; Pizar-Quiroz, A.M.; Rodríguez-Hernández, A.A.; Rodríguez-Monroy, M.; Sepúlveda-Jiménez, G. Trichoderma longibrachiatum, a biological control agent of Sclerotium cepivorum on onion plants under salt stress. *Biol. Control* **2023**, *180*, 105168. [CrossRef]
156. El-Komy, M.H.; Al-Qahtani, R.M.; Ibrahim, Y.E.; Almasrahi, A.A.; Al-Saleh, M.A. Soil application of *Trichoderma asperellum* strains significantly improves Fusarium root and stem rot disease management and promotes growth in cucumbers in semi-arid regions. *Eur. J. Plant Pathol.* **2022**, *162*, 637–653. [CrossRef]
157. Wang, H.Y.; Zhang, R.; Duan, Y.N.; Jiang, W.T.; Chen, X.S.; Shen, X.; Yin, C.M.; Mao, Z.Q. The endophytic strain *Trichoderma asperellum* 6S-2: An efficient biocontrol agent against apple replant disease in China and a potential plant-growth-promoting fungus. *J. Fungi* **2021**, *7*, 1050. [CrossRef] [PubMed]
158. John, R.P.; Tyagi, R.D.; Prévost, D.; Brar, S.K.; Pouleur, S.; Surampalli, R.Y. Mycoparasitic Trichoderma viride as a biocontrol agent against *Fusarium oxysporum* f. sp. *adzuki* and *Pythium arrhenomanes* and as a growth promoter of soybean. *Crop Prot.* **2010**, *29*, 1452–1459. [CrossRef]
159. Zegeye, E.D.; Santhanam, A.; Gorfu, D.; Tessera, M.; Kassa, B. Biocontrol activity of Trichoderma viride and Pseudomonas fluorescens against Phytophthora infestans under greenhouse conditions. *J. Agric. Technol.* **2011**, *7*, 1589–1602.
160. Zaharia, R.; Petrisor, C.; Fatu, V.; Leveanu, I.; Botea, M.C.; Chireceanu, C. Biocontrol potential of Trichoderma viride against main phytopathogenic fungi associated with Capsicum peppers cultivated in IPM system. In Proceedings of the IX South-Eastern Europe Symposium on Vegetables and Potatoes, Bucharest, Romania, 5–9 September 2023; ISHS: Leuven, Belgium, 2023; Volume 1391, pp. 401–406.
161. Djonovic, S.; Pozo, M.J.; Kenerley, C.M. Tvbgn3, a β-1, 6-glucanase from the biocontrol fungus Trichoderma virens, is involved in mycoparasitism and control of Pythium ultimum. *Appl. Environ. Microbiol.* **2006**, *72*, 7661–7670. [CrossRef]
162. Etebarian, H.R.; Scott, E.S.; Wicks, T.J. Trichoderma harzianum T39 and T. virens DAR 74290 as potential biological control agents for Phytophthora erythroseptica. *Eur. J. Plant Pathol.* **2000**, *106*, 329–337. [CrossRef]

163. Galli, V.; Romboli, Y.; Barbato, D.; Mari, E.; Venturi, M.; Guerrini, S.; Granchi, L. Indigenous *Aureobasidium pullulans* Strains as Biocontrol Agents of *Botrytis cinerea* on Grape Berries. *Sustainability* **2021**, *13*, 9389. [CrossRef]
164. Taping, J.M.F.; Borja, B.T.; Bretaña, B.L.P.; Tanabe, M.E.N.; Cabasan, M.T.N. Fungal endophytes as potential biocontrol agent of Panama disease of banana. *Egypt. J. Biol. Pest Control* **2023**, *33*, 84. [CrossRef]
165. Aggarwal, R. *Chaetomium globosum*: A potential biocontrol agent and its mechanism of action. *Indian Phytopathol.* **2015**, *68*, 8–24.
166. Shanthiyaa, V.; Saravanakumar, D.; Rajendran, L.; Karthikeyan, G.; Prabakar, K.; Raguchander, T. Use of *Chaetomium globosum* for biocontrol of potato late blight disease. *Crop Prot.* **2013**, *52*, 33–38. [CrossRef]
167. Di Pietro, A.; Gut-Rella, M.; Pachlatko, J.P.; Schwinn, F.J. Role of antibiotics produced by *Chaetomium* globosum in biocontrol of *Pythium ultimum*, a causal agent of damping-off. *Phytopathology* **1992**, *82*, 131–135. [CrossRef]
168. Alexander, B.J.R.; Stewart, A. Glasshouse screening for biological control agents of Phytophthora cactorum on apple (Malus domestica). *N. Z. J. Crop Hortic. Sci.* **2001**, *29*, 159–169. [CrossRef]
169. Utkhede, R.S. Biological Control of Apple Crown Rot and Replant Disease. In *Biological Control of Plant Diseases*; Tjamos, E.C., Papavizas, G.C., Cook, R.J., Eds.; NATO ASI Series; Springer: Boston, MA, USA, 1992; Volume 230, pp. 410–415.
170. Nakayama, T.; Sayama, M. Suppression of potato powdery scab caused by *Spongospora subterranea* using an antagonistic fungus *Aspergillus versicolor* isolated from potato roots [Conference poster]. In Proceedings of the Ninth Symposium of the International Working Group on Plant Viruses with Fungal Vectors, Obihiro, Japan, 19–22 August 2013.
171. Chin-A-Woeng, T.F.; Bloemberg, G.V.; Lugtenberg, B.J.J. Phenazines and their role in biocontrol by Pseudomonas bacteria. *New Phytol.* **2003**, *157*, 503–523. [CrossRef] [PubMed]
172. Mouloud, G.; Daoud, H.; Bassem, J.; Laribi Atef, I.; Hani, B. New bacteriocin from Bacillus clausii strainGM17: Purification, characterization, and biological activity. *Appl. Biochem. Biotechnol.* **2013**, *171*, 2186–2200. [CrossRef]
173. Bonaterra, A.; Badosa, E.; Daranas, N.; Francés, J.; Roselló, G.; Montesinos, E. Bacteria as Biological Control Agents of Plant Diseases. *Microorganisms* **2022**, *10*, 1759. [CrossRef]
174. Legein, M.; Smets, W.; Vandenheuvel, D.; Eilers, T.; Muyshondt, B.; Prinsen, E.; Samson, R.; Lebeer, S. Modes of action of microbial biocontrol in the phyllosphere. *Front. Microbiol.* **2020**, *11*, 1619. [CrossRef]
175. Kalia, V.C.; Patel, S.K.S.; Kang, Y.C.; Lee, J.K. Quorum sensing inhibitors as antipathogens: Biotechnological applications. *Biotechnol. Adv.* **2019**, *37*, 68–90. [CrossRef]
176. Savini, V.; Fazii, P. Bacillus thuringiensis insecticide properties. In *The Diverse Faces of Bacillus Cereus*; Academic Press: Cambridge, MA, USA, 2016; pp. 139–155.
177. Lobanovska, M.; Pilla, G. Focus: Drug development: Penicillin's discovery and antibiotic resistance: Lessons for the future? *Yale J. Biol. Med.* **2017**, *90*, 135.
178. Ongena, M.; Jourdan, E.; Adam, A.; Paquot, M.; Brans, A.; Joris, B.; Arpigny, J.L.; Thonart, P. Surfactin and fengycin lipopeptides of *Bacillus subtilis* as elicitors of induced systemic resistance in plants. *Environ. Microbiol.* **2007**, *9*, 1084–1090. [CrossRef]
179. Farag, M.A.; Zhang, H.; Ryu, C.M. Dynamic chemical communication between plants and bacteria through airborne signals: Induced resistance by bacterial volatiles. *J. Chem. Ecol.* **2013**, *39*, 1007–1018. [CrossRef]
180. Raza, W.; Ling, N.; Liu, D.; Wei, Z.; Huang, Q.; Shen, Q. Volatile organic compounds produced by *Pseudomonas fluorescens* WR-1 restrict the growth and virulence traits of *Ralstonia solanacearum*. *Microbiol. Res.* **2016**, *192*, 103–113. [CrossRef]
181. Kamilova, F.; Validov, S.; Azarova, T.; Mulders, I.; Lugtenberg, B. Enrichment for enhanced competitive plant root tip colonizers selects for a new class of biocontrol bacteria. *Environ. Microbiol.* **2005**, *7*, 1809–1817. [CrossRef]
182. Schuhegger, R.; Ihring, A.; Gantner, S.; Bahnweg, G.; Knappe, C.; Vogg, G.; Hutzler, P.; Schmid, M.; Van Breusegem, F.; Eberl, L.; et al. Induction of systemic resistance in tomato by N-acyl-L-homoserine lactone-producing rhizosphere bacteria. *Plant Cell Environ.* **2006**, *29*, 909–918. [CrossRef] [PubMed]
183. Wu, Y.; Tan, Y.; Peng, Q.; Xiao, Y.; Xie, J.; Li, Z.; Ding, H.; Pan, H.; Wei, L. Biocontrol potential of endophytic bacterium Bacillus altitudinis GS-16 against tea anthracnose caused by Colletotrichum gloeosporioides. *Peer J.* **2024**, *12*, 16761. [CrossRef]
184. Gupta, R.; Keppanan, R.; Leibman-Markus, M.; Matveev, S.; Rav-David, D.; Shulhani, R.; Elad, Y.; Ment, D.; Bar, M. Bacillus thuringiensis promotes systemic immunity in tomato, controlling pests and pathogens and promoting yield. *Food Secur.* **2024**, *16*, 675–690. [CrossRef]
185. Heo, Y.; Lee, Y.; Balaraju, K.; Jeon, Y. Characterization and evaluation of Bacillus subtilis GYUN-2311 as a biocontrol agent against *Colletotrichum* spp. on apple and hot pepper in Korea. *Front Microbiol.* **2024**, *14*, 1322641. [CrossRef]
186. Wockenfuss, A.; Chan, K.; Cooper, J.G.; Chaya, T.; Mauriello, M.A.; Yannarell, S.M.; Maresca, J.A.; Donofrio, N.M. A Bacillus velezensis strain shows antimicrobial activity against soilborne and foliar fungi and oomycetes. *Front Fung Biol.* **2024**, *5*, 1332755. [CrossRef]
187. Sumera Yasmin, S.Y.; Hafeez, F.Y.; Mirza, M.S.; Rasul, M.; Arshad, H.M.I.; Muhammad Zubair, M.Z.; Mazhar Iqbal, M.I. Biocontrol of Bacterial Leaf Blight of rice and profiling of secondary metabolites produced by rhizospheric Pseudomonas aeruginosa BRp3. *Front. Microbiol.* **2017**, *8*, 1895.
188. Xie, J.; Singh, P.; Qi, Y.; Singh, R.K.; Qin, Q.; Jin, C.; Wang, B.; Fang, W. Pseudomonas aeruginosa Strain 91: A Multifaceted Biocontrol Agent against Banana Fusarium Wilt. *J. Fungi* **2023**, *9*, 1047. [CrossRef] [PubMed]
189. Wang, X.; Zhou, X.; Cai, Z.; Guo, L.; Chen, X.; Chen, X.; Liu, J.; Feng, M.; Qiu, Y.; Zhang, Y.; et al. A biocontrol strain of Pseudomonas aeruginosa CQ-40 promote growth and control Botrytis cinerea in tomato. *Pathogens* **2020**, *10*, 22. [CrossRef] [PubMed]

190. Amaradasa, B.S.; Mei, C.; He, Y.; Chretien, R.L.; Doss, M.; Durham, T.; Lowman, S. Biocontrol potential of endophytic Pseudomonas strain IALR1619 against two Pythium species in cucumber and hydroponic lettuce. *PLoS ONE* **2024**, *19*, 0298514. [CrossRef] [PubMed]
191. Cao, Y.; Pi, H.; Chandrangsu, P.; Li, Y.; Wang, Y.; Zhou, H.; Xiong, H.; Helmann, J.D.; Cai, Y. Antagonism of two plant-growth promoting Bacillus velezensis isolates against Ralstonia solanacearum and Fusarium oxysporum. *Sci Rep.* **2018**, *8*, 4360. [CrossRef]
192. Yi, Y.; Luan, P.; Liu, S.; Shan, Y.; Hou, Z.; Zhao, S.; Jia, S.; Li, R. Efficacy of Bacillus subtilis XZ18-3 as a Biocontrol Agent against Rhizoctonia cerealis on Wheat. *Agriculture* **2022**, *12*, 258. [CrossRef]
193. Myers, J.M.; James, T.Y. Mycoviruses. *Curr. Biol.* **2022**, *32*, 150–155. [CrossRef]
194. Di Giallonardo, F.; Holmes, E.C. Viral biocontrol: Grand experiments in disease emergence and evolution. *Trends Microbiol.* **2015**, *23*, 83–90. [CrossRef]
195. Wagemans, J.; Holtappels, D.; Vainio, E.; Rabiey, M.; Marzachì, C.; Herrero, S.; Turina, M. Going viral: Virus-based biological control agents for plant protection. *Ann. Rev. Phytopathol.* **2022**, *60*, 21–42. [CrossRef]
196. Mallmann, W.; Hemstreet, C. Isolation of an inhibitory substance from plants. *Agric. Res.* **1924**, *28*, 599–602.
197. Nakayinga, R.; Makumi, A.; Tumuhaise, V.; Tinzaara, W. *Xanthomonas* bacteriophages: A review of their biology and biocontrol applications in agriculture. *BMC Microbiol.* **2021**, *21*, 291. [CrossRef] [PubMed]
198. Myers, J.M. Check for Mycoviruses Jillian M. Myers and Timothy Y. James. *Evol. Fungi Fungal-Like Org.* **2023**, *14*, 151.
199. Moscardi, F. Assessment of the application of baculoviruses for control of Lepidoptera. *Annu. Rev. Entomol.* **1999**, *44*, 257–289. [CrossRef]
200. Pechinger, K.; Choo, K.M.; MacDiarmid, R.M.; Harper, S.J.; Ziebell, H. A new era for mild strain crossprotection. *Viruses* **2019**, *11*, 670. [CrossRef] [PubMed]
201. Ruiz-Padilla, A.; Rodriguez-Romero, J.; Gomez-Cid, I.; Pacifico, D.; Ayllón, M.A. Novel mycoviruses discovered in the mycovirome of a necrotrophic fungus. *mBio* **2021**, *12*, 3705–3720. [CrossRef]
202. Vainio, E.J.; Hantula, J. Fungal viruses. In *Viruses of Microorganisms*; Hyman, P., Abedon, S.T., Eds.; Caster Acad.: Norfolk, UK, 2018; pp. 193–209.
203. Xie, J.; Jiang, D. New insights into mycoviruses and exploration for the biological control of crop fungal diseases. *Annu. Rev. Phytopathol.* **2014**, *52*, 45–68. [CrossRef]
204. Chen, B.; Chen, C.H.; Bowman, B.H.; Nuss, D.L. Phenotypic changes associated with wild-type and mutant hypovirus RNA transfection of plant pathogenic fungi phylogenetically related to *Cryphonectria parasitica*. *Phytopathology* **1996**, *86*, 301–310. [CrossRef]
205. van Heerden, S.W.; Geletka, L.M.; Preisig, O.; Nuss, D.L.; Wingfield, B.D.; Wingfield, M.J. Characterization of South African Cryphonectria cubensis isolates infected with a C. parasitica hypovirus. *Phytopathology* **2001**, *91*, 628–632. [CrossRef] [PubMed]
206. Sasaki, A.; Onoue, M.; Kanematsu, S.; Suzaki, K.; Miyanishi, M.; Suzuki, N.; Nuss, D.L.; Yoshida, K. Extending chestnut blight hypovirus host range within diaporthales by biolistic delivery of viral cDNA. *Mol. Plant. Microbe Interact.* **2002**, *15*, 780–789. [CrossRef]
207. Kanematsu, S.; Sasaki, A.; Onoue, M.; Oikawa, Y.; Ito, T. Extending the fungal host range of a partitivirus and a mycoreovirus from Rosellinia necatrix by inoculation of protoplasts with virus particles. *Phytopathology* **2010**, *100*, 922–930. [CrossRef]
208. Xiao, X.; Cheng, J.; Tang, J.; Fu, Y.; Jiang, D.; Baker, T.S.; Ghabrial, S.A.; Xie, J. A novel partitivirus that confers hypovirulence on plant pathogenic fungi. *J. Virol.* **2014**, *88*, 10120–10133. [CrossRef] [PubMed]
209. Lee, K.M.; Yu, J.; Son, M.; Lee, Y.W.; Kim, K.H. Transmission of Fusarium boothii mycovirus via protoplast fusion causes hypovirulence in other phytopathogenic fungi. *PLoS ONE* **2011**, *6*, e21629. [CrossRef] [PubMed]
210. Islam, T.; Haque, M.A.; Barai, H.R.; Istiaq, A.; Kim, J.J. Antibiotic Resistance in Plant Pathogenic Bacteria: Recent Data and Environmental Impact of Unchecked Use and the Potential of Biocontrol Agents as an Eco-Friendly Alternative. *Plants* **2024**, *13*, 1135. [CrossRef] [PubMed]
211. Vu, N.T.; Oh, C.S. Bacteriophage usage for bacterial disease management and diagnosis in plants. *Plant Pathol. J.* **2020**, *36*, 204–217. [CrossRef] [PubMed]
212. Balogh, B.; Jones, J.B.; Momol, M.T.; Olson, S.M.; Obradovic, A. Improved efficacy of newly formulated bacteriophages for management of bacterial spot on tomato. *Plant Dis.* **2003**, *87*, 949–954. [CrossRef]
213. Balogh, B.; Canteros, B.I.; Stall, K.E.; Jones, J.B. Control of citrus canker and citrus bacterial spot with bacteriophages. *Plant Dis.* **2008**, *92*, 1048–1052. [CrossRef] [PubMed]
214. Meczker, K.; Doemoetoer, D.; Vass, J.; Rakhely, G.; Schneider, G.; Kovacs, T. The genome of the *Erwinia amylovora* phage PhiEaH1 reveals greater diversity and broadens the applicability of phages for the treatment of fire blight. *FEMS Microbiol. Lett.* **2014**, *350*, 25–27. [CrossRef] [PubMed]
215. Saponari, M.; Loconsole, G.; Cornara, D.; Yokomi, R.K.; De Stradis, A.; Boscia, D.; Bosco, D.; Martelli, G.P.; Krugner, R.; Porcelli, F. Infectivity and transmission of Xylella fastidiosa by Philaenus spumarius (Hemiptera: Aphrophoridae) in Apulia, Italy. *J. Econ. Entomol.* **2014**, *107*, 1316–1319. [CrossRef]
216. Ahern, S.J.; Das, M.; Bhowmick, T.S.; Young, R.; Gonzalez, C.F. Characterization of novel virulent broad.host.range phages of Xylella fastidiosa and Xanthomonas. *J. Bacteriol.* **2014**, *196*, 459–471. [CrossRef]
217. Ahmad, A.A.; Askora, A.; Kawasaki, T.; Fujie, M.; Yamada, T. The filamentous phage XacF1 causes loss of virulence in *Xanthomonas axonopodis* pv. *citri*, the causative agent of citrus canker disease. *Front. Microbiol.* **2014**, *5*, 321. [CrossRef]

218. Álvarez, B.; López, M.M.; Biosca, E.G. Biocontrol of the major plant pathogen Ralstonia solanacearum in irrigation water and host plants by novel waterborne lytic bacteriophages. *Front. Microbiol.* **2019**, *10*, 2813. [CrossRef]
219. Doemoetoer, D.; Becsagh, P.; Rakhely, G.; Schneider, G.; Kovacs, T. Complete genomic sequence of *Erwinia amylovora* phage PhiEaH2. *J. Virol.* **2012**, *86*, 10899. [CrossRef] [PubMed]
220. Mckinney, H.H. Mosaic diseases in the Canary Islands, West Africa, and Gibraltar. *J. Agric. Res.* **1929**, *39*, 557–578.
221. Agrios, G.N. *Plant Pathology*; Elsevier: Amsterdam, The Netherlands, 2005.
222. Gal On, A.; Shiboleth, Y.M. Cross protection. In *Natural Resistance Mechanisms of Plants to Viruses*; Loebenstein, G., Carr, J.P., Eds.; Springer: Berlin, Germany, 2006; pp. 261–288.
223. Tomitaka, Y.; Shimomoto, Y.; Ryang, B.S.; Hayashi, K.; Oki, T.; Matsuyama, M.; Sekine, K.T. Development and Application of Attenuated Plant Viruses as Biological Control Agents in Japan. *Viruses* **2024**, *16*, 517. [CrossRef]
224. Salaman, R.N. Protective inoculation against a plant virus. *Nature* **1933**, *131*, 468. [CrossRef]
225. Holmes, F.O. A masked strain of tobacco mosaic virus. *Phytopathology* **1934**, *24*, 845–873.
226. Grant, T.J.; Costa, A.S. A mild strain of tristeza virus of citurs. *Phytopathology* **1951**, *41*, 114–122.
227. Posnette, A.F.; Todd, J.M. Viruse diseases of Cacao in West Africa IX. Strain variation and interference in virus 1A. *Ann. Appl. Biol.* **1955**, *43*, 433–453. [CrossRef]
228. Aguero, J.; Gomez-Aix, C.; Sempere, R.N.; Garcia-Villalba, J.; Garcia-Nunez, J. Stable and broad spectrum cross-protection against Pepino mosaic virus attained by mixed infection. *Front. Plant Sci.* **2018**, *9*, 1810. [CrossRef]
229. Yeh, S.D.; Gonsalves, D.; Wang, H.L.; Namba, R.; Chiu, R.J. Control of papaya ringspot virus by cross protection. *Plant Dis.* **1988**, *72*, 375–380. [CrossRef]
230. Folimonova, S.Y. Developing an understanding of cross-protection by Citrus tristeza virus. *Front. Microbiol.* **2013**, *4*, 76. [CrossRef] [PubMed]
231. Cook, G.; Breytenbach, J.H.; Steyn, C.; de Bruyn, R.; van Vuuren, S.P.; Burger, J.T.; Maree, H.J. Grapefruit field trial evaluation of Citrus tristeza virus T68-strain sources. *Plant Dis.* **2021**, *105*, 361–367. [CrossRef] [PubMed]
232. Fravel, D.R.; Rhodes, D.J.; Larkin, R.P. Production and commercialization of biocontrol products. In *Integrated Pest and Disease Management in Greenhouse Crops*; Kluwer Academic Publishers: Dordrecht, The Netherlands, 1999; pp. 365–376.
233. Nunes, C. Biological control of postharvest diseases of fruit. *Eur. J. Plant Pathol.* **2012**, *133*, 181–196. [CrossRef]
234. Whipps, J.M. Developments in the biological control for soilborne plant pathogens. *Adv. Bot. Res.* **1997**, *26*, 1–134.
235. Larena, I.; De Cal, A.; Melgarejo, P. Solid substrate production of *Epicoccum nigrum* conidia for biological control of brown rot on stone fruits. *Int. J. Food Microbiol.* **2004**, *94*, 161–167. [CrossRef] [PubMed]
236. Prasad, R.D.; Rangeshwaran, R.; Anuroop, C.P.; Phanikumar, P.R. Bioefficacy and shelf life of conidial and chlamydospore formulations of *Trichoderma harzianum* Rifai. *J. Biol. Control* **2002**, *16*, 145–148.
237. Jeyarajan, R. Prospects of indigenous mass production and formulation of Trichoderma. In *Current Status of Biological Control of Plant Diseases Using Antagonistic Organisms in India, Proceedings of the Group Meeting on Antagonistic Organisms in Plant Disease Management Held at Project Directorate of Biological Control, Bangalore, India, 10–11 July 2003*; Project Directorate of Biological Control, Indian Council of Agricultural Research: Bangalore, India, 2006; pp. 10–11.
238. Angeli, D.; Saharan, K.; Segarra, G.; Sicher, C.; Pertot, I. Production of *Ampelomyces quisqualis* conidia in submerged fermentation and improvements in the formualtion for increases shelf-lefe. *Crop Prot.* **2017**, *97*, 135–144. [CrossRef]
239. Carbó, A.; Torres, R.; Usall, J.; Ballesta, J.; Teixidó, N. Biocontrol potential of *Ampelomyces quisqualis* strain CPA-9 against powdery mildew: Conidia production in liquid medium and efficacy on zucchini leaves. *Sci. Hortic.* **2020**, *267*, 109337. [CrossRef]
240. Hynes, R.K.; Boyetchko, S.M. Research initiatives in the art and science of biopesticide formulations. *Soil Biol. Biochem.* **2006**, *38*, 845–849. [CrossRef]
241. Gotor-Vila, A.; Usall, J.; Torres, R.; Abadias, M.; Teixidó, N. Formulation of the biocontrol agent *Bacillus amyloliquefaciens* CPA-8 using different approaches: Liquid, freeze-drying and fluid-bed spray-drying. *BioControl* **2017**, *62*, 545–555. [CrossRef]
242. Keswani, C.; Bisen, K.; Singh, V.; Sarma, B.K.; Singh, H.B. Formulation technology of biocontrol agents: Present status and future prospects. In *Bioformulations: For Sustainable Agriculture*; Springer: New Delhi, India, 2016; pp. 35–52.
243. Fravel, D.R.; Connick, W.J.; Lewis, J.A. Formulation of microorganisms to control plant diseases. In *Formulation of Microbial Biopesticides*; Kluwer Academic Publishers: Dordrecht, The Netherlands, 1998; pp. 187–202.
244. Droby, S.; Wisniewski, M.; Teixidó, N.; Spadaro, D.; Jijakli, M.H. The science, development, and commercialization of postharvest biocontrol products. *Postharvest Biol. Technol.* **2016**, *122*, 22–29. [CrossRef]
245. Usall, J.; Torres, R.; Teixidó, N. Biological control of postharvest diseases on fruit: A suitable alternative. *Curr. Opin. Food Sci.* **2016**, *11*, 51–55. [CrossRef]
246. Bejarano, A.; Puopolo, G. Bioformulation of microbial biocontrol agents for a sustainable agriculture. In *How Research Can Stimulate the Development of Commercial Biological Control against Plant Diseases*; Springer: Berlin/Heidelberg, Germany, 2020; pp. 275–293.
247. Brar, S.K.; Verma, M.; Tyagi, R.G.; Valéro, J.R. Recent advances in dowstream processing and formulations of *Bacillus thuringiensis* based biopestices. *Process Biochem.* **2006**, *41*, 323–342. [CrossRef]
248. Batta, Y.A. Postharvest biological control of apple gray mold by *Trichoderma harzianum* Rifai formulated in an invert emulsion. *Crop Prot.* **2004**, *23*, 19–26. [CrossRef]

249. Nandhini, M.; Harish, S.; Aiyanathan, K.E.A.; Durgadevi, D.; Beaulah, A. Glycerol-based liquid formulation of the epiphytic yeast *Hanseniaspora guilliermondii* isolate YBB3 with multiple modes of action controls postharvest Aspergillus rot in grapes. *J. Plant Pathol.* **2021**, *103*, 1253–1264. [CrossRef]
250. Abadias, M.; Teixidó, N.; Usall, J.; Solsona, C.; Viñas, I. Survival of the postharvest biocontrol yeast Candida sake CPA-1 after dehydration by spray-drying. *Biocontrol Sci. Technol.* **2005**, *15*, 835–846. [CrossRef]
251. Adams, G. The principles of Freeze-Drying. In *Cryopreservation and Freeze-Drying Protocols*, 2nd ed.; Day, J.G., Stacey, G.N., Eds.; Humana Press Inc.: Totowa, NJ, USA, 2007; pp. 15–38.
252. Prakash, O.; Nimonkar, Y.; Shouche, Y.S. Practice and prospects of microbial preservation. *FEMS Microbiol. Lett.* **2013**, *339*, 1–9. [CrossRef] [PubMed]
253. Costa, S.S.; Machado, B.A.S.; Martin, A.R.; Bagnara, F.; Ragadalli, S.A.; Alves, A.R.C. Drying by spray drying in the food industry:Micro-encapsulation, process parameters and main carriers used. *Afr. J. Food Sci.* **2015**, *9*, 462–470.
254. Yánez-Mendizábal, V.; Viñas, I.; Usall, J.; Cañamás, T.; Teixidó, N. Endospore production allows using spray-drying as a possible formulation system of the biocontrol agent Bacillus subtilis CPA-8. *Biotechnol. Lett.* **2012**, *34*, 729–735. [CrossRef]
255. Strasser, S. Innovative Product Formulations Applying the Fluidised Bed Technology. Ph.D. Thesis, University of Natural Resources and Life Sciences, Vienna, Austria, 2008.
256. Larena, I.; De Cal, A.; Linan, M.; Melgarejo, P. Drying of *Epicoccum nigrum* conidia for obtaining a shelf-stable biological product against brown rot disease. *J. Appl. Microbiol.* **2003**, *94*, 508–514. [CrossRef]
257. Vemmer, M.; Patel, A.V. Review of encapsulation methods suitable for microbial biological control agents. *Biol. Control* **2013**, *67*, 380–389. [CrossRef]
258. González, L.E.; Bashan, Y. Increased growth of the microalga Chlorella vulgaris when coimmobilized and cocultured in alginate beads with the plant-growth-promoting bacterium *Azospirillum brasilense*. *Appl. Environ. Microbiol.* **2000**, *66*, 1527–1531. [CrossRef] [PubMed]
259. Lumsden, R.D.; Walter, J.F. Development of the biocontrol fungus *Gliocladium virens*: Risk assessment and approval for horticultural use. In *Biological Control: Benefits and Risks*; Hokkanen, H.M.T., Lynch, J.M., Eds.; Cambridge University Press: Cambridge, UK, 1995; pp. 263–269.
260. Kruitwagen, A.; Beukeboom, L.W.; Wertheim, B. Optimization of native biocontrol agents, with parasitoids of the invasive pest Drosophila suzukii as an example. *Evol. Appl.* **2018**, *11*, 1473–1497. [CrossRef]
261. Bielza, P.; Balanza, V.; Cifuentes, D.; Mendoza, J.E. Challenges facing arthropod biological control: Identifying traits for genetic improvement of predators in protected crops. *Pest Manag. Sci.* **2020**, *76*, 3517–3526. [CrossRef]
262. Leung, K.; Ras, E.; Ferguson, K.B.; Ariëns, S.; Babendreier, D.; Bijma, P.; Bourtzis, K.; Brodeur, J.; Bruins, M.A.; Centurión, A.; et al. Next-generation biological control: The need for integrating genetics and genomics. *Biol. Rev.* **2020**, *95*, 1838–1854. [CrossRef] [PubMed]
263. Rutledge, P.J.; Challis, G.L. Discovery of Microbial Natural Products by Activation of Silent Biosynthetic Gene Clusters. *Nat. Rev. Microbiol.* **2015**, *13*, 509–523. [CrossRef] [PubMed]
264. Nogueira, T.; Botelho, A. Metagenomics and Other Omics Approaches to Bacterial Communities and Antimicrobial Resistance Assessment in Aquacultures. *Antibiotics* **2021**, *10*, 787. [CrossRef]
265. Lahlali, R.; Ibrahim, D.S.S.; Belabess, Z.; Roni, M.Z.K.; Radouane, N.; Vicente, C.S.L.; Menendez, E.; Mokrini, F.; Barka, E.A.; de Melo e Mota, M.G.; et al. High-Throughput Molecular Technologies for Unraveling the Mystery of Soil Microbial Community: Challenges and Future Prospects. *Heliyon* **2021**, *7*, e08142. [CrossRef] [PubMed]
266. Barahona, E.; Navazo, A.; Martínez-Granero, F.; Zea-Bonilla, T.; Pérez-Jiménez, R.M.; Martín, M.; Rivilla, R. Pseudomonas fluorescens F113 mutant with enhanced competitive colonization ability and improved biocontrol activity against fungal root pathogens. *Appl. Environ. Microbiol.* **2011**, *77*, 5412–5419. [CrossRef]
267. Leclère, V.; Béchet, M.; Adam, A.; Guez, J.-S.; Wathelet, B.; Ongena, M.; Thonart, P.; Gancel, F.; Chollet-Imbert, M.; Jacques, P. Mycosubtilin overproduction by Bacillus subtilis BBG100 enhances the organism's antagonistic and biocontrol activities. *Appl. Environ. Microbiol.* **2005**, *71*, 4577–4584. [CrossRef]
268. Okay, S.; Tefon, B.; Özkan, M.; Özcengiz, G. Expression of chitinase A (chiA) gene from a local isolate of Serratia marcescens in Coleoptera-specific Bacillus thuringiensis. *J. Appl. Microbiol.* **2008**, *104*, 161–170. [CrossRef] [PubMed]
269. Hofmann, D.; Kempf, H.J.; van Pée, K.H. Natural product with antimicrobial activity from Pseudomonas biocontrol bacteria. In *Pesticide Chemistry and Bioscience—The Food-Environment Challenge*; Brooks, G.T., Roberts, T.R., Eds.; The Royal Society of Chemistry: Cambridge, UK, 1999; Volume 22, pp. 179–189.
270. Kota, M.; Daniell, H.; Varma, S.; Garczynski, S.F.; Gould, F.; Moar, W.J. Overexpression of the Bacillus thuringiensis (Bt) Cry2Aa2 protein in chloroplasts confers resistance to plants against susceptible and Bt-resistant insects. *Proc. Natl. Acad. Sci. USA* **1999**, *96*, 1840–1845. [CrossRef] [PubMed]
271. Chen, X.H.; Koumoutsi, A.; Scholz, R.; Schneider, K.; Vater, J.; Süssmuth, R.; Piel, J.; Borriss, R. Genome analysis of Bacillus amyloliquefaciens FZB42 reveals its potential for biocontrol of plant pathogens. *J. Biotechnol.* **2009**, *140*, 27–37. [CrossRef] [PubMed]
272. Clermont, N.; Lerat, S.; Beaulieu, C. Genome shuffling enhances biocontrol abilities of Streptomyces strains against two potato pathogens. *J. Appl. Microbiol.* **2011**, *111*, 671–682. [CrossRef]

273. Zhao, P.; Quan, C.; Wang, Y.; Wang, J.; Fan, S. Bacillus amyloliquefaciens Q-426 as a potential biocontrol agent against *Fusarium oxysporum* f. sp. spinaciae. *J. Basic Microbiol.* **2014**, *54*, 448–456. [CrossRef] [PubMed]
274. Côrtes, M.V.; de Sousa Oliveira, M.I.; Mateus, J.R.; Seldin, L.; Silva-Lobo, V.L.; Freire, D.M. A pipeline for the genetic improvement of a biological control agent enhances its potential for controlling soil-borne plant pathogens. *Biol. Control* **2021**, *152*, 104460. [CrossRef]
275. Downing, K.J.; Thomson, J.A. Introduction of the Serratia marcescens chiA gene into an endophytic Pseudomonas fluorescens for the biocontrol of phytopathogenic fungi. *Can. J. Microbiol.* **2000**, *46*, 363–369. [CrossRef]
276. Baek, J.M.; Howell, C.R.; Kenerley, C.M. The role of an extracellular chitinase from Trichoderma virens Gv29-8 in the biocontrol of Rhizoctonia solani. *Curr. Genet.* **1999**, *35*, 41–50. [CrossRef] [PubMed]
277. Sun, Z.B.; Sun, M.H.; Zhou, M.; Li, S.D. Transformation of the endochitinase gene Chi67-1 in Clonostachys rosea 67-1 increases its biocontrol activity against *Sclerotinia sclerotiorum*. *AMB Express* **2017**, *7*, 1. [CrossRef]
278. Bakker, P.A.; Glandorf, D.; Viebahn, M.; Ouwens, T.W.; Smit, E.; Leeflang, P.; van Loon, L.C. Effects of Pseudomonas putida modified to produce phenazine-1-carboxylic acid and 2, 4-diacetylphloroglucinol on the microflora of field grown wheat. *Antonie Leeuwenhoek* **2002**, *81*, 617–624. [CrossRef]
279. Mendoza-Mendoza, A.; Pozo, M.J.; Grzegorski, D.; Martínez, P.; García, J.M.; Olmedo-Monfil, V.; Cortés, C.; Kenerley, C.; Herrera-Estrella, A. Enhanced biocontrol activity of Trichoderma through inactivation of a mitogen-activated protein kinase. *Proc. Natl. Acad. Sci. USA* **2003**, *100*, 15965–15970. [CrossRef] [PubMed]
280. Deveau, H.; Garneau, J.E.; Moineau, S. CRISPR/Cas system and its role in phage-bacteria interactions. *Annu. Rev. Microbiol.* **2010**, *64*, 475–493. [CrossRef] [PubMed]
281. Wang, Y.; Chen, H.; Ma, L.; Gong, M.; Wu, Y.; Bao, D.; Zou, G. Use of CRISPR-Cas tools to engineer Trichoderma species. *Microb. Biotechnol.* **2022**, *10*, 2521–2532. [CrossRef] [PubMed]
282. Zhao, D.; Zhu, X.; Zhou, H.; Sun, N.; Wang, T.; Bi, C.; Zhang, X. CRISPR-based metabolic pathway engineering. *Metab. Eng.* **2021**, *63*, 148–159. [CrossRef] [PubMed]
283. Song, Y.P.; Fang, S.T.; Miao, F.P.; Yin, X.L.; Ji, N.Y. Diterpenes and sesquiterpenes from the marine algicolous fungus Trichoderma harzianum X-5. *J. Nat. Prod.* **2018**, *81*, 2553–2559. [CrossRef] [PubMed]
284. Kai, K.; Mine, K.; Akiyama, K.; Ohki, S.; Hayashi, H. Anti-plant viral activity of peptaibols, trichorzins HA II, HA V, and HA VI, isolated from Trichoderma harzianum HK-61. *J. Pestic. Sci.* **2018**, *43*, 283–286. [CrossRef]
285. van Bohemen, A.I.; Ruiz, N.; Zalouk-Vergnoux, A.; Michaud, A.; Robiou du Pont, T.; Druzhinina, I.; Atanasova, L.; Prado, S.; Bodo, B.; Pouchus, Y.F.; et al. Pentadecaibins I-V: 15-residue peptaibols produced by a marine-derived *Trichoderma* sp. of the harzianum clade. *J. Nat. Prod.* **2021**, *84*, 1271–1282. [CrossRef]

MDPI AG
Grosspeteranlage 5
4052 Basel
Switzerland
Tel.: +41 61 683 77 34

Horticulturae Editorial Office
E-mail: horticulturae@mdpi.com
www.mdpi.com/journal/horticulturae

www.ingramcontent.com/pod-product-compliance
Lightning Source LLC
Chambersburg PA
CBHW041351210926
51850CB00013B/3044